D1246022

R for Everyone

Second Edition

The Addison-Wesley Data and Analytics Series

Addison-Wesley

Visit **informit.com/awdataseries** for a complete list of available publications.

The **Addison-Wesley Data and Analytics Series** provides readers with practical knowledge for solving problems and answering questions with data. Titles in this series primarily focus on three areas:

1. **Infrastructure:** how to store, move, and manage data
2. **Algorithms:** how to mine intelligence or make predictions based on data
3. **Visualizations:** how to represent data and insights in a meaningful and compelling way

The series aims to tie all three of these areas together to help the reader build end-to-end systems for fighting spam; making recommendations; building personalization; detecting trends, patterns, or problems; and gaining insight from the data exhaust of systems and user interactions.

Make sure to connect with us!
informit.com/socialconnect

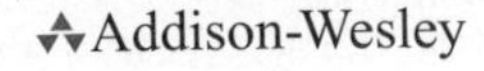

R for Everyone

Advanced Analytics and Graphics

Second Edition

Jared P. Lander

Addison-Wesley

Boston • Columbus • Indianapolis • New York • San Francisco • Amsterdam • Cape Town
Dubai • London • Madrid • Milan • Munich • Paris • Montreal • Toronto • Delhi • Mexico City
São Paulo • Sydney • Hong Kong • Seoul • Singapore • Taipei • Tokyo

Many of the designations used by manufacturers and sellers to distinguish their products are claimed as trademarks. Where those designations appear in this book, and the publisher was aware of a trademark claim, the designations have been printed with initial capital letters or in all capitals.

The author and publisher have taken care in the preparation of this book, but make no expressed or implied warranty of any kind and assume no responsibility for errors or omissions. No liability is assumed for incidental or consequential damages in connection with or arising out of the use of the information or programs contained herein.

For information about buying this title in bulk quantities, or for special sales opportunities (which may include electronic versions; custom cover designs; and content particular to your business, training goals, marketing focus, or branding interests), please contact our corporate sales department at corpsales@pearsoned.com or (800) 382-3419.

For government sales inquiries, please contact governmentsales@pearsoned.com.

For questions about sales outside the U.S., please contact intlcs@pearson.com.

Visit us on the Web: informit.com/aw

Library of Congress Control Number: 2017934582

Copyright © 2017 Pearson Education, Inc.

All rights reserved. Printed in the United States of America. This publication is protected by copyright, and permission must be obtained from the publisher prior to any prohibited reproduction, storage in a retrieval system, or transmission in any form or by any means, electronic, mechanical, photocopying, recording, or likewise. For information regarding permissions, request forms and the appropriate contacts within the Pearson Education Global Rights & Permissions Department, please visit www.pearsoned.com/permissions/.

ISBN-13: 978-0-13-454692-6
ISBN-10: 0-13-454692-X

1 17

❖

To Becky

❖

Contents

Foreword

R has had tremendous growth in popularity over the last five years. Based on that, you'd think that it was a new, up-and-coming language. But surprisingly, R has been around since 1993. Why the sudden uptick in popularity? The somewhat obvious answer seems to be the emergence of data science as a career and field of study. But the underpinnings of data science have been around for many decades. Statistics, linear algebra, operations research, artificial intelligence and machine learning all contribute parts to the tools that a modern data scientist uses. R, more than most languages, has been built to make most of these tools only a single function call away.

That's why I'm excited that Jared has chosen to revisit his bestselling first edition and provide us with this updated second edition that brings in many of the recent innovations in the R community. R is indispensable for many data science tasks. Many algorithms useful for prediction and analysis can be accessed through only a few lines of code, which makes it a great fit for solving modern data challenges. Data science as a field isn't just about math and statistics, and it isn't just about programming and infrastructure. This book provides a well-balanced introduction to the power and expressiveness of R that is aimed at a general audience.

I can't think of a better author to provide an introduction to R than Jared Lander. Jared and I first met through the NYC machine learning community in late 2009. Back then, the NYC data community was small enough to fit in a single conference room, and many of the other data meetups had yet to be formed. Over the last seven years Jared has been at the forefront of the emerging data science profession.

Through running the Open Statistical Programming Meetup, speaking at events, and teaching a course on R at Columbia University, Jared has helped grow the community by educating programmers, data scientists, journalists and statisticians alike. Jared's expertise isn't limited to teaching. As an everyday practitioner he puts these tools to use while consulting for clients big and small. In the time since the first edition of this book was published Jared has continued to do great work in the R community: from organizing the New York R Conference, to speaking at many meetups and conferences, to evaluating the 2016 NFL Draft with R.

This book provides both an introduction to programming in R and the various statistical methods and tools an everyday R programmer uses. This second edition adds new material, making it current with the latest in the R community. This includes sections on data munging with libraries from the Tidyverse, as well as new chapters on RMarkdown, Shiny and others. Examples use publicly available datasets that Jared has helpfully cleaned and made accessible through his Web site. By using real data and setting up interesting problems, this book stays engaging to the end.

—*Paul Dix*
Series Editor

Preface

With the increasing prevalence of data in our daily lives, new and better tools are needed to analyze the deluge. Traditionally there have been two ends of the spectrum: lightweight, individual analysis using tools like Excel or SPSS, and heavy duty, high-performance analysis built with C++ and the like. With the increasing strength of personal computers grew a middle ground that was both interactive and robust. Analysis done by an individual on his or her own computer in an exploratory fashion could quickly be transformed into something destined for a server, underpinning advanced business processes. This area is the domain of R, Python and other scripted languages.

R, invented by Robert Gentleman and Ross Ihaka of the University of Auckland in 1993, grew out of S, which was invented by John Chambers at Bell Labs. It is a high-level language that was originally intended to be run interactively, where the user runs a command, gets a result and then runs another command. It has since evolved into a language that can also be embedded in systems and tackle complex problems.

In addition to transforming and analyzing data, R can produce amazing graphics and reports with ease. It is now being used as a full stack for data analysis, extracting and transforming data, fitting models, drawing inferences and making predictions, plotting and reporting results.

R's popularity has skyrocketed since the late 2000s as it has stepped out of academia and into banking, marketing, pharmaceuticals, politics, genomics and many other fields. Its new users are often shifting from low-level, compiled languages like C++, other statistical packages such as SAS or SPSS and from the 800-pound gorilla, Excel. This time period also saw a rapid surge in the number of add-on packages, libraries of prewritten code that extend R's functionality.

While R can sometimes be intimidating to beginners, especially for those without programming experience, I find that programming analysis, instead of pointing and clicking, soon becomes much easier, more convenient and more reliable. It is my goal to make that learning process easier and quicker.

This book lays out information in a way I wish I were taught when learning R in graduate school. Coming full circle, the content of this book was developed in conjunction with the data science course I teach at Columbia University. It is not meant to cover every minute detail of R but rather the 20% of functionality needed to accomplish 80% of the work. The content is organized into self-contained chapters as follows.

The second edition has been updated to cover many tools that have been developed or improved since the publication of the first edition. Primary among the new additions are **dplyr**, **tidyr** and **purrr** from the Tidyverse for munging data. Model fitting gained more attention with discussion of boosted trees and **caret** for parameter tuning. The **knitr** chapter was split in two, with one covering **knitr** and LaTeX and the other devoted to RMarkdown, which has been significantly improved in the past few years, including the creation of **htmlwidgets** that allow for the inclusion of JavaScript into documents. An entire chapter is dedicated to Shiny, a new tool for creating interactive Web-based dashboards in R. The chapter on writing R packages has been updated to include code testing, and the chapter on reading data has been updated to cover new ways of reading data, including using **readr**, **readxl** and **jsonlite**. The new content reflects many of the new practices in the R community.

Chapter 1, "Getting R," covers where to download R and how to install it. This deals with the various operating systems and 32-bit versus 64-bit versions. It also gives advice on where to install R.

Chapter 2, "The R Environment," provides an overview of using R, particularly from within RStudio. RStudio projects and Git integration are covered, as is customizing and navigating RStudio.

Chapter 3, "Packages," is concerned with how to locate, install and load R packages.

Chapter 4, "Basics of R," is about using R for math. Variable types such as `numeric`, `character` and `Date` are detailed as are `vectors`. There is a brief introduction to calling functions and finding documentation on functions.

Chapter 5, "Advanced Data Structures," is about the most powerful and commonly used data structure, `data.frames`, along with `matrices` and `lists`, are introduced.

Chapter 6, "Reading Data into R," is about getting data into R. Before data can be analyzed, it must be read into R. There are numerous ways to ingest data, including reading from CSVs and databases.

Chapter 7, "Statistical Graphics," makes it clear why graphics are a crucial part of preliminary data analysis and communicating results. R can make beautiful plots using its powerful plotting utilities. Base graphics and **ggplot2** are introduced and detailed here.

Chapter 8, "Writing R Functions," shows that repeatable analysis is often made easier with user defined functions. The structure, arguments and return rules are discussed.

Chapter 9, "Control Statements," covers controlling the flow of programs using **if**, **ifelse** and complex checks.

Chapter 10, "Loops, the Un-R Way to Iterate," introduces iterating using **for** and **while** loops. While these are generally discouraged, they are important to know.

Chapter 11, "Group Manipulations," provides a better alternative to loops—vectorization. Vectorization does not quite iterate through data so much as operate on all elements at once. This is more efficient and is primarily performed with the **apply** family of functions and **plyr** package.

Chapter 12, "Faster Group Manipulation with `dplyr`," covers the next evolution in group manipulation, **dplyr**. This new package has been optimized to work with `data.frames` and takes advantage of pipes for efficient coding that is easier to read.

Chapter 13, "Iterating with `purrr`," provides another alternative to loops with `purrr`, for iterating over `lists` and `vectors`. This represents a return to the functional roots of R.

Chapter 14, "Data Reshaping," is about the fact that combining multiple datasets, whether by stacking or joining, is commonly necessary as is changing the shape of data. The **plyr** and **reshape2** packages offer good functions for accomplishing this in addition to base tools such as **rbind**, **cbind** and **merge**.

Chapter 15, "Reshaping Data in the Tidyverse," showcases another example of package evolution as **dplyr** and **tidyr** replace **plyr** and **reshape2** for combining, reshaping and joining data.

Chapter 16, "Manipulating Strings," is about text. Most people do not associate character data with statistics, but it is an important form of data. R provides numerous facilities for working with strings, including combining them and extracting information from within. Regular expressions are also detailed.

Chapter 17, "Probability Distributions," provides a thorough look at the normal, binomial and Poisson distributions. The formulas and functions for many distributions are noted.

Chapter 18, "Basic Statistics," covers the first statistics most people are taught, such as mean, standard deviation and t-tests.

Chapter 19, "Linear Models," extensively details the most powerful and common tool in statistics—linear models.

Chapter 20, "Generalized Linear Models," shows how linear models are extended to include logistic and Poisson regression. Survival analysis is also covered.

Chapter 21, "Model Diagnostics," establishes the methods for determining the quality of models and variable selection using residuals, AIC, cross-validation, the bootstrap and stepwise variable selection.

Chapter 22, "Regularization and Shrinkage," covers prevention of overfitting using the Elastic Net and Bayesian methods.

Chapter 23, "Nonlinear Models," covers those cases where linear models are inappropriate and nonlinear models are a good solution. Nonlinear least squares, splines, generalized additive models, decision trees, boosted trees and random forests are discussed.

Chapter 24, "Time Series and Autocorrelation," covers methods for the analysis of univariate and multivariate time series data.

Chapter 25, "Clustering," shows how clustering, the grouping of data, is accomplished by various methods such as K-means and hierarchical clustering.

Chapter 26, "Model Fitting with Caret," introduces the **caret** package for automatic model tuning. The package also provides a uniform interface for hundreds of models, easing the analysis process.

Chapter 27, "Reproducibility and Reports with `knitr`," gets into integrating R code and results into reports from within R. This is made easy with **knitr** and LaTeX.

Chapter 28, "Rich Documents with RMarkdown," showcases how to generate reproducible reports, slide shows and Web pages from within R with RMarkdown. Interactivity is accomplished using **htmlwidgets** such as **leaflet** and **dygraphs**.

Chapter 29, "Interactive Dashboards with Shiny," introduces interactive dashboards using Shiny which can generate Web-based dashboards with the full power of R as a backend.

Chapter 30, "Building R Packages," is about how R packages are great for portable, reusable code. Building these packages has been made incredibly easy with the advent of **devtools** and **Rcpp**.

Appendix A, "Real-Life Resources," is a listing of our favorite resources for learning more about R and interacting with the community.

Appendix B, "Glossary," is a glossary of terms used throughout this book.

A good deal of the text in this book is either R code or the results of running code. Code and results are most often in a separate block of text and set in a distinctive font, as shown in the following example. The different parts of code also have different colors. Lines of code start with >, and if code is continued from one line to another, the continued line begins with +.

```
> # this is a comment
>
> # now basic math
> 10 * 10

[1] 100

> # calling a function
> sqrt(4)

[1] 2
```

Certain Kindle devices do not display color, so the digital edition of this book will be viewed in grayscale on those devices.

There are occasions where code is shown inline and looks like `sqrt(4)`.

In the few places where math is necessary, the equations are indented from the margin and are numbered.

$$e^{i\pi} + 1 = 0 \tag{1}$$

Within equations, normal variables appear as italic text (x), vectors are bold lowercase letters ($\mathbf{x}$) and matrices are bold uppercase letters ($\mathbf{X}$). Greek letters, such as α and β, follow the same convention.

Function names are written as **join** and package names as **plyr**. Objects generated in code that are referenced in text are written as `object1`.

Learning R is a gratifying experience that makes life so much easier for so many tasks. I hope you enjoy learning with me.

Register your copy of *R for Everyone*, Second Edition, at informit.com/register for convenient access to downloads, updates, and corrections as they become available (you must log-in or create a new account). Enter the product ISBN (9780134546926) and click Submit. Once the process is complete, you will find any available bonus content under "Registered Products." If you would like to be notified of exclusive offers on new editions and updates, please check the box to receive eMail from us.

Acknowledgments

Acknowledgments for the Second Edition

First and foremost, I am most appreciative of my wife-to-be, Rebecca Martin. Writing this second edition meant playing in R for hours at a time, which is fun on its own, but was greatly enhanced by her presence. She is amazing in so many ways, not least of which is that she uses R. She even indulged my delusions of writing like Orwell and Kipling while cruising up the Irrawaddy on the road to Mandalay.

As before, my family has always supported me in so many ways. My parents, Gail and Howard Lander, encouraged me on this path to math and data science. When this book was first published they said it would be too boring for them to enjoy and have since kept their promise of never reading it. It sits similarly unread, yet proudly displayed, in the homes of my grandmother and all my aunts and uncles. My sister and brother-in-law, Aimee and Eric Schechterman, always humor my antics with their kids, Noah and Lila, whom I am beginning to teach to program.

There are many people in the open-source community, particularly those who attend and contribute to the New York Open Statistical Computing Meetup, whose work and encouragement have been great motivators. Principal among them is Drew Conway, the early leader of the meetup who provided a place for my love of R to grow and eventually turned the meetup over to my stewardship. The friendship of Paul Puglia, Saar Golde, Jay Emerson, Adam Hogan, John Mount, Nina Zumel, Kirk Mettler, Max Kuhn, Bryan Lewis, David Smith, Dirk Eddelbuettel, JD Long, Ramnath Vaidyanathan, Hilary Parker and David Robinson has made the experience incredibly entertaining. I even enjoy my time with Python users Wes McKinney, Ben Lerner and James Powell.

The Work-Bench family, my coorganizers for the New York R Conference, are fantastic people. Jon Lehr, Jess Lin, Stephanie Manning, Kelley Mak, Vipin Chamakkala, Laurel Woerner, Michael Yamnitsky and Mickey Graham (despite his obsession with the Oxford comma) are great to be around.

As my business has grown in recent years, many people have helped, either as employees and clients or by providing valuable advice. Among these are Joseph Sherman, Jeff Horner, Lee Medoff, Jeroen Janssens, Jonathan Hersh, Matt Sheridan, Omar De La Cruz Cabrera, Benjamin De Groot, Vinny Saulys, Rick Spielman, Scott Kuhn, Mike Band, Nate Shea-Han, Greg Fuller, Mark Barry and Lenn Robbins. The teachings of Andrew Gelman, David Madigan and Richard Garfield have stayed with me far beyond the university.

This book is largely possible due to the tireless efforts of the RStudio team. The efforts of JJ Allaire, Winston Chang, Joe Cheng, Garrett Grolemund, Hadley Wickham and Yihui Xie provide the tools that make this book, and much of what we do in R technically feasible. Tareef Kawaf, Pete Knast, Sean Lopp, Roger Oberg, Joe Rickert,

Nathan Stephens, Jim Clemens, Anne Carome, Bill Carney and many others support and spur the growth of the R community.

The material for this book was largely drawn from the class I taught at Columbia University with Rachel Schutt, Introduction to Data Science. The students in that class largely shaped the material and tone of the book, including how it was presented. Vivian Peng, Dan Chen, Michael Piccirilli, Adam Obeng, Eurry Kim and Kaz Sakamoto all inspired my writing.

Numerous people helped with the writing, validating, proofreading and editing of this book. Michael Beigelmacher ensured the code works while Chris Zahn did the same with the prose. Paul Dix introduced me to Pearson, enabling the whole process. My editor, Debra Williams Cauley, has now gone through two editions and three videos of my work patterns and is the driving force that has made the book succeed. Without her, this would not exist.

This second edition is built upon all the help of those mentioned in the acknowledgments from my original book, who are still very dear to me.

Acknowledgments for the First Edition

To start, I must thank my mother, Gail Lander, for encouraging me to become a math major. Without that I would never have followed the path that led me to statistics and data science. In a similar vein I have to thank my father, Howard Lander, for paying all those tuition bills. He has been a valuable source of advice and guidance throughout my life and someone I have aspired to emulate in many ways. While they both insist they do not understand what I do, they love that I do it and have helped me all along the way. Staying with family, I should thank my sister and brother-in-law, Aimee and Eric Schechterman, for letting me teach math to Noah, their five-year-old son.

There are many teachers that have helped shape me over the years. The first is Rochelle Lecke who tutored me in middle school math even when my teacher told me I did not have worthwhile math skills.

Then there is Beth Edmondson, my precalc teacher at Princeton Day School. After wasting the first half of high school as a mediocre student she told me I had "some nerve signing up for next year's AP Calc given my grades." She agreed to let me in AP Calc if I went from a C to an A+ in her class, never thinking I stood a chance. Three months later she stood in stood in disbelief as I not only got the A+ but turned around my entire academic career and became an excellent student. She changed my life. Without her I do not know where I would be. I am forever grateful that she was my teacher.

For the first two years at Muhlenberg College I was determined to be a Business and Communications major yet took math classes because they just came naturally to me. Penny Dunham, Bill Dunham and Linda McGuire all convinced me to become a math major, a decision that has certainly improved my life. Greg Cicconetti gave me my first glimpse of rigorous statistics, my first research opportunity and planted the idea in my head that I should go to grad school for statistics. Fortunately, I eventually listened to him.

My time at Columbia University was spent surrounded by brilliant minds in statistics and programming. David Madigan opened my eyes to modern machine learning and Bodhi Sen got thinking about statistical programming. I had the privilege to do research with Andrew Gelman whose insights have been immeasurably important to me. Richard

Garfield showed me how to use statistics to help people in disaster and war zones. His advice and friendship over the years have been dear to me. Jingchen Liu allowed and encouraged me to write my thesis on New York City pizza,[1] which has brought me an inordinate amount of attention.

While at Columbia University I met my good friend—and one time TA—Ivor Cribben who filled in so many gaps in my knowledge. Through him I met Rachel Schutt, who was a source of great advice in grad school and who I am now honored to teach with at Columbia University.

Grad school might never have happened without the encouragement and support of Shanna Lee. She took good care of me and helped maintain my sanity while I was incredibly overcommited to two jobs, classes and Columbia University's hockey team. I am not sure I would have made it through without her.

Steve Czetty gave me my first job in analytics at Sky IT Group and taught me about databases while letting me run wild, programming anything I wanted. This sparked my interest in statistics and data. Joe DeSiena, Philip DuPlessis and Ed Bobrin at the Bardess Group are some of the finest people I have ever had the pleasure to work with and the work they gave me helped put me through grad school. I am proud to be able to do statistics work with them to this day. Mike Minelli, Rich Kittler, Mark Barry, David Smith, Joseph Rickert, Norman Nie, James Peruvankal, Neera Talbert and Dave Rich at Revolution Analytics let me do one of the best jobs I could possibly imagine: Explaining to people in industry why they should be using R. Kirk Mettler, Richard Schultz, Bryan Lewis and Jim Winfield at Big Computing encourage me to have fun, tackling interesting problems in R. Vinny Saulys and Saar Golde were a lot of fun to work with at Goldman Sachs and also very educational.

Throughout the course of writing this book so many people helped, or rather put up with, me. First and foremost is Yin Cheung who saw all the stress I constantly felt. There were many nights and days ruined when I had to work or write and she suffered through those.

My editor, Debra Williams, knew just how to handle me when I was churning out pages, and more frequently, when I was letting time slip by. Her guiding hand has been invaluable. Paul Dix, the series editor and friend of mine, is the person who suggested I write this book, so without him none of this would have happened. Thanks to Caroline Senay and Andrea Fox I realized quite how many mistakes I made as a writer. Without them, this book would not be nearly as well put together. Robert Mauriello's technical review was incredibly useful in honing the presentation of the included material. The folks at RStudio, particularly JJ Allaire and Josh Paulson, make an amazing product, which made the writing process far easier than it would have been otherwise. Yihui Xie, the author of the `knitr` package, put up with a long series of personal feature requests that I desperately needed to write this book. His software, and his speed at implementing my requests, allowed me to make this book look and feel just the way I felt was right.

Numerous people have looked over parts of this book and given me valuable feedback, including some of those already mentioned. Others who have greatly helped me are

1. `http://slice.seriouseats.com/archives/2010/03/the-moneyball-of-pizza-statistician-uses-statistics-to-find-nyc-best-pizza.html`

Chris Bethel, Dirk Eddelbuettel, Ramnath Vaidyanathan, Eran Bellin, Avi Fisher, Brian Ezra, Paul Puglia, Nicholas Galasinao, Aaron Schumaker, Adam Hogan, Jeffrey Arnold and John Houston.

Last fall was my first time teaching and I am thankful to the students from the Fall 2012 Introduction to Data Science class at Columbia University for being the guinea pigs for the material that ultimately ended up in this book.

There are many people who have helped me along the way and I am grateful to them all.

About the Author

Jared P. Lander is the Chief Data Scientist of Lander Analytics, a New York-based data science firm that specializes in statistical consulting and training services, the Organizer of the New York Open Statistical Programming Meetup—the world's largest R meetup—and the New York R Conference and an Adjunct Professor of Statistics at Columbia University. He is also a tour guide for Scott's Pizza Tours. With a masters from Columbia University in statistics and a bachelors from Muhlenberg College in mathematics, he has experience in both academic research and industry. Very active in the data community, Jared is a frequent speaker at conferences such as Strata and the MIT Sloan Sports Analytics Conference, universities and meetups around the world. His writings on statistics can be found at jaredlander.com, and his work has been featured in many outlets, in particular CBS and the *Wall Street Journal*.

1 Getting R

R is a wonderful tool for statistical analysis, visualization and reporting. Its usefulness is best seen in the wide variety of fields where it is used. We alone have used R for projects with banks, political campaigns, tech startups, food startups, international development and aid organizations, hospitals and real estate developers. Other areas where we have seen it used are online advertising, insurance, ecology, genetics and pharmaceuticals. R is used by statisticians with advanced machine learning training and by programmers familiar with other languages and also by people who are not necessarily trained in advanced data analysis but are tired of using Excel.

Before it can be used it needs to be downloaded and installed, a process that is no more complicated than installing any other program.

1.1 Downloading R

The first step in using R is getting it on the computer. Unlike with languages such as C++, R must be installed in order to run.[1] The program is easily obtainable from the Comprehensive R Archive Network (CRAN), the maintainer of R, at `http://cran.r-project.org/`. At the top of the page are links to download R for Windows, Mac OS X and Linux.

There are prebuilt installations available for Windows and Mac OS X, while those for Linux usually compile from source. Installing R on any of these platforms is just like installing any other program.

Windows users should click the link "Download R for Windows," then "base" and then "Download R 3.x.x for Windows"; the x's indicate the version of R. This changes periodically as improvements are made.

Similarly, Mac users should click "Download R for (Mac) OS X" and then "R-3.x.x.pkg"; again, the x's indicate the current version of R. This will also install both 32- and 64-bit versions.

Linux users should download R using their standard distribution mechanism whether that is apt-get (Ubuntu and Debian), yum (Red Hat), zypper (SUSE) or another source. This will also build and install R.

---

1. Technically C++ cannot be set up on its own without a compiler, so something would still need to be installed anyway.

1.2 R Version

As of this writing, R is at version 3.4.0 and has seen a lot of improvements since the first edition of this book when the version was 3.0.1. CRAN follows a one-year release cycle where each major version change increases the middle of the three numbers in the version. For instance, version 3.2.0 was released in 2015. In 2016 the version was incremented to 3.3.0 with 3.4.0 released in 2017. The last number in the version is for minor updates to the current major version.

Most R functionality is usually backward compatible with previous versions.

1.3 32-bit vs. 64-bit

The choice between using 32-bit and using 64-bit comes down to whether the computer supports 64-bit—most new machines do—and the size of the data to be worked with. The 64-bit versions can address arbitrarily large amounts of memory (or RAM), so it might as well be used.

This is especially important starting with version 3.0.0, as that adds support for 64-bit integers, meaning far greater amounts of data can be stored in R objects.

In the past, certain packages required the 32-bit version of R, but that is exceedingly rare these days. The only reason for installing the 32-bit version now is to support some legacy analysis or for use on a machine with a 32-bit processor such as Intel's low-power Atom chip.

1.4 Installing

Installing R on Windows and Mac is just like installing any other program.

1.4.1 Installing on Windows

Find the appropriate installer where it was downloaded. For Windows users, it will look like Figure 1.1.

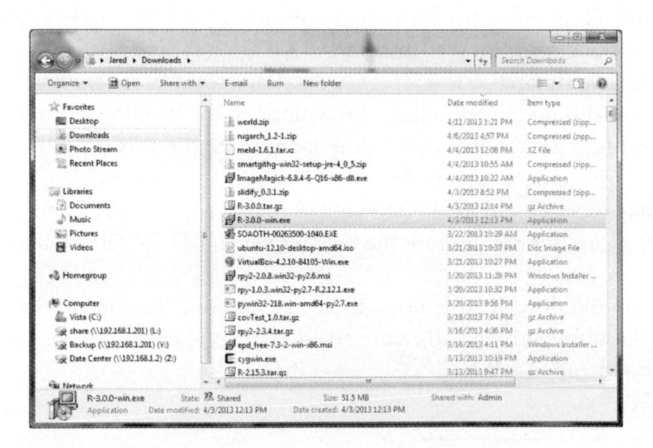

Figure 1.1 Location of R installer.

R should be installed using administrator privileges. This means right-clicking the installer and then selecting Run as Administrator. This brings up a prompt where the administrator password should be entered.

The first dialog, shown in Figure 1.2, offers a choice of language, defaulted at English. Choose the appropriate language and click OK.

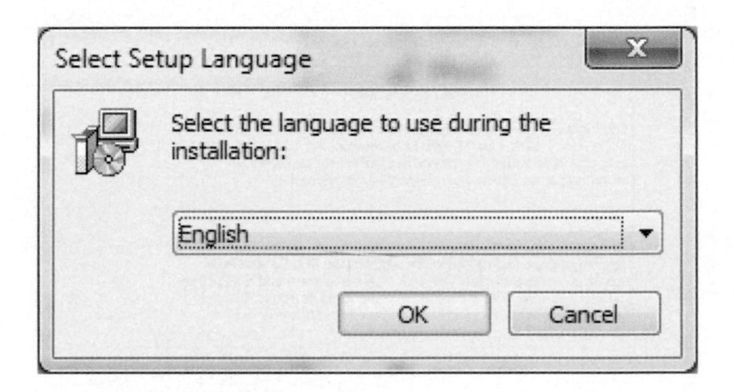

Figure 1.2 Language selection for Windows.

Next, the caution shown in Figure 1.3 recommends that all other programs be closed. This advice is rarely followed or necessary anymore, so clicking Next is appropriate.

Figure 1.3 With modern versions of Windows, this suggestion can be safely ignored.

The software license is then displayed, as in Figure 1.4. R cannot be used without agreeing to this (important) license, so the only recourse is to click Next.

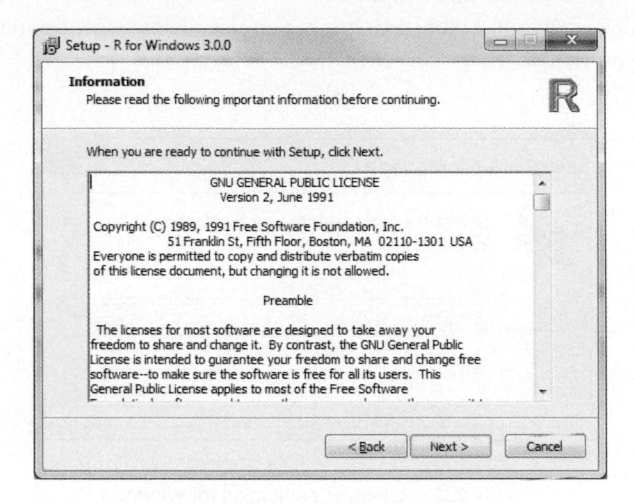

Figure 1.4 The license agreement must be acknowledged to use R.

The installer then asks for a destination location. Even though the official advice from CRAN is that R should be installed in a directory with no spaces in the name, half the time the default installation directory is `Program Files\R`, which causes trouble if we try to build packages that require compiled code such as C++ for FORTRAN. Figure 1.5 shows this dialog. It is important to choose a directory with no spaces, even if the default installation says otherwise.

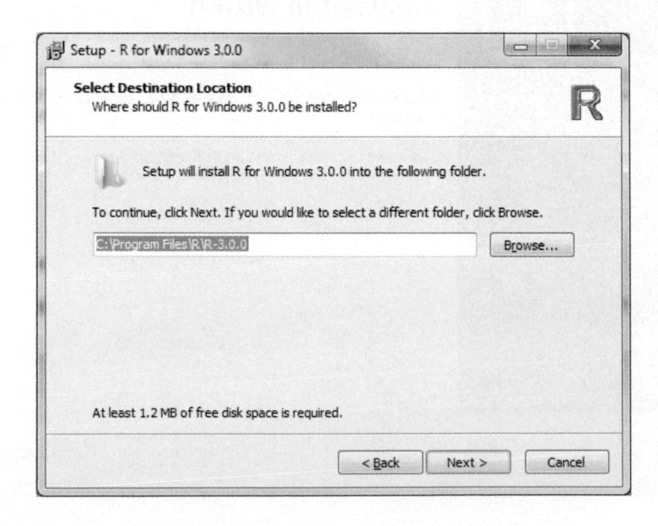

Figure 1.5 It is important to choose a destination folder with no spaces in the name.

If that is the case, click the Browse button to bring up folder options like the ones shown in Figure 1.6.

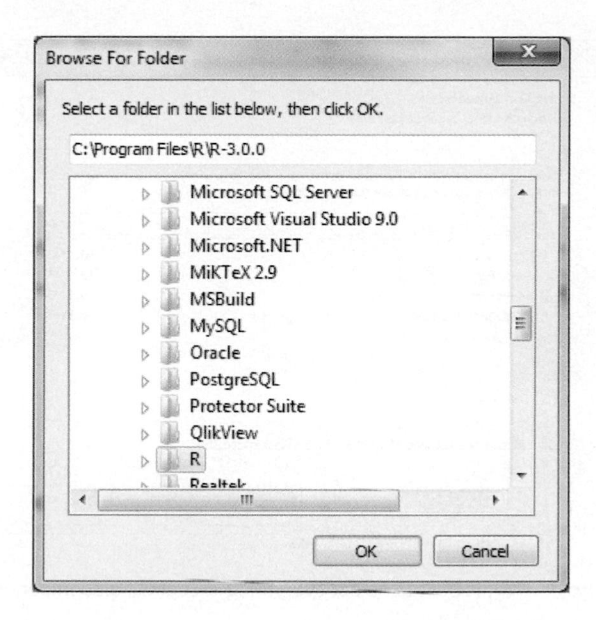

Figure 1.6 This dialog is used to choose the destination folder.

It is best to choose a destination folder that is on the `C:` drive (or another hard disk drive) or inside `My Documents`, which despite that user-friendly name is actually located at `C:\Users\UserName\Documents`, which contains no spaces. Figure 1.7 shows a proper destination for the installation.

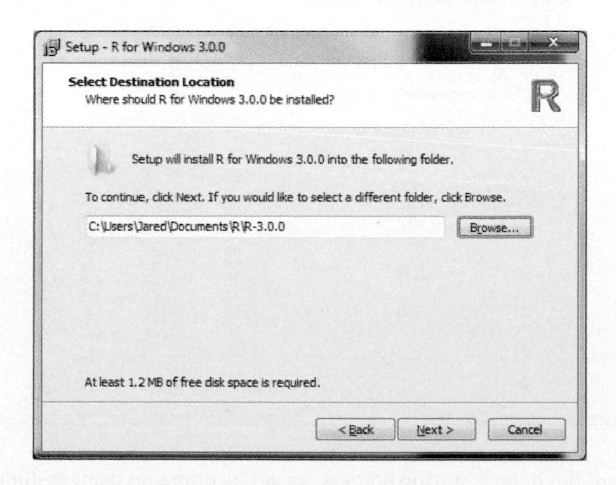

Figure 1.7 This is a proper destination, with no spaces in the name.

Next, Figure 1.8, shows a list of components to install. Unless there is a specific need for 32-bit files, that option can be unchecked. Everything else should be selected.

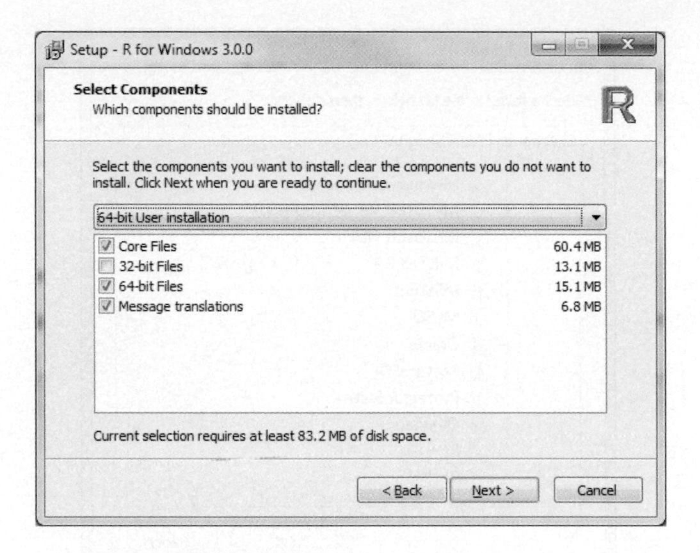

Figure 1.8 It is best to select everything except 32-bit components.

The startup options should be left at the default, No, as in Figure 1.9, because there are not a lot of options and we recommend using RStudio as the front end anyway.

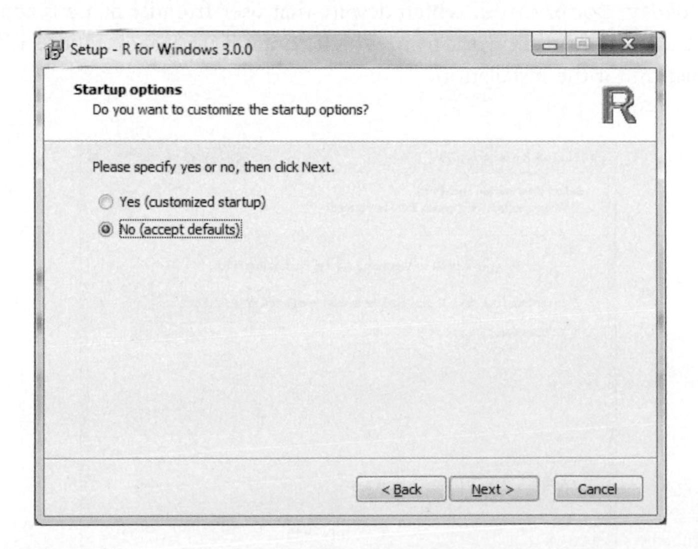

Figure 1.9 Accept the default startup options, as we recommend using RStudio as the front end, and these will not be important.

Next, choose where to put the start menu shortcuts. We recommend simply using R and putting every version in there as shown in Figure 1.10.

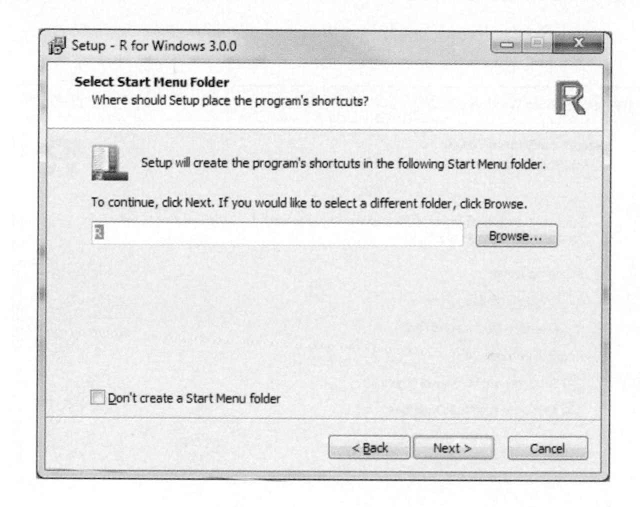

Figure 1.10 Choose the Start Menu folder where the shortcuts will be installed.

We have many versions of R, all inside the same Start Menu folder, which allows code to be tested in different versions. This is illustrated in Figure 1.11.

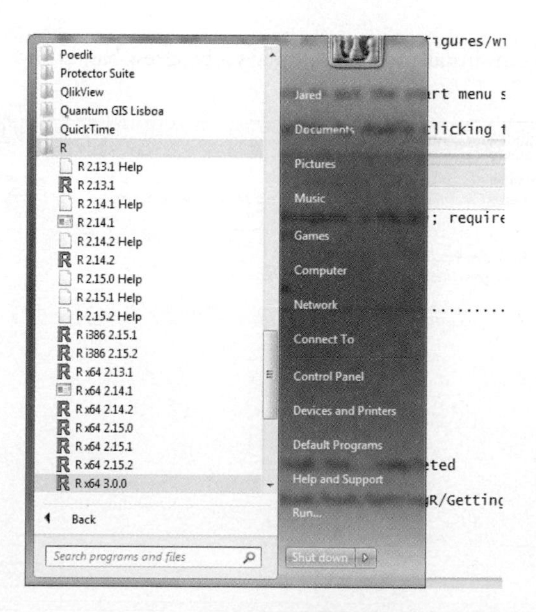

Figure 1.11 We have multiple versions of R installed to allow development and testing with different versions.

The last option is choosing whether to complete some additional tasks such as creating a desktop icon (not too useful if using RStudio). We highly recommend saving the version number in the registry and associating R with `RData` files. These options are shown in Figure 1.12.

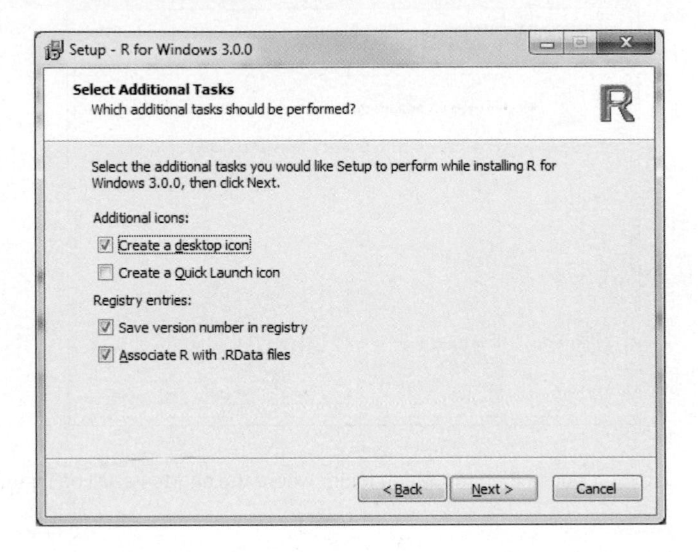

Figure 1.12 We recommend saving the version number in the registry and associating R with `RData` files.

Clicking Next begins installation and displays a progress bar, as shown in Figure 1.13.

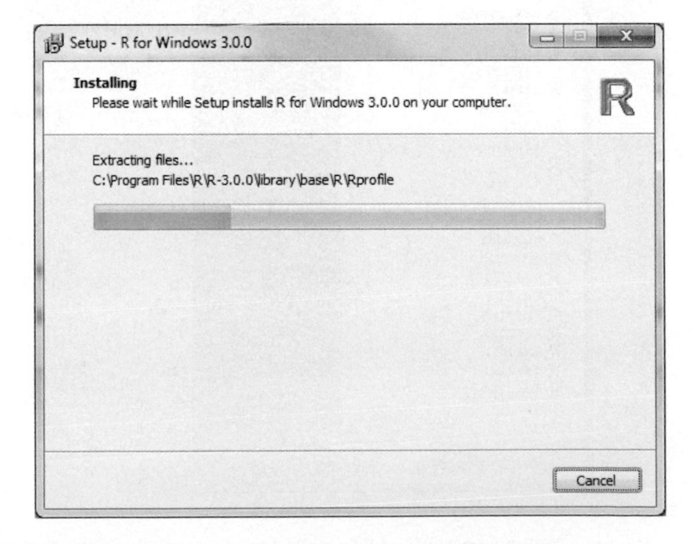

Figure 1.13 A progress bar is displayed during installation.

The last step, shown in Figure 1.14, is to click Finish, confirming the installation is complete.

Figure 1.14 Confirmation that installation is complete.

1.4.2 Installing on Mac OS X

Find the appropriate installer, which ends in `.pkg`, and launch it by double-clicking. This brings up the introduction, shown in Figure 1.15. Click Continue to begin the installation process.

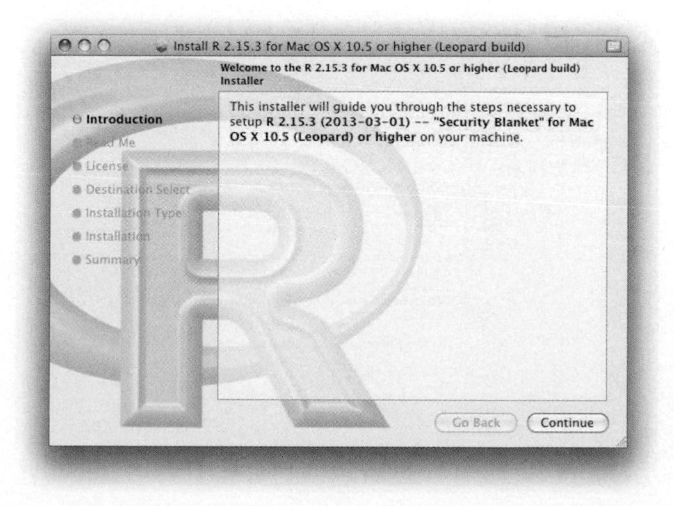

Figure 1.15 Introductory screen for installation on a Mac.

This brings up some information about the version of R being installed. There is nothing to do except click Continue, as shown in Figure 1.16.

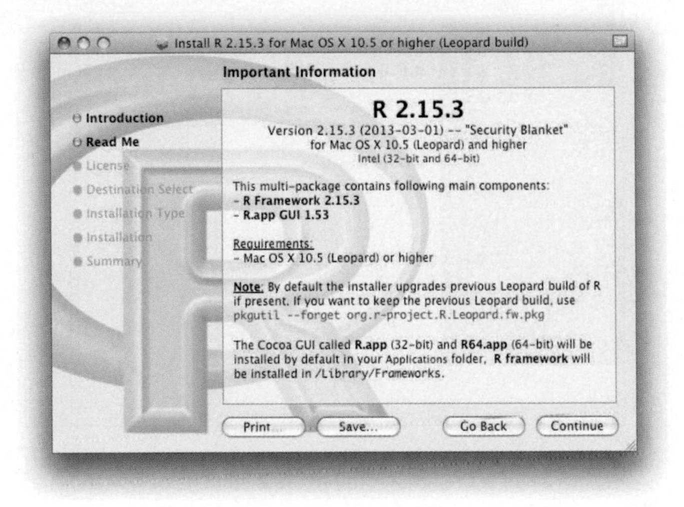

Figure 1.16 Version selection.

Then the license information is displayed, as in Figure 1.17. Click Continue to proceed, the only viable option in order to use R.

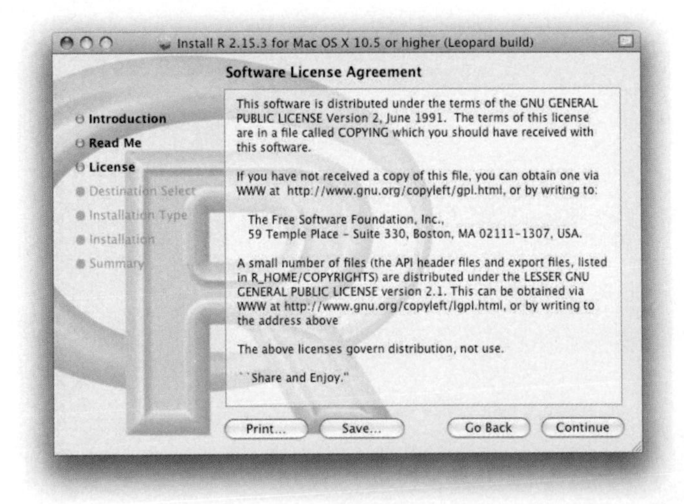

Figure 1.17 The license agreement, which must be acknowledged to use R.

Click Agree to confirm that the license is agreed to, which is mandatory to use R, as is evidenced in Figure 1.18.

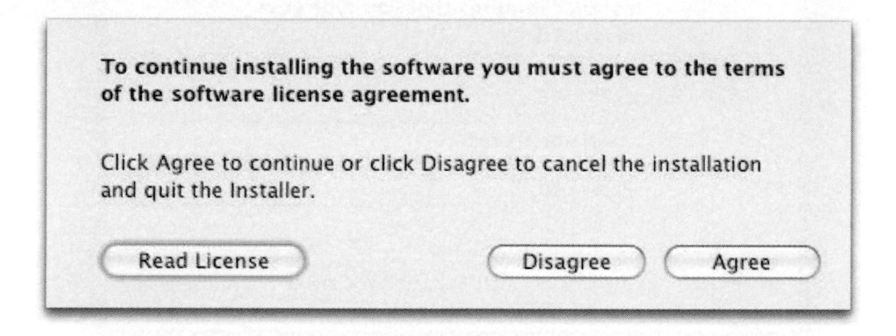

Figure 1.18 The license agreement must also be agreed to.

To install R for all users, click Install; otherwise, click Change Install Location to pick a different location. This is shown in Figure 1.19.

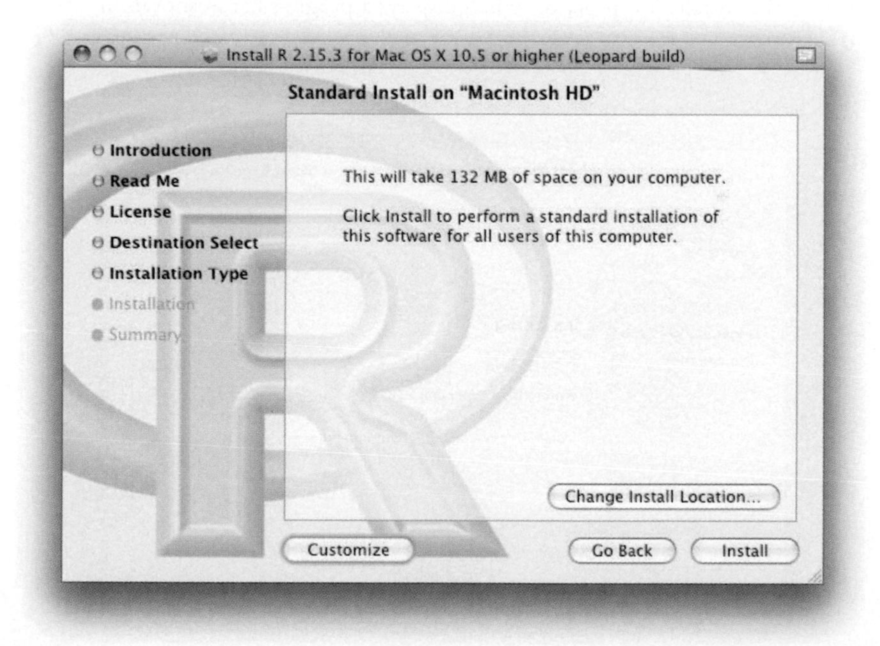

Figure 1.19 By default R is installed for all users, although there is the option to choose a specific location.

If prompted, enter the necessary password as shown in Figure 1.20.

Figure 1.20 The administrator password might be required for installation.

This starts the installation process, which displays a progress bar as shown in Figure 1.21.

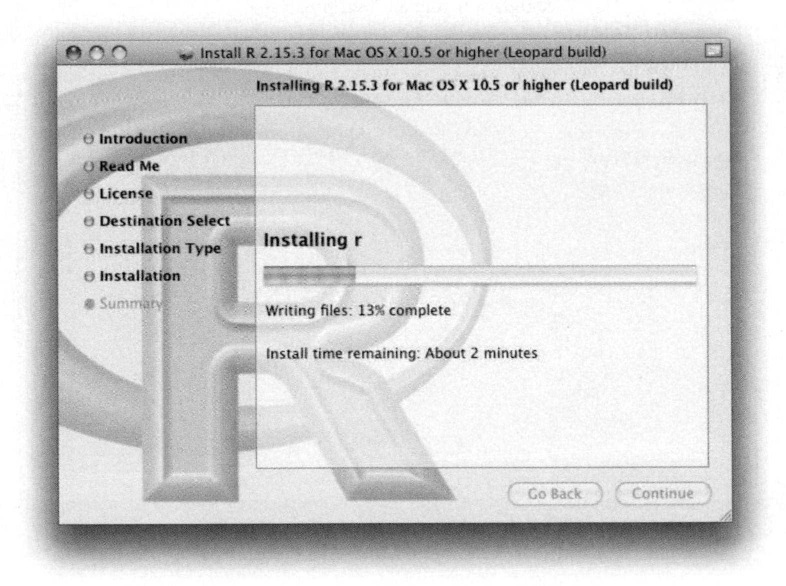

Figure 1.21 A progress bar is displayed during installation.

When done, the installer signals success as Figure 1.22 shows. Click Close to finish the installation.

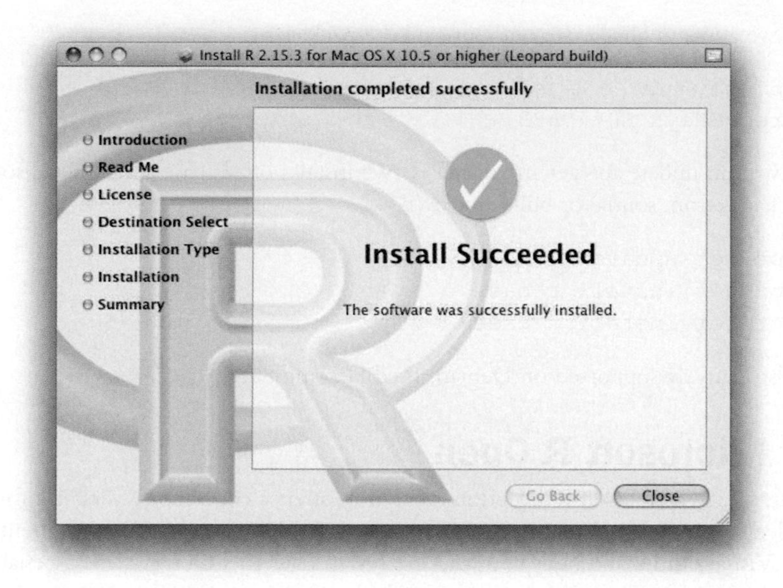

Figure 1.22 This signals a successful installation.

1.4.3 Installing on Linux

Retrieving R from its standard distribution mechanism will download, build and install R in one step. We will focus on Ubuntu, which uses apt-get.

The first step is to edit the file `/etc/apt/sources.list`, which lists package sources. Two pieces of information need to be included: the CRAN mirror and the version of Ubuntu or Debian.

Any CRAN mirror can be used, so we choose the RStudio mirror at `http://cran.rstudio.com/bin/linux/ubuntu`.

The supported versions of Ubuntu, as of early 2017, are Yakkety Yak (16.10), Xenial Xerus (16.04), Wily Werewolf (15.10), Vivid Vervet (15.04), Trusty Tahr (14.04; LTS) and Precise Pangolin (12.04; LTS).[2]

To install R from the RStudio CRAN mirror on Ubuntu 16.04, we need to add the line

```
deb http://cran.rstudio.com/bin/linux/ubuntu xenial/
```

to `/etc/apt/sources.list`. This can be done manually or by running the following command in the terminal.

```
sudo sh -c \
'echo "deb http://cran.rstudio.com/bin/linux/ubuntu xenial/" \
>> /etc/apt/sources.list'
```

2. According to `https://cran.r-project.org/bin/linux/ubuntu/README`

Then we add a public key to authenticate the packages.

```
sudo apt-key adv --keyserver keyserver.ubuntu.com
    --recv-keys E084DAB9
```

Now we can update apt-get and install R. We install both R base and R devel so we can build packages from source or build our own.

```
sudo apt-get update
sudo apt-get install r-base
sudo apt-get install r-base-dev
```

R is also natively supported on Debian, Red Hat and SuSE.

1.5 Microsoft R Open

Microsoft, which purchased Revolution Analytics, offers a community version of its build of R, called Microsoft R Open, featuring an Integrated Development Environment based on Visual Studio and built with the Intel Matrix Kernel Library (MKL), enabling faster matrix computations. It is available for free at `https://mran.microsoft.com/download/`. They also offer a paid version—Microsoft R Server—that provides specialized algorithms to work on large data and greater operability with Microsoft SQL Server and Hadoop. More information is available at `https://www.microsoft.com/en-us/server-cloud/products/r-server/`.

1.6 Conclusion

At this point R is fully usable and comes with a crude GUI. However, it is best to install RStudio and use its interface, which is detailed in Section 2.2. The process involves downloading and launching an installer, just as with any other program.

2

The R Environment

Now that R is downloaded and installed, it is time to get familiar with how to use R. The basic R interface on Windows is fairly Spartan, as seen in Figure 2.1. The Mac interface (Figure 2.2) has some extra features and Linux has far fewer, being just a terminal.

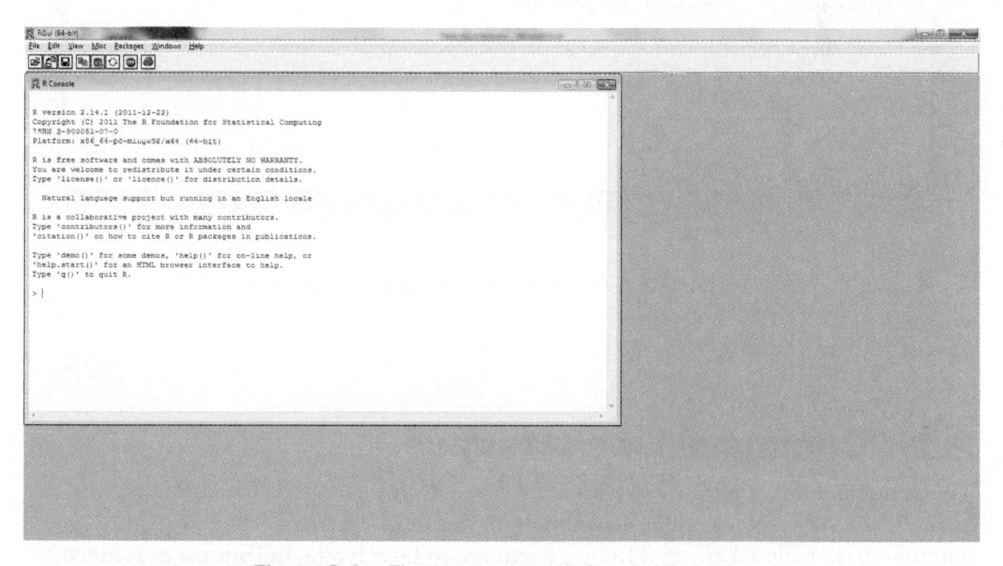

Figure 2.1 The standard R interface in Windows.

Unlike other languages, R is very interactive. That is, results can be seen one command at a time. Languages such as C++ require that an entire section of code be written, compiled and run in order to see results. The state of objects and results can be seen at any point in R. This interactivity is one of the most amazing aspects of working with R.

There have been numerous Integrated Development Environments (IDEs) built for R. For the purposes of this book we will assume that RStudio is being used, which is discussed in Section 2.2.

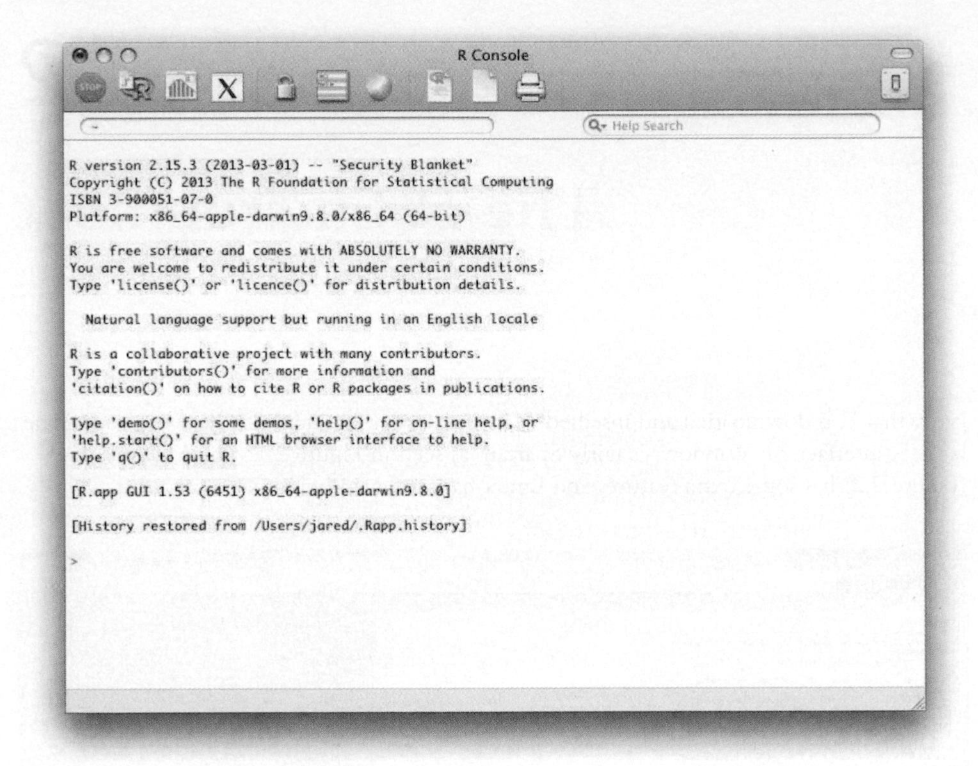

Figure 2.2 The standard R interface on Mac OS X.

2.1 Command Line Interface

The command line interface is what makes R so powerful, and also frustrating to learn. There have been attempts to build point-and-click interfaces for R, such as Rcmdr, but none have truly taken off. This is a testament to how typing in commands is much better than using a mouse. That might be hard to believe, especially for those coming from Excel, but over time it becomes easier and less error prone.

For instance, fitting a regression in Excel takes at least seven mouse clicks, often more: `Data >> Data Analysis >> Regression >> OK >> Input Y Range >> Input X Range >> OK`. Then it may need to be done all over again to make one little tweak or because there are new data. Even harder is walking a colleague through those steps via email. In contrast, the same command is just one line in R, which can easily be repeated and copied and pasted. This may be hard to believe initially, but after some time the command line makes life much easier.

To run a command in R, type it into the console next to the > symbol and press the Enter key. Entries can be as simple as the number 2 or complex functions, such as those seen in Chapter 8.

To repeat a line of code, simply press the Up Arrow key and hit Enter again. All previous commands are saved and can be accessed by repeatedly using the Up and Down Arrow keys to cycle through them.

Interrupting a command is done with `Esc` in Windows and Mac and `Ctrl-C` in Linux.

Often when working on a large analysis it is good to have a file of the code used. Until a few years ago, the most common way to handle this was to use a text editor[1] such as Sublime Text or Notepad++ to write code and then copy and paste it into the R console. While this worked, it was sloppy and led to a lot of switching between programs. Thankfully, there is now RStudio, which is a game changer and detailed in Section 2.2.

2.2 RStudio

While there are a number of IDEs available, the best right now is RStudio, created by a team led by JJ Allaire whose previous products include ColdFusion and Windows Live Writer. It is available for Windows, Mac and Linux and looks identical in all of them. Even more impressive is the RStudio Server, which runs an R instance on a Linux server and enables the user to run commands through the standard RStudio interface in a Web browser. It works with any version of R (greater than 2.11.1), including Microsoft R Open and Microsoft R Server from Microsoft. RStudio has so many options that it can be a bit overwhelming. We cover some of the most useful or frequently used features.

RStudio is highly customizable, but the basic interface looks roughly like Figure 2.3. In this case the lower left pane is the R console, which can be used just like the standard R console. The upper left pane takes the place of a text editor but is far more powerful. The upper right pane holds information about the workspace, command history, files in the current folder and Git version control. The lower right pane displays plots, package information and help files.

There are a number of ways to send and execute commands from the editor to the console. To send one line, place the cursor at the desired line and press `Ctrl+Enter` (`Command+Enter` on Mac). To insert a selection, simply highlight the selection and press `Ctrl+Enter`. To run an entire file of code, press `Ctrl+Shift+S`.

When typing code, such as an object name or function name, hitting `Tab` will autocomplete the code. If more than one object or function matches the letters typed so far, a dialog will pop up giving the matching options, as shown in Figure 2.4.

Typing `Ctrl+1` moves the cursor to the text editor area and `Ctrl+2` moves it to the console. To move to the previous tab in the text editor, press `Ctrl+Alt+Left` on Windows, `Ctrl+PageUp` in Linux and `Ctrl+Option+Left` on Mac. To move to the next tab in the text editor, press `Ctrl+Alt+Right` in Windows, `Ctrl+PageDown` in Linux and `Ctrl+Option+Right` on Mac. On some Windows machines these shortcuts can cause the screen to rotate, so `Ctrl+F11` and `Ctrl+F12` also move between tabs as does `Ctrl+Alt+Left` and `Ctrl+Alt+Right`, though only in the desktop client. For an almost-complete list of shortcuts, click `Help >> Keyboard Shortcuts` or use the keyboard shortcut `Alt+Shift+K` on Windows and Linux and `Option+Shift+K` on

1. This means a programming text editor as opposed to a word processor such as Microsoft Word. A text editor preserves the structure of the text, whereas word processors may add formatting that makes it unsuitable for insertion into the console.

Mac. A more complete list is available at `https://support.rstudio.com/hc/en-us/articles/200711853-Keyboard-Shortcuts`.

Figure 2.3 The general layout of RStudio.

```
14                          Yes        1
15                          yes!       1
16                          yes.       1
17                          Yes.       1
18                         Yes.        1
> 5+2+37+77+1+1
[1] 123
> 130*2/8
[1] 32.5
> 150*2/8
[1] 37.5
> 35*20
[1] 700
> 2.75*8
[1] 22
> 200*4.55
[1] 910
> 437/29
[ eventID
> eventSize
[
>
[
>
[          "eventID" "rsvp"    "theRSVP"
>           10
> even
```

Figure 2.4 Object Name Autocomplete in RStudio.

2.2.1 RStudio Projects

A primary feature of RStudio is projects. A project is a collection of files—and possibly data, results and graphs—that are all related to each other.[2] Each package even has its own working directory. This is a great way to keep organized.

The simplest way to start a new project is to click `File >> New Project`, as in Figure 2.5.

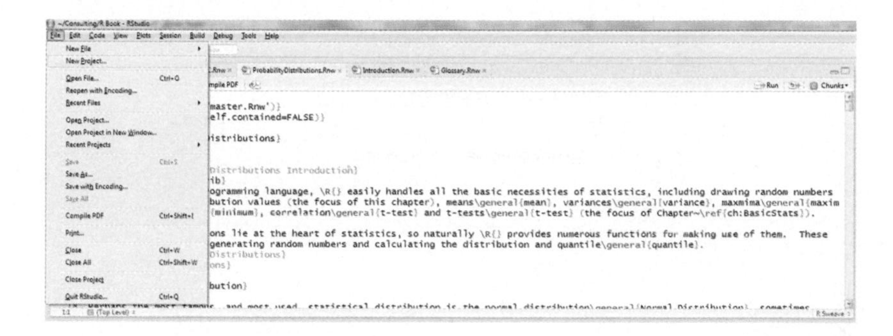

Figure 2.5 Clicking `File >> New Project` begins the project creation process.

Three options are available, shown in Figure 2.6: starting a new project in a new directory, associating a project with an existing directory or checking out a project from a version control repository such as Git or SVN[3]. In all three cases a `.Rproj` file is put into the resulting directory and keeps track of the project.

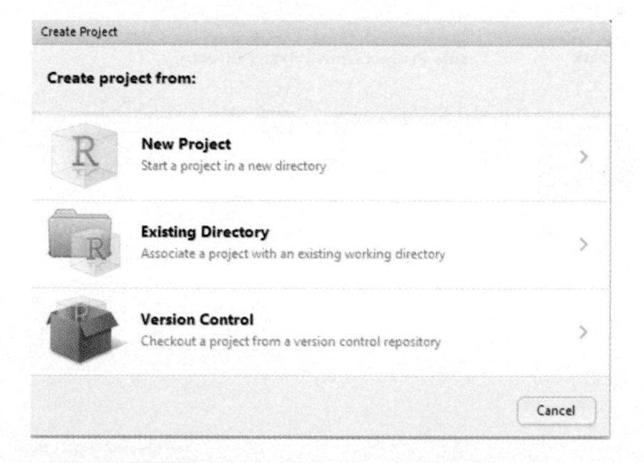

Figure 2.6 Three options are available to start a new project: a new directory, associating a project with an existing directory or checking out a project from a version control repository.

2. This is different from an R session, which is all the objects and work done in R and kept in memory for the current usage period, and usually resets upon restarting R.

3. Using version control requires that the version control program is installed on the computer.

Choosing to create a new directory brings up a dialog, shown in Figure 2.7, that requests a project name and where to create a new directory.

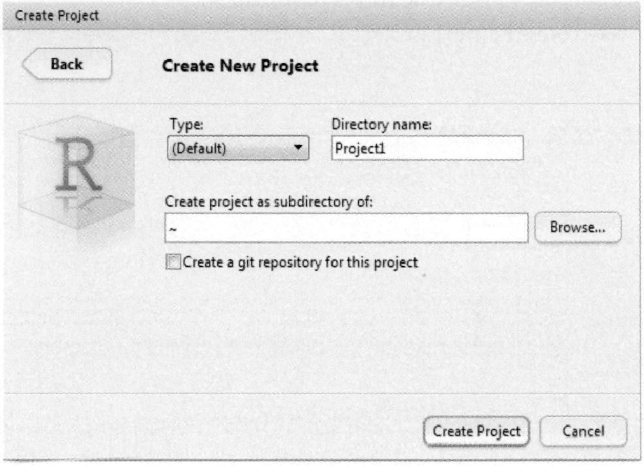

Figure 2.7 Dialog to choose the location of a new project directory.

Choosing an existing directory asks for the name of the directory, as shown in Figure 2.8.

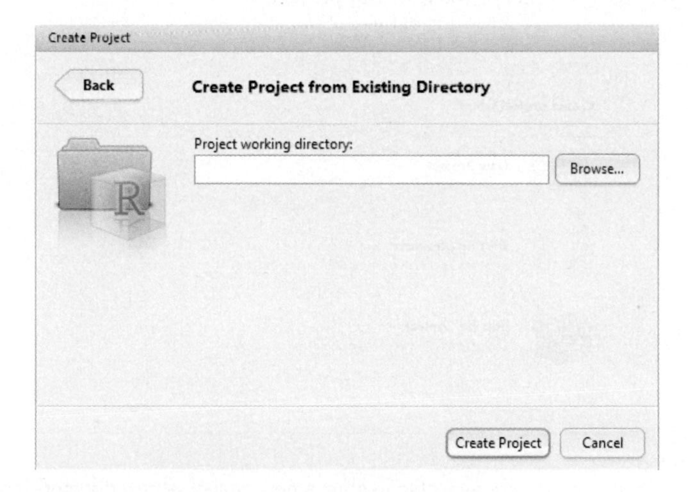

Figure 2.8 Dialog to choose an existing directory in which to start a project.

Choosing to use version control (we prefer Git) firsts asks whether to use Git or SVN as in Figure 2.9.

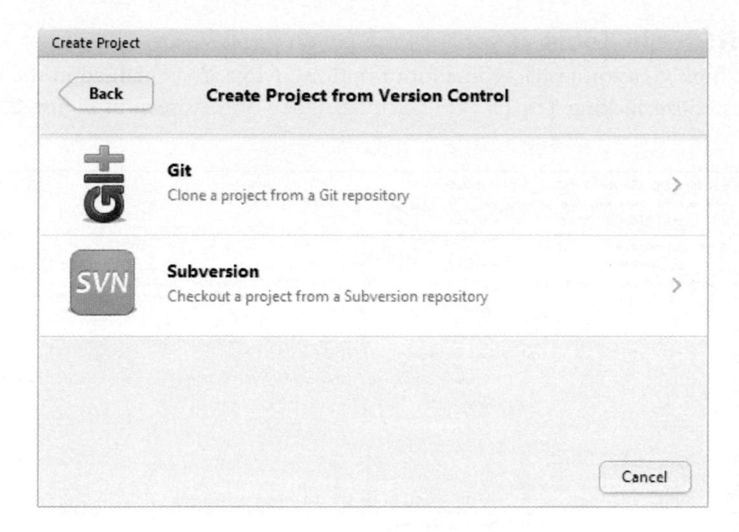

Figure 2.9 Here is the option to choose which type of repository to start a new project from.

Selecting Git asks for a repository URL, such as `git@github.com:jaredlander/coefplot.git`, which will then fill in the project directory name, as shown in Figure 2.10. As with creating a new directory, this will ask where to put this new directory.

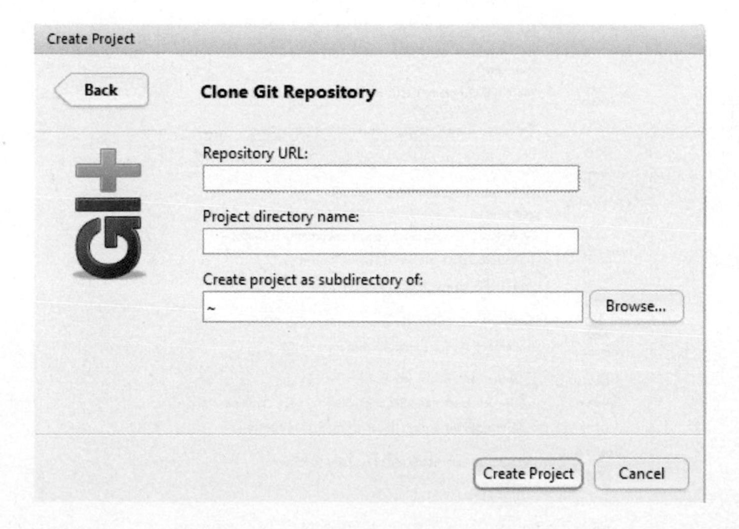

Figure 2.10 Enter the URL for a Git repository, as well as the folder where this should be cloned to.

2.2.2 RStudio Tools

RStudio is highly customizable with a lot of options. Most are contained in the Options dialog accessed by clicking `Tools >> Global Options`, as seen in Figure 2.11.

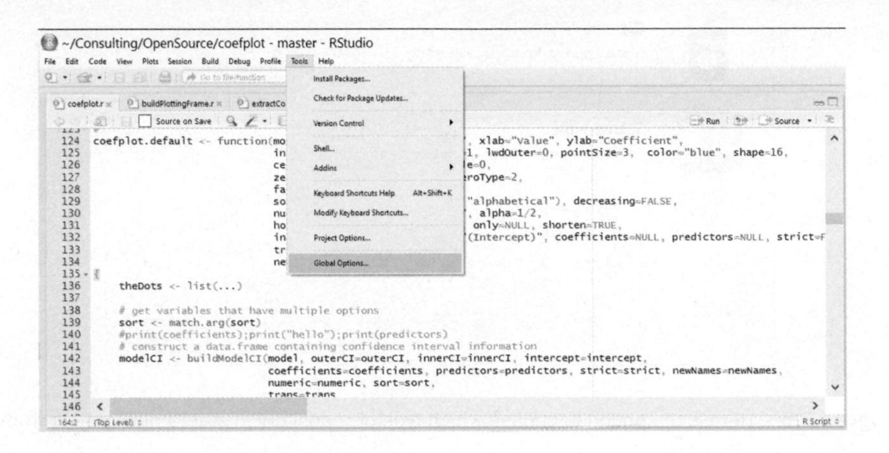

Figure 2.11 Clicking `Tools >> Options` brings up RStudio options.

First are the General options, shown in Figure 2.12. On Windows there is a control for selecting which version of R to use. This is a powerful tool when a computer has a

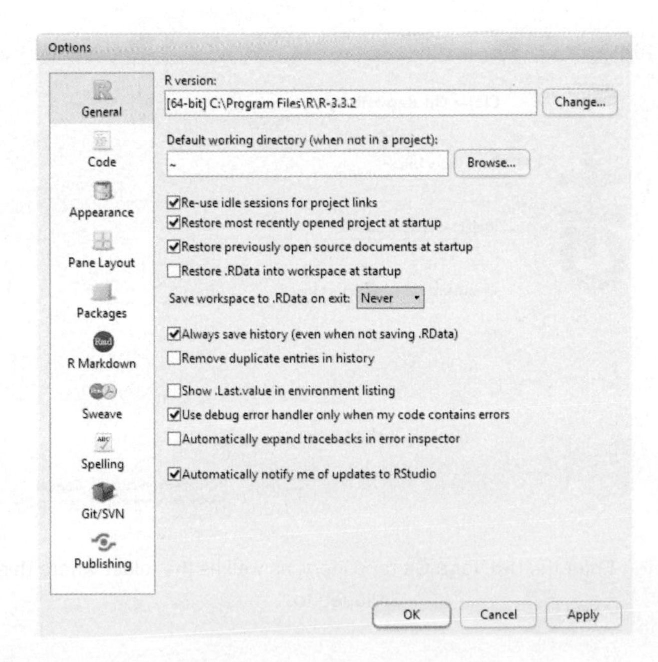

Figure 2.12 General options in RStudio.

number of versions of R. However, RStudio must be restarted after changing the R version. In the future, RStudio is slated to offer the capability to set different versions of R for each project. It is also a good idea to not restore or save `.RData` files on startup and exiting.[4] This way each time R is started it is a fresh session without potential variable corruptions or unnecessary data occupying memory.

Code editing options, shown in Figure 2.13, control the way code is entered and displayed in the text editor. It is generally considered good practice to replace tabs with spaces, either two or four,[5] as tabs can sometimes be interpreted differently by different text editors. Some hard-core programmers will appreciate vim and Emacs modes.

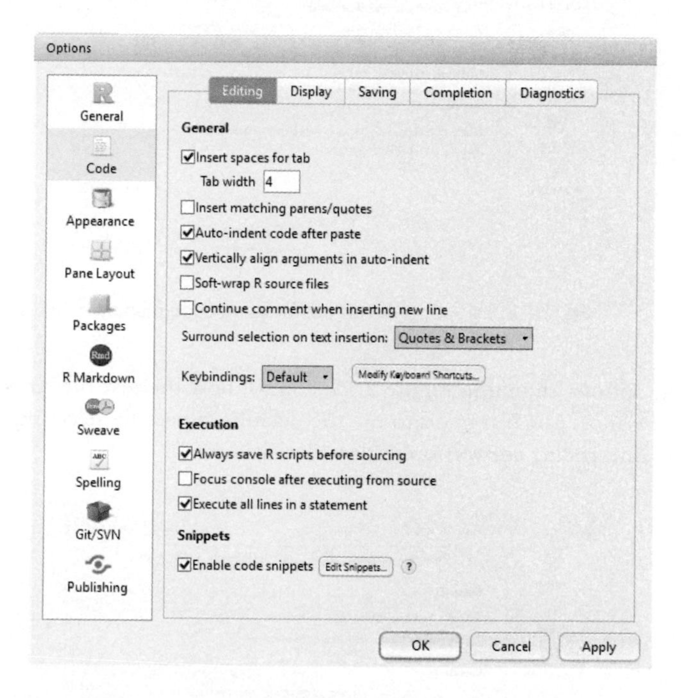

Figure 2.13 Options for customizing code editing.

Code display options, shown in Figure 2.14, control visual cues in the text editor and console. Highlighting selected words makes it easy to spot multiple occurrences. Showing line numbers are a must to ease code navigation. Showing a margin column gives a good indication of when a line of code is too long to be easily read.

4. `RData` files are a convenient way of saving and sharing R objects and are discussed in Section 6.5.
5. Four is better for working with Markdown documents.

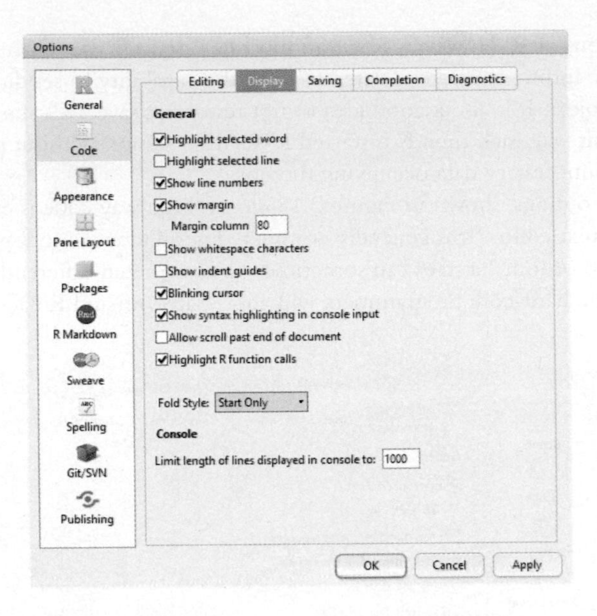

Figure 2.14 Options for customizing code display.

Code saving options, shown in Figure 2.15, control how the text files containing code are saved. For the most part it is good to use the defaults, especially selecting "Platform Native" for the line ending conversion under "Serialization."

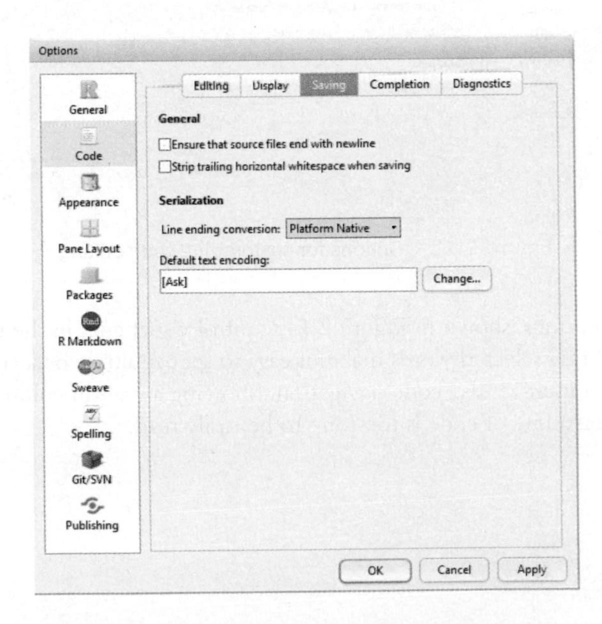

Figure 2.15 Options for customizing code saving.

Code Completion options, shown in Figure 2.16, control how code is completed while programming. Some people like having parentheses added automatically after typing a function, and others do not. One particularly divisive setting is whether to put spaces around the equals sign for named function arguments.

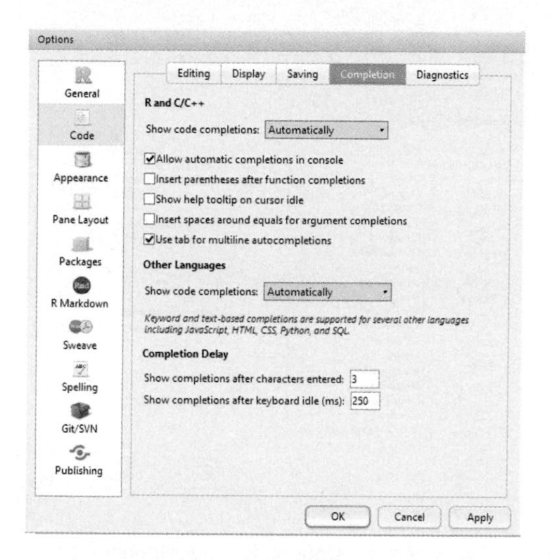

Figure 2.16 Options for customizing code completion.

Code diagnostics options, shown in Figure 2.17, enable code checking. These can be very helpful in identifying mistyped object names, poor style and general mistakes.

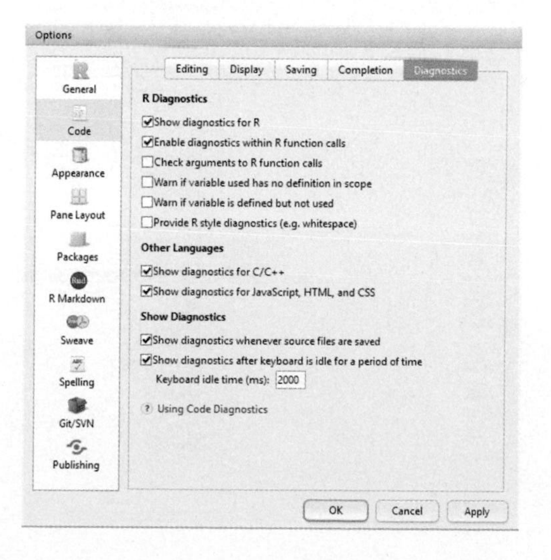

Figure 2.17 Options for customizing code diagnostics.

Appearance options, shown in Figure 2.18, change the way code looks, aesthetically. The font, size and color of the background and text can all be customized here.

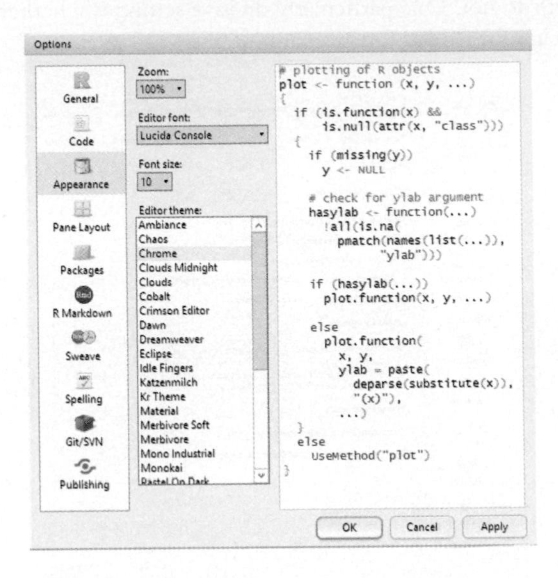

Figure 2.18 Options for code appearance.

The Pane Layout options, shown in Figure 2.19, simply rearrange the panes that make up RStudio.

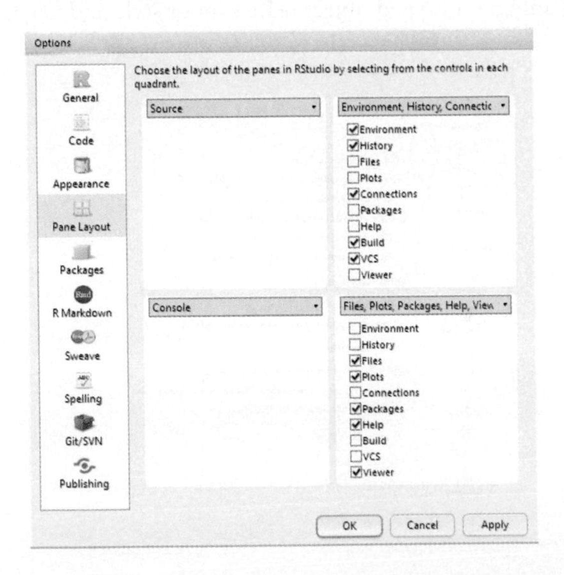

Figure 2.19 These options control the placement of the various panes in RStudio.

The Packages options, shown in Figure 2.20, set options regarding packages, although the most important is the CRAN mirror. While this is changeable from the console, this is the default setting. It is best to pick the mirror that is geographically the closest.

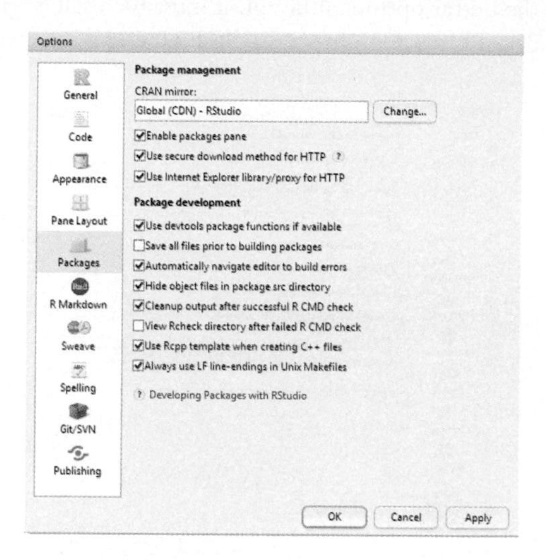

Figure 2.20 Options related to packages. The most important is the CRAN mirror selection.

The RMarkdown options, seen in Figure 2.21, control settings for working with RMarkdown documents. This allows rendered documents to be previewed in an external window or in the Viewer pane. It also lets RMarkdown files be treated like notebooks, rendering results, images and equations inline.

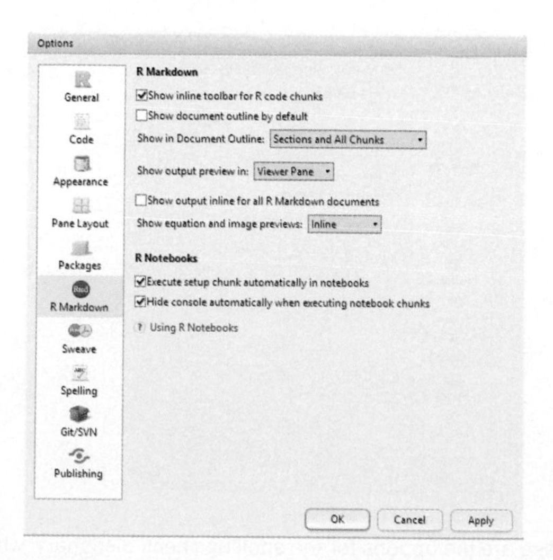

Figure 2.21 Options for RMarkdown, including whether to treat them like notebooks.

The Sweave options, seen in Figure 2.22, may be a bit misnamed, as this is where to choose between using Sweave or **knitr**. Both are used for the generation of PDF documents with **knitr** also enabling the creation of `HTML` documents. **knitr**, detailed in Chapter 27, is by far the better option, although it must be installed first, which is explained in Section 3.1. This is also where the PDF viewer is selected.

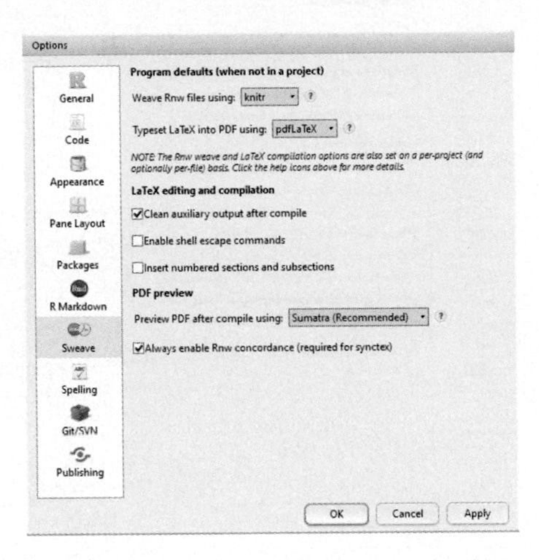

Figure 2.22 This is where to choose whether to use Sweave or knitr and select the PDF viewer.

RStudio contains a spelling checker for writing LaTeX and Markdown documents (using **knitr**, preferably), which is controlled from the Spelling options, shown in Figure 2.23. Not much needs to be set here.

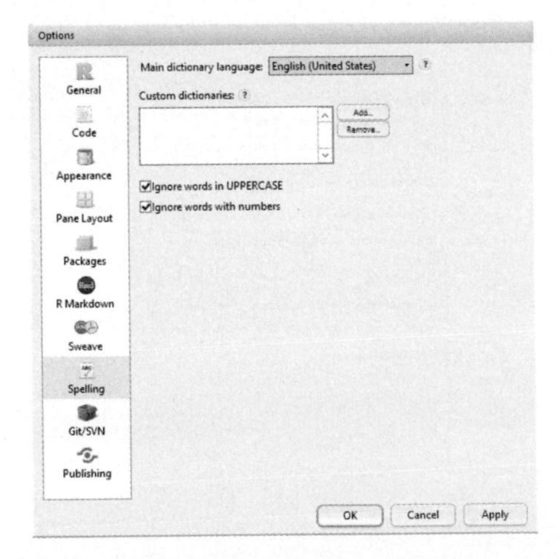

Figure 2.23 These are the options for the spelling check dictionary, which allows language selection and the custom dictionaries.

The Git/SVN options, shown in Figure 2.24, indicates where the executables for Git and SVN exist. This needs to be set only once but is necessary for version control.

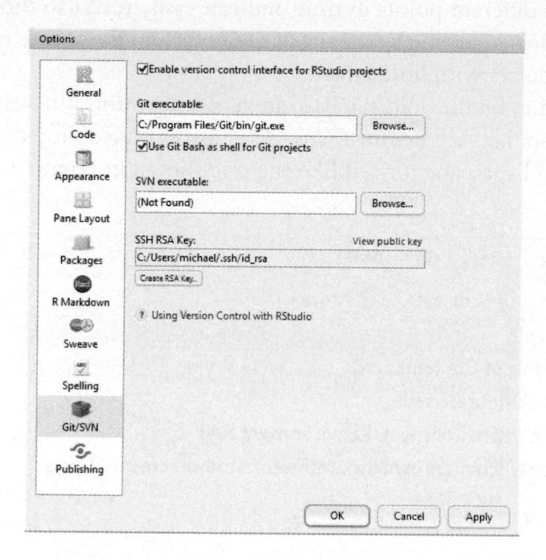

Figure 2.24 This is where to set the location of Git and SVN executables so they can be used by RStudio.

The last option, Publishing, Figure 2.25, sets connections for publishing documents to ShinyApps.io or RStudio Connect.

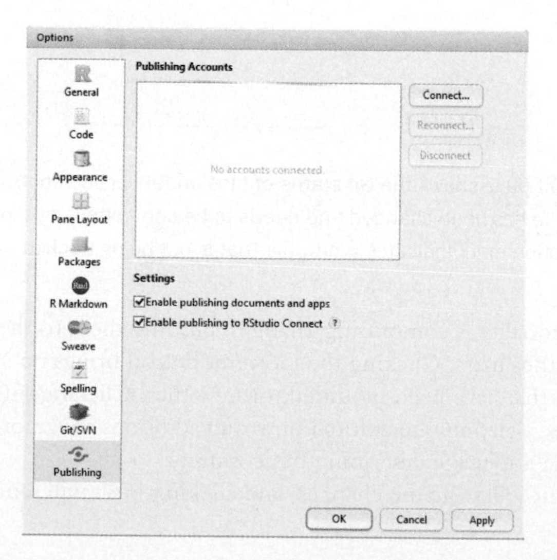

Figure 2.25 This is where to set connections to ShinyApps.io or RStudio Connect.

2.2.3 Git Integration

Using version control is a great idea for many reasons. First and foremost it provides snapshots of code at different points in time and can easily revert to those snapshots. Ancillary benefits include having a backup of the code and the ability to easily transfer the code between computers with little effort.

While SVN used to be the gold standard in version control, it has since been superseded by Git, so that will be our focus. After associating a project with a Git repository[6] RStudio has a pane for Git like the one shown in Figure 2.26.

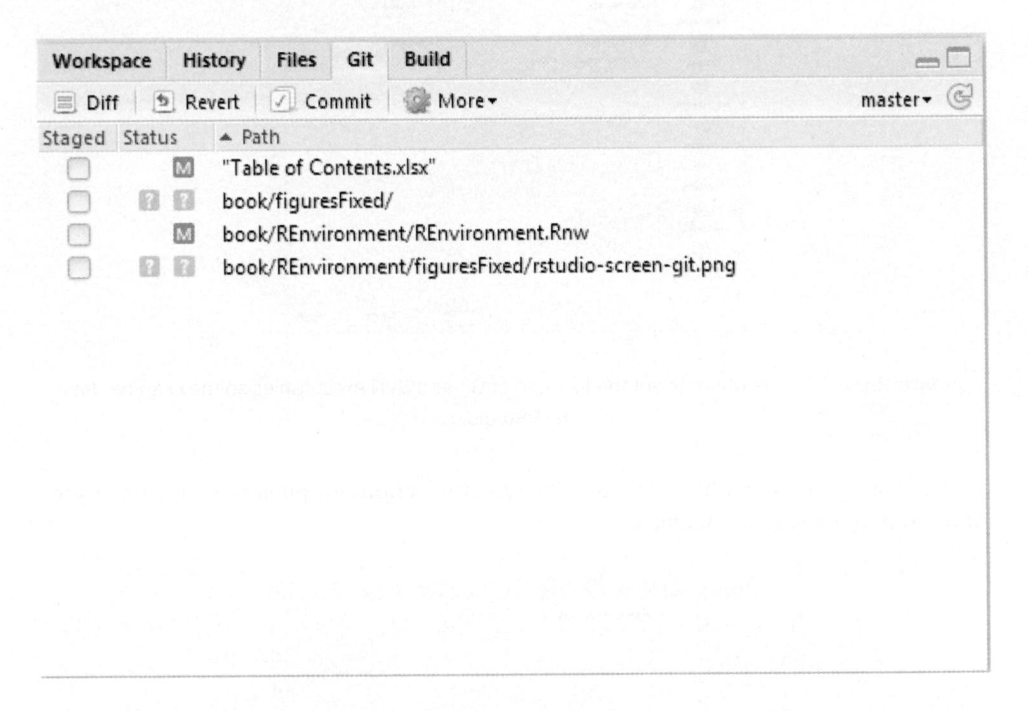

Figure 2.26 The Git pane shows the Git status of files under version control. A blue square with a white M indicates a file has been changed and needs to be committed. A yellow square with a white question mark indicates a new file that is not being tracked by Git.

The main functionality is committing changes, pushing them to the server and pulling changes made by other users. Clicking the Commit button brings up a dialog, Figure 2.27, which displays files that have been modified, or new files. Clicking on one of these files displays the changes; deletions are colored pink and additions are colored green. There is also a space to write a message describing the commit.

Clicking Commit will stage the changes, and clicking Push will send them to the server.

---

6. A Git account should be set up with either GitHub (`https://github.com/`) or Bitbucket (`https://bitbucket.org/`) beforehand.

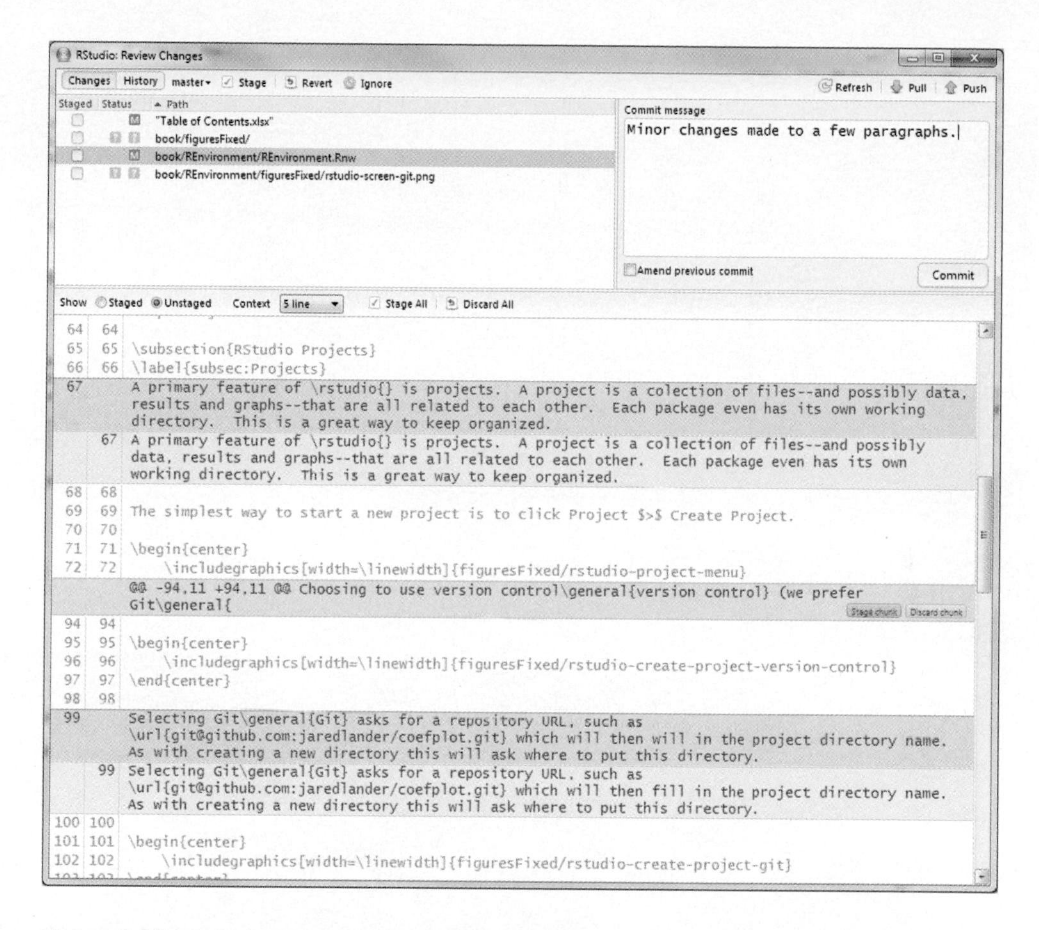

Figure 2.27 This displays files and the changes made to the files, with green being additions and pink being deletions. The upper right contains a space for writing commit messages.

2.3 Microsoft Visual Studio

Microsoft Visual Studio provides IDE tools for working with R. While most R users will be more comfortable using RStudio, this is a nice option for those familiar with Visual Studio.

2.4 Conclusion

R's usability has greatly improved over the past few years, mainly thanks to RStudio. Using an IDE can greatly improve proficiency, and change working with R from merely tolerable to actually enjoyable.[7] RStudio's code completion, text editor, Git integration and projects are indispensable for a good programming work flow.

7. One of our students relayed that he preferred Matlab to R until he used RStudio.

3

R Packages

Perhaps the biggest reason for R's phenomenally ascendant popularity is its collection of user contributed packages. As of early February 2017 there were over 10,000 packages available on CRAN[1] written by more than 2,000 different people. Odds are good that if a statistical technique exists, it has been written in R and contributed to CRAN. Not only is there an incredibly large number of packages; many are written by the authorities in the field, such as Andrew Gelman, Trevor Hastie, Dirk Eddelbuettel and Hadley Wickham.

A package is essentially a library of prewritten code designed to accomplish some task or a collection of tasks. The **survival** package is used for survival analysis, **ggplot2** is used for plotting and **sp** is for dealing with spatial data.

It is important to remember that not all packages are of the same quality. Some are built to be very robust and are well-maintained, while others are built with good intentions but can fail with unforeseen errors, and others still are just plain poor. Even with the best packages, it is important to remember that most were written by statisticians for statisticians, so they may differ from what a computer engineer would expect.

This book will not attempt to be an exhaustive list of good packages to use, because that is constantly changing. However, there are some packages that are so pervasive that they will be used in this book as if they were part of base R. Some of these are **ggplot2**, **tidyr** and **dplyr** by Hadley Wickham; **glmnet** by Trevor Hastie, Robert Tibshirani and Jerome Friedman; **Rcpp** by Dirk Eddelbuettel and **knitr** by Yihui Xie. We have written a package on CRAN, **coefplot**, **useful** and **resumer** with more to follow.

3.1 Installing Packages

As with many tasks in R, there are multiple ways to install packages. The simplest is to install them using the GUI provided by RStudio and shown in Figure 3.1. Access the Packages pane shown in this figure either by clicking its tab or by pressing `Ctrl+7` on the keyboard.

1. `http://cran.r-project.org/web/packages/`

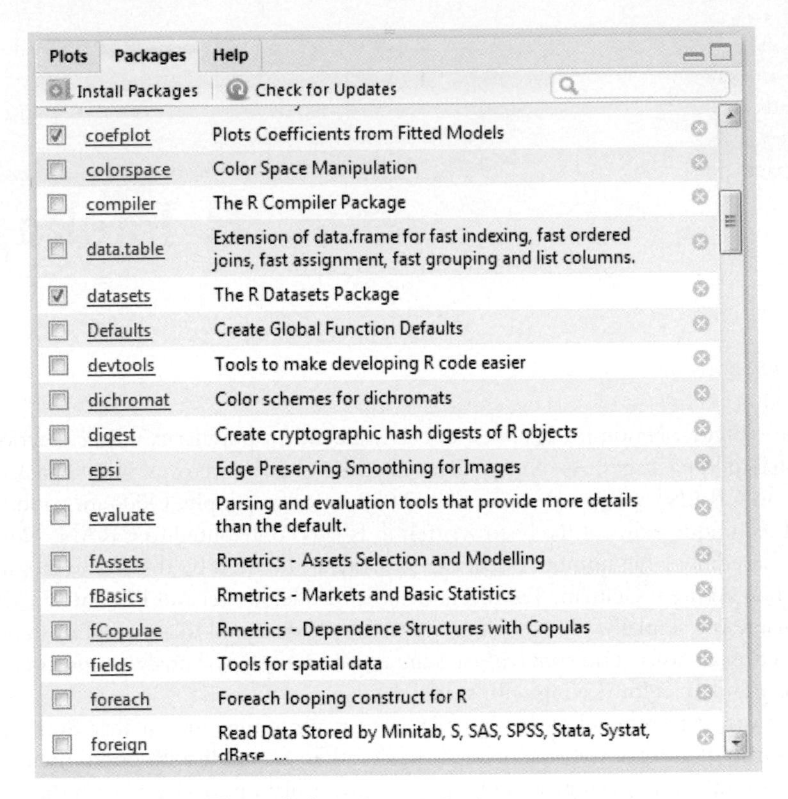

Figure 3.1 RStudio's Packages pane.

In the upper-left corner, click the Install Packages button to bring up the dialog in Figure 3.2.

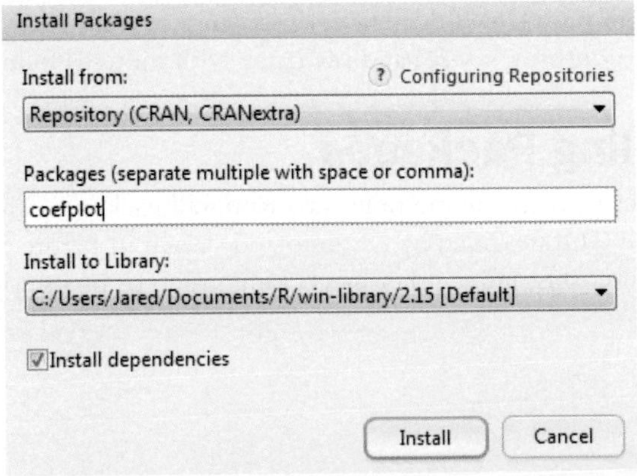

Install Packages

Install from:

Configuring Repositories

Repository (CRAN, CRANextra)

Packages (separate multiple with space or comma):

coefplot

Install to Library:

C:/Users/Jared/Documents/R/win-library/2.15 [Default]

Install dependencies

Install

Cancel

Figure 3.2 RStudio's package installation dialog.

From here simply type the name of a package (RStudio has a nice autocomplete feature for this) and click Install. Multiple packages can be specified, separated by commas. This downloads and installs the desired package, which is then available for use. Selecting the Install dependencies checkbox will automatically download and install all packages that the desired package requires to work. For example, our **coefplot** package depends on **ggplot2**, **plyr**, **dplyr**, **useful**, **stringr** and **reshape2**, and each of those may have further dependencies.

An alternative is to type a very simple command into the console:

```
> install.packages("coefplot")
```

This will accomplish the same thing as working in the GUI.

There has been a movement recently to install packages directly from GitHub or BitBucket repositories, especially to get the development versions of packages. This can be accomplished using **devtools**.

```
> library(devtools)
> install_github(repo="coefplot/jaredlander")
```

In order to use functions in the **devtools** package we first needed to load the package using **library** which is explained in Section 3.2.

If the package being installed from a repository contains source code for a compiled language—generally C++ or FORTRAN—then the proper compilers must be installed. More information is in Section 30.7.

Sometimes there is a need to install a package from a local file, either a zip of a prebuilt package or a tar.gz of package code. This can be done using the installation dialog mentioned before but switching the Install from: option to Package Archive File as shown in Figure 3.3. Then browse to the file and install. Note that this will not install dependencies, and if they are not present, the installation will fail. Be sure to install dependencies first.

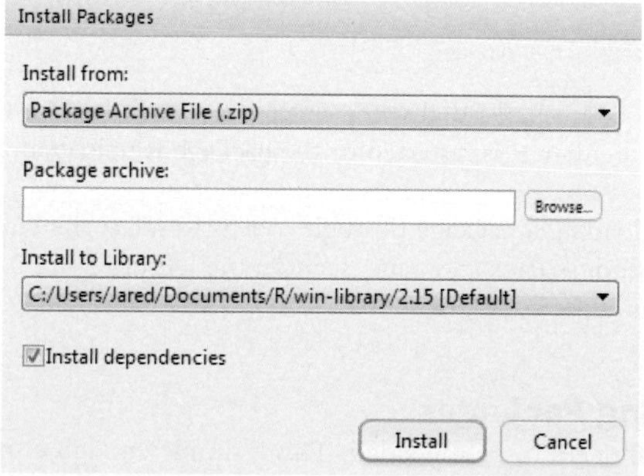

Figure 3.3 RStudio's package installation dialog to install from an archive file.

Similarly to before, this can be accomplished using **install.packages**.

```
> install.packages("coefplot_1.1.7.zip")
```

3.1.1 Uninstalling Packages

In the rare instance when a package needs to be uninstalled, it is easiest to click the white X inside a gray circle on the right of the package descriptions in RStudio's Packages pane shown in Figure 3.1. Alternatively, this can be done with **remove.packages**, where the first argument is a `character vector` naming the packages to be removed.

3.2 Loading Packages

Now that packages are installed they are almost ready to use and just need to be loaded first. There are two commands that can be used, either **library** or **require**. They both accomplish the same thing: Loading the package. Using **require** will return `TRUE` if it succeeds and `FALSE` with a warning if it cannot find the package. This returned value is useful when loading a package from within a function, a practice considered acceptable to some, improper to others. Calling **library** on a package that is not installed will cause an error which can be advantageous when running code in scripts. In interactive usage there is not much of a difference but it is preferable, when writing scripts, to use **library**. The argument to either function is the name of the desired package, with or without quotes. So loading the **coefplot** package would look like:

```
> library(coefplot)

Loading required package: ggplot2
```

It also prints out the dependent packages that get loaded as well. This can be suppressed by setting the argument `quietly` to `TRUE`.

```
> library(coefplot, quietly=TRUE)
```

A package only needs to be loaded when starting a new R session. Once loaded, it remains available until either R is restarted or the package is unloaded, as described in Section 3.2.1.

An alternative to loading a package through code is to select the checkbox next to the package name in RStudio's Packages pane, seen on the left of Figure 3.1. This will load the package by running the code just shown.

3.2.1 Unloading Packages

Sometimes a package needs to be unloaded. This is simple enough either by clearing the checkbox in RStudio's Packages pane or by using the **detach** function. The function takes the package name preceded by `package:` all in quotes.

```
> detach("package:coefplot")
```

It is not uncommon for functions in different packages to have the same name. For example, **coefplot** is in both **arm** (by Andrew Gelman) and **coefplot**.[2] If both packages are loaded, the function in the package loaded last will be invoked when calling that function. A way around this is to precede the function with the name of the package, separated by two colons (`::`).

```
> arm::coefplot(object)
> coefplot::coefplot(object)
```

Not only does this call the appropriate function; it also allows the function to be called without even loading the package beforehand.

3.3 Building a Package

Building a package is one of the more rewarding parts of working with R, especially sharing that package with the community through CRAN. Chapter 30 discusses this process in detail.

3.4 Conclusion

Packages make up the backbone of the R community and experience. They are often considered what makes working with R so desirable. This is how the community makes its work, and so many of the statistical techniques, available to the world. With such a large number of packages, finding the right one can be overwhelming. CRAN Task Views (`http://cran.r-project.org/web/views/`) offers a curated listing of packages for different needs. However, the best way to find a new package might just be to ask the community. Appendix A gives some resources for doing just that.

2. This particular instance is because we built `coefplot` as an improvement on the one available in `arm`. There are other instances where the names have nothing in common.

[illegible]

[illegible]

[illegible]

3.3 Building a Package

[illegible]

3.4 Conclusion

[illegible]

[illegible]

4

Basics of R

R is a powerful tool for all manner of calculations, data manipulation and scientific computations. Before getting to the complex operations possible in R we must start with the basics. Like most languages R has its share of mathematical capability, variables, functions and data types.

4.1 Basic Math

Being a statistical programming language, R can certainly be used to do basic math and that is where we will start.

We begin with the "hello, world!" of basic math: $1 + 1$. In the console there is a right angle bracket (>) where code should be entered. Simply test R by running

```
> 1 + 1

[1] 2
```

If this returns 2, then everything is great; if not, then something is very, very wrong. Assuming it worked, let's look at some slightly more complicated expressions:

```
> 1 + 2 + 3

[1] 6

> 3 * 7 * 2

[1] 42

> 4 / 2

[1] 2

> 4 / 3

[1] 1.333333
```

These follow the basic order of operations: Parenthesis, Exponents, Multiplication, Division, Addition and Subtraction (PEMDAS). This means operations inside parentheses take priority over other operations. Next on the priority list is exponentiation. After that, multiplication and division are performed, followed by addition and subtraction.

This is why the first two lines in the following code have the same result, while the third is different.

```
> 4 * 6 + 5

[1] 29

> (4 * 6) + 5

[1] 29

> 4 * (6 + 5)

[1] 44
```

So far we have put white space in between each operator, such as * and /. This is not necessary but is encouraged as good coding practice.

4.2 Variables

Variables are an integral part of any programming language and R offers a great deal of flexibility. Unlike statically typed languages such as C++, R does not require variable types to be declared. A variable can take on any available data type as described in Section 4.3. It can also hold any R object such as a function, the result of an analysis or a plot. A single variable can at one point hold a number, then later hold a character and then later a number again.

4.2.1 Variable Assignment

There are a number of ways to assign a value to a variable, and again, this does not depend on the type of value being assigned.

The valid assignment operators are <- and =, with the first being preferred.

For example, let's save 2 to the variable x and 5 to the variable y.

```
> x <- 2
> x

[1] 2

> y = 5
> y

[1] 5
```

The arrow operator can also point in the other direction.

```
> 3 -> z
> z

[1] 3
```

The assignment operation can be used successively to assign a value to multiple variables simultaneously.

```
> a <- b <- 7
> a

[1] 7

> b

[1] 7
```

A more laborious, though sometimes necessary, way to assign variables is to use the **assign** function.

```
> assign("j", 4)
> j

[1] 4
```

Variable names can contain any combination of alphanumeric characters along with periods (.) and underscores (_). However, they cannot *start* with a number or an underscore.

The most common form of assignment in the R community is the left arrow (<-), which may seem awkward to use at first but eventually becomes second nature. It even seems to make sense, as the variable is sort of pointing to its value. There is also a particularly nice benefit for people coming from languages like SQL, where a single equal sign (=) tests for equality.

It is generally considered best practice to use actual names, usually nouns, for variables instead of single letters. This provides more information to the person reading the code. This is seen throughout this book.

4.2.2 Removing Variables

For various reasons a variable may need to be removed. This is easily done using **remove** or its shortcut **rm**.

```
> j

[1] 4

> rm(j)
> # now it is gone
> j

Error in eval(expr, envir, enclos): object 'j' not found
```

This frees up memory so that R can store more objects, although it does not necessarily free up memory for the operating system. To guarantee that, use **gc**, which performs garbage collection, releasing unused memory to the operating system. R automatically does garbage collection periodically, so this function is not essential.

Variable names are case sensitive, which can trip up people coming from a language like SQL or Visual Basic.

```
> theVariable <- 17
> theVariable

[1] 17

> THEVARIABLE

Error in eval(expr, envir, enclos): object 'THEVARIABLE' not found
```

4.3 Data Types

There are numerous data types in R that store various kinds of data. The four main types of data most likely to be used are `numeric`, `character` (string), `Date/POSIXct` (time-based) and `logical` (`TRUE/FALSE`).

The type of data contained in a variable is checked with the **class** function.

```
> class(x)

[1] "numeric"
```

4.3.1 Numeric Data

As expected, R excels at running numbers, so numeric data is the most common type in R. The most commonly used numeric data is `numeric`. This is similar to a `float` or `double` in other languages. It handles integers and decimals, both positive and negative, and of course, zero. A numeric value stored in a variable is automatically assumed to be `numeric`. Testing whether a variable is `numeric` is done with the function **is.numeric**.

```
> is.numeric(x)

[1] TRUE
```

Another important, if less frequently used, type is `integer`. As the name implies this is for whole numbers only, no decimals. To set an integer to a variable it is necessary to append the value with an `L`. As with checking for a `numeric`, the **is.integer** function is used.

```
> i <- 5L
> i

[1] 5

> is.integer(i)

[1] TRUE
```

Do note that, even though i is an integer, it will also pass a numeric check.

```
> is.numeric(i)

[1] TRUE
```

R nicely promotes integers to numeric when needed. This is obvious when multiplying an integer by a numeric, but importantly it works when dividing an integer by another integer, resulting in a decimal number.

```
> class(4L)

[1] "integer"

> class(2.8)

[1] "numeric"

> 4L * 2.8

[1] 11.2

> class(4L * 2.8)

[1] "numeric"

> class(5L)

[1] "integer"

> class(2L)

[1] "integer"

> 5L / 2L

[1] 2.5

> class(5L / 2L)

[1] "numeric"
```

4.3.2 Character Data

Even though it is not explicitly mathematical, the character (string) data type is very common in statistical analysis and must be handled with care. R has two primary ways of handling character data: character and factor. While they may seem similar on the surface, they are treated quite differently.

```
> x <- "data"
> x

[1] "data"
```

```
> y <- factor("data")
> y

[1] data
Levels: data
```

Notice that x contains the word “data” encapsulated in quotes, while y has the word “data” without quotes and a second line of information about the `levels` of y. That is explained further in Section 4.4.2 about `vectors`.

Characters are case sensitive, so “Data” is different from “data” or “DATA”.

To find the length of a `character` (or `numeric`) use the **nchar** function.

```
> nchar(x)

[1] 4

> nchar("hello")

[1] 5

> nchar(3)

[1] 1

> nchar(452)

[1] 3
```

This will not work for `factor` data.

```
> nchar(y)

Error in nchar(y): 'nchar()' requires a character vector
```

4.3.3 Dates

Dealing with dates and times can be difficult in any language, and to further complicate matters R has numerous different types of dates. The most useful are `Date` and `POSIXct`. `Date` stores just a date while `POSIXct` stores a date and time. Both objects are actually represented as the number of days (`Date`) or seconds (`POSIXct`) since January 1, 1970.

```
> date1 <- as.Date("2012-06-28")
> date1

[1] "2012-06-28"

> class(date1)

[1] "Date"

> as.numeric(date1)

[1] 15519
```

```
> date2 <- as.POSIXct("2012-06-28 17:42")
> date2
```

```
[1] "2012-06-28 17:42:00 EDT"
```

```
> class(date2)
```

```
[1] "POSIXct" "POSIXt"
```

```
> as.numeric(date2)
```

```
[1] 1340919720
```

Easier manipulation of date and time objects can be accomplished using the **lubridate** and **chron** packages.

Using functions such as **as.numeric** or **as.Date** does not merely change the formatting of an object but actually changes the underlying type.

```
> class(date1)
```

```
[1] "Date"
```

```
> class(as.numeric(date1))
```

```
[1] "numeric"
```

4.3.4 Logical

`Logicals` are a way of representing data that can be either `TRUE` or `FALSE`. Numerically, `TRUE` is the same as 1 and `FALSE` is the same as 0. So `TRUE` $*$ 5 equals 5 while `FALSE` $*$ 5 equals 0.

```
> TRUE * 5
```

```
[1] 5
```

```
> FALSE * 5
```

```
[1] 0
```

Similar to other types, `logicals` have their own test, using the **is.logical** function.

```
> k <- TRUE
> class(k)
```

```
[1] "logical"
```

```
> is.logical(k)
```

```
[1] TRUE
```

R provides T and F as shortcuts for TRUE and FALSE, respectively, but it is best practice not to use them, as they are simply variables storing the values TRUE and FALSE and can be overwritten, which can cause a great deal of frustration as seen in the following example.

```
> TRUE

[1] TRUE

> T

[1] TRUE

> class(T)

[1] "logical"

> T <- 7
> T

[1] 7

> class(T)

[1] "numeric"
```

Logicals can result from comparing two numbers, or characters.

```
> # does 2 equal 3?
> 2 == 3

[1] FALSE

> # does 2 not equal three?
> 2 != 3

[1] TRUE

> # is two less than three?
> 2 < 3

[1] TRUE

> # is two less than or equal to three?
> 2 <= 3

[1] TRUE

> # is two greater than three?
> 2 > 3

[1] FALSE

> # is two greater than or equal to three?
> 2 >= 3
```

```
[1] FALSE

> # is "data" equal to "stats"?
> "data" == "stats"

[1] FALSE

> # is "data" less than "stats"?
> "data" < "stats"

[1] TRUE
```

4.4 Vectors

A vector is a collection of elements, all of the same type. For instance, `c(1, 3, 2, 1, 5)` is a vector consisting of the numbers 1, 3, 2, 1, 5, in that order. Similarly, `c("R", "Excel", "SAS", "Excel")` is a vector of the character elements, "R", "Excel", "SAS", and "Excel". A vector cannot be of mixed type.

Vectors play a crucial, and helpful, role in R. More than being simple containers, vectors in R are special in that R is a vectorized language. That means operations are applied to each element of the vector automatically, without the need to loop through the vector. This is a powerful concept that may seem foreign to people coming from other languages, but it is one of the greatest things about R.

Vectors do not have a dimension, meaning there is no such thing as a column vector or row vector. These vectors are not like the mathematical vector, where there is a difference between row and column orientation.[1]

The most common way to create a vector is with **c**. The "c" stands for combine because multiple elements are being combined into a vector.

```
> x <- c(1, 2, 3, 4, 5, 6, 7, 8, 9, 10)
> x

 [1]  1  2  3  4  5  6  7  8  9 10
```

4.4.1 Vector Operations

Now that we have a vector of the first ten numbers, we might want to multiply each element by 3. In R this is a simple operation using just the multiplication operator (*).

```
> x * 3

 [1]  3  6  9 12 15 18 21 24 27 30
```

No loops are necessary. Addition, subtraction and division are just as easy. This also works for any number of operations.

```
> x + 2
```

1. Column or row vectors can be represented as one-dimensional matrices, which are discussed in Section 5.3.

```
 [1]  3  4  5  6  7  8  9 10 11 12

> x - 3

 [1] -2 -1  0  1  2  3  4  5  6  7

> x / 4

 [1] 0.25 0.50 0.75 1.00 1.25 1.50 1.75 2.00 2.25 2.50

> x^2

 [1]   1   4   9  16  25  36  49  64  81 100

> sqrt(x)

 [1] 1.000000 1.414214 1.732051 2.000000 2.236068 2.449490 2.645751
 [8] 2.828427 3.000000 3.162278
```

Earlier we created a `vector` of the first ten numbers using the **c** function, which creates a `vector`. A shortcut is the **:** operator, which generates a sequence of consecutive numbers, in either direction.

```
> 1:10

 [1]  1  2  3  4  5  6  7  8  9 10

> 10:1

 [1] 10  9  8  7  6  5  4  3  2  1

> -2:3

[1] -2 -1  0  1  2  3

> 5:-7

 [1]  5  4  3  2  1  0 -1 -2 -3 -4 -5 -6 -7
```

Vector operations can be extended even further. Let's say we have two vectors of equal length. Each of the corresponding elements can be operated on together.

```
> # create two vectors of equal length
> x <- 1:10
> y <- -5:4
> # add them
> x + y

 [1] -4 -2  0  2  4  6  8 10 12 14

> # subtract them
> x - y

 [1] 6 6 6 6 6 6 6 6 6 6
```

```
> # multiply them
> x * y

 [1] -5 -8 -9 -8 -5  0  7 16 27 40

> # divide them--notice division by 0 results in Inf
> x / y

 [1] -0.2 -0.5 -1.0 -2.0 -5.0  Inf  7.0  4.0  3.0  2.5

> # raise one to the power of the other
> x^y

 [1] 1.000000e+00 6.250000e-02 3.703704e-02 6.250000e-02 2.000000e-01
 [6] 1.000000e+00 7.000000e+00 6.400000e+01 7.290000e+02 1.000000e+04

> # check the length of each
> length(x)

[1] 10

> length(y)

[1] 10

> # the length of them added together should be the same
> length(x + y)

[1] 10
```

In the preceding code block, notice the hash # symbol. This is used for comments. Anything following the hash, on the same line, will be commented out and not run.

Things get a little more complicated when operating on two `vectors` of unequal length. The shorter `vector` gets recycled—that is, its elements are repeated, in order, until they have been matched up with every element of the longer `vector`. If the longer one is not a multiple of the shorter one, a warning is given.

```
> x + c(1, 2)

 [1]  2  4  4  6  6  8  8 10 10 12

> x + c(1, 2, 3)

Warning in x + c(1, 2, 3): longer object length is not a
multiple of shorter object length

 [1]  2  4  6  5  7  9  8 10 12 11
```

Comparisons also work on `vectors`. Here the result is a vector of the same length containing `TRUE` or `FALSE` for each element.

```
> x <= 5

 [1]  TRUE  TRUE  TRUE  TRUE  TRUE FALSE FALSE FALSE FALSE FALSE
```

```
> x > y

 [1] TRUE TRUE TRUE TRUE TRUE TRUE TRUE TRUE TRUE TRUE

> x < y

 [1] FALSE FALSE FALSE FALSE FALSE FALSE FALSE FALSE FALSE FALSE
```

To test whether all the resulting elements are TRUE, use the **all** function. Similarly, the **any** function checks whether any element is TRUE.

```
> x <- 10:1
> y <- -4:5
> any(x < y)

[1] TRUE

> all(x < y)

[1] FALSE
```

The **nchar** function also acts on each element of a vector.

```
> q <- c("Hockey", "Football", "Baseball", "Curling", "Rugby",
+        "Lacrosse", "Basketball", "Tennis", "Cricket", "Soccer")
> nchar(q)

 [1]  6  8  8  7  5  8 10  6  7  6

> nchar(y)

 [1] 2 2 2 2 1 1 1 1 1 1
```

Accessing individual elements of a vector is done using square brackets ([ ]). The first element of x is retrieved by typing x[1], the first two elements by x[1:2] and nonconsecutive elements by x[c(1, 4)].

```
> x[1]

[1] 10

> x[1:2]

[1] 10  9

> x[c(1, 4)]

[1] 10  7
```

This works for all types of vectors whether they are numeric, logical, character and so forth.

It is possible to give names to a vector either during creation or after the fact.

```
> # provide a name for each element of an array using a name-value pair
> c(One="a", Two="y", Last="r")

 One  Two Last
 "a"  "y"  "r"

> # create a vector
> w <- 1:3
> # name the elements
> names(w) <- c("a", "b", "c")
> w

a b c
1 2 3
```

4.4.2 Factor Vectors

Factors are an important concept in R, especially when building models. Let's create a simple vector of text data that has a few repeats. We will start with the vector q we created earlier and add some elements to it.

```
> q2 <- c(q, "Hockey", "Lacrosse", "Hockey", "Water Polo",
+         "Hockey", "Lacrosse")
```

Converting this to a factor is easy with **as.factor**.

```
> q2Factor <- as.factor(q2)
> q2Factor

 [1] Hockey     Football   Baseball   Curling    Rugby      Lacrosse
 [7] Basketball Tennis     Cricket    Soccer     Hockey     Lacrosse
[13] Hockey     Water Polo Hockey     Lacrosse
11 Levels: Baseball Basketball Cricket Curling Football ... Water Polo
```

Notice that after printing out every element of q2Factor, R also prints the levels of q2Factor. The levels of a factor are the unique values of that factor variable. Technically, R is giving each unique value of a factor a unique integer, tying it back to the character representation. This can be seen with **as.numeric**.

```
> as.numeric(q2Factor)

 [1]  6  5  1  4  8  7  2 10  3  9  6  7  6 11  6  7
```

In ordinary factors the order of the levels does not matter and one level is no different from another. Sometimes, however, it is important to understand the order of a factor, such as when coding education levels. Setting the ordered argument to TRUE creates an ordered factor with the order given in the levels argument.

```
> factor(x=c("High School", "College", "Masters", "Doctorate"),
+        levels=c("High School", "College", "Masters", "Doctorate"),
+        ordered=TRUE)
```

```
[1] High School College     Masters     Doctorate
Levels: High School < College < Masters < Doctorate
```

`Factors` can drastically reduce the size of the variable because they are storing only the unique values, but they can cause headaches if not used properly. This will be discussed further throughout the book.

4.5 Calling Functions

Earlier we briefly used a few basic functions like **nchar**, **length** and **as.Date** to illustrate some concepts. Functions are very important and helpful in any language because they make code easily repeatable. Almost every step taken in R involves using functions, so it is best to learn the proper way to call them. R function calling is filled with a good deal of nuance, so we are going to focus on the gist of what is needed to know. Of course, throughout the book there will be many examples of calling functions.

Let's start with the simple **mean** function, which computes the average of a set of numbers. In its simplest form it takes a `vector` as an argument.

```
> mean(x)

[1] 5.5
```

More complicated functions have multiple arguments that can either be specified by the order they are entered or by using their name with an equal sign. We will see further use of this throughout the book.

R provides an easy way for users to build their own functions, which we will cover in more detail in Chapter 8.

4.6 Function Documentation

Any function provided in R has accompanying documentation, with varying quality, of course. The easiest way to access that documentation is to place a question mark in front of the function name, like this: `?mean`.

To get help on binary operators like +, * or == surround them with back ticks (`).

```
> ?`+`
> ?`*`
> ?`==`
```

There are occasions when we have only a sense of the function we want to use. In that case we can look up the function by using part of the name with **apropos**.

```
> apropos("mea")

 [1] ".colMeans"            ".rowMeans"            "colMeans"
 [4] "influence.measures"   "kmeans"               "mean"
 [7] "mean.Date"            "mean.default"         "mean.difftime"
[10] "mean.POSIXct"         "mean.POSIXlt"         "mean_cl_boot"
[13] "mean_cl_normal"       "mean_sdl"             "mean_se"
[16] "rowMeans"             "weighted.mean"
```

4.7 Missing Data

Missing data plays a critical role in both statistics and computing, and R has two types of missing data, NA and NULL. While they are similar, they behave differently and that difference needs attention.

4.7.1 NA

Often we will have data that has missing values for any number of reasons. Statistical programs use various techniques to represent missing data such as a dash, a period or even the number 99. R uses NA. NA will often be seen as just another element of a vector. **is.na** tests each element of a vector for missingness.

```
> z <- c(1, 2, NA, 8, 3, NA, 3)
> z

[1]  1  2 NA  8  3 NA  3

> is.na(z)

[1] FALSE FALSE  TRUE FALSE FALSE  TRUE FALSE
```

NA is entered simply by typing the letters "N" and "A" as if they were normal text. This works for any kind of vector.

```
> zChar <- c("Hockey", NA, "Lacrosse")
> zChar

[1] "Hockey"   NA         "Lacrosse"

> is.na(zChar)

[1] FALSE  TRUE FALSE
```

If we calculate the mean of z, the answer will be NA since **mean** returns NA if even a single element is NA.

```
> mean(z)

[1] NA
```

When the na.rm is TRUE, **mean** first removes the missing data, then calculates the mean.

```
> mean(z, na.rm=TRUE)

[1] 3.4
```

There is similar functionality with **sum**, **min**, **max**, **var**, **sd** and other functions as seen in Section 18.1.

Handling missing data is an important part of statistical analysis. There are many techniques depending on field and preference. One popular technique is multiple imputation, which is discussed in detail in Chapter 25 of Andrew Gelman and Jennifer Hill's book *Data Analysis Using Regression and Multilevel/Hierarchical Models*, and is implemented in the **mi**, **mice** and **Amelia** packages.

4.7.2 NULL

NULL is the absence of anything. It is not exactly missingness, it is nothingness. Functions can sometimes return NULL and their arguments can be NULL. An important difference between NA and NULL is that NULL is atomical and cannot exist within a vector. If used inside a vector, it simply disappears.

```
> z <- c(1, NULL, 3)
> z

[1] 1 3
```

Even though it was entered into the vector z, it did not get stored in z. In fact, z is only two elements long.

The test for a NULL value is **is.null**.

```
> d <- NULL
> is.null(d)

[1] TRUE

> is.null(7)

[1] FALSE
```

Since NULL cannot be a part of a vector, **is.null** is appropriately not vectorized.

4.8 Pipes

A new paradigm for calling functions in R is the pipe. The pipe from the **magrittr** package works by taking the value or object on the left-hand side of the pipe and inserting it into the first argument of the function that is on the right-hand side of the pipe. A simple example example would be using a pipe to feed x to the **mean** function.

```
> library(magrittr)
> x <- 1:10
> mean(x)

[1] 5.5

> x %>% mean

[1] 5.5
```

The result is the same but they are written differently. Pipes are most useful when used in a pipeline to chain together a series of function calls. Given a vector z that contains numbers and NAs, we want to find out how many NAs are present. Traditionally, this would be done by nesting functions.

```
> z <- c(1, 2, NA, 8, 3, NA, 3)
> sum(is.na(z))

[1] 2
```

This can also be done using pipes.

```
> z %>% is.na %>% sum

[1] 2
```

Pipes read more naturally in a left-to-right fashion, making the code easier to comprehend. Using pipes is negligibly slower than nesting function calls, though as Hadley Wickham notes, pipes will not be a major bottleneck in code.

When piping an object into a function and not setting any additional arguments, no parentheses are needed. However, if additional arguments are used, then they should be named and included inside the parentheses after the function call. The first argument is not used, as the pipe already inserted the left-hand object into the first argument.

```
> z %>% mean(na.rm=TRUE)

[1] 3.4
```

Pipes are used extensively in a number of modern packages after being popularized by Hadley Wickham in the **dplyr** package, as detailed in Chapter 14.

4.9 Conclusion

Data come in many types, and R is well equipped to handle them. In addition to basic calculations, R can handle numeric, character and time-based data. One of the nicer parts of working with R, although one that requires a different way of thinking about programming, is vectorization. This allows operating on multiple elements in a `vector` simultaneously, which leads to faster and more mathematical code.

4.9 Conclusion

5

Advanced Data Structures

Sometimes data require more complex storage than simple vectors and thankfully R provides a host of data structures. The most common are the data.frame, matrix and list, followed by the array. Of these, the data.frame will be most familiar to anyone who has used a spreadsheet, the matrix to people familiar with matrix math and the list to programmers.

5.1 data.frames

Perhaps one of the most useful features of R is the data.frame. It is one of the most often cited reasons for R's ease of use.

On the surface a data.frame is just like an Excel spreadsheet in that it has columns and rows. In statistical terms, each column is a variable and each row is an observation.

In terms of how R organizes data.frames, each column is actually a vector, each of which has the same length. That is very important because it lets each column hold a different type of data (see Section 4.3). This also implies that within a column each element must be of the same type, just like with vectors.

There are numerous ways to construct a data.frame, the simplest being to use the **data.frame** function. Let's create a basic data.frame using some of the vectors we have already introduced, namely x, y and q.

```
> x <- 10:1
> y <- -4:5
> q <- c("Hockey", "Football", "Baseball", "Curling", "Rugby",
+        "Lacrosse", "Basketball", "Tennis", "Cricket", "Soccer")
> theDF <- data.frame(x, y, q)
> theDF

   x  y          q
1 10 -4     Hockey
2  9 -3   Football
3  8 -2   Baseball
4  7 -1    Curling
5  6  0      Rugby
6  5  1   Lacrosse
7  4  2 Basketball
```

```
8   3  3      Tennis
9   2  4     Cricket
10  1  5      Soccer
```

This creates a 10x3 `data.frame` consisting of those three `vectors`. Notice the names of `theDF` are simply the variables. We could have assigned names during the creation process, which is generally a good idea.

```
> theDF <- data.frame(First=x, Second=y, Sport=q)
> theDF

   First Second      Sport
1     10     -4     Hockey
2      9     -3   Football
3      8     -2   Baseball
4      7     -1    Curling
5      6      0      Rugby
6      5      1   Lacrosse
7      4      2 Basketball
8      3      3     Tennis
9      2      4    Cricket
10     1      5     Soccer
```

`data.frames` are complex objects with many attributes. The most frequently checked attributes are the number of rows and columns. Of course there are functions to do this for us: **nrow** and **ncol**. And in case both are wanted at the same time there is the **dim** function.

```
> nrow(theDF)

[1] 10

> ncol(theDF)

[1] 3

> dim(theDF)

[1] 10  3
```

Checking the column names of a `data.frame` is as simple as using the **names** function. This returns a `character vector` listing the columns. Since it is a `vector` we can access individual elements of it just like any other `vector`.

```
> names(theDF)

[1] "First"  "Second" "Sport"

> names(theDF)[3]

[1] "Sport"
```

We can also check and assign the row names of a data.frame.

```
> rownames(theDF)

 [1] "1"  "2"  "3"  "4"  "5"  "6"  "7"  "8"  "9"  "10"

> rownames(theDF) <- c("One", "Two", "Three", "Four", "Five", "Six",
+                     "Seven", "Eight", "Nine", "Ten")
> rownames(theDF)

 [1] "One"   "Two"   "Three" "Four"  "Five"  "Six"   "Seven" "Eight"
 [9] "Nine"  "Ten"

> # set them back to the generic index
> rownames(theDF) <- NULL
> rownames(theDF)

 [1] "1"  "2"  "3"  "4"  "5"  "6"  "7"  "8"  "9"  "10"
```

Usually a data.frame has far too many rows to print them all to the screen, so thankfully the **head** function prints out only the first few rows.

```
> head(theDF)

  First Second    Sport
1    10     -4   Hockey
2     9     -3 Football
3     8     -2 Baseball
4     7     -1  Curling
5     6      0    Rugby
6     5      1 Lacrosse

> head(theDF, n=7)

  First Second      Sport
1    10     -4     Hockey
2     9     -3   Football
3     8     -2   Baseball
4     7     -1    Curling
5     6      0      Rugby
6     5      1   Lacrosse
7     4      2 Basketball

> tail(theDF)

   First Second      Sport
5      6      0      Rugby
6      5      1   Lacrosse
7      4      2 Basketball
8      3      3     Tennis
```

```
9       2     4   Cricket
10      1     5    Soccer
```

As we can with other variables, we can check the `class` of a `data.frame` using the **class** function.

```
> class(theDF)

[1] "data.frame"
```

Since each column of the `data.frame` is an individual `vector`, it can be accessed individually and each has its own `class`. Like many other aspects of R, there are multiple ways to access an individual column. There is the `$` operator and also the square brackets. Running `theDF$Sport` will give the third column in `theDF`. That allows us to specify one particular column by name.

```
> theDF$Sport

 [1] Hockey     Football   Baseball   Curling    Rugby      Lacrosse
 [7] Basketball Tennis     Cricket    Soccer
10 Levels: Baseball Basketball Cricket Curling Football ... Tennis
```

Similar to `vectors`, `data.frames` allow us to access individual elements by their position using square brackets, but instead of having one position, two are specified. The first is the row number and the second is the column number. So to get the third row from the second column we use `theDF[3, 2]`.

```
> theDF[3, 2]

[1] -2
```

To specify more than one row or column, use a `vector` of indices.

```
> # row 3, columns 2 through 3
> theDF[3, 2:3]

  Second    Sport
3     -2 Baseball

> # rows 3 and 5, column 2
> # since only one column was selected it was returned as a vector
> # hence the column names will not be printed
> theDF[c(3, 5), 2]

[1] -2  0

> # rows 3 and 5, columns 2 through 3
> theDF[c(3, 5), 2:3]

  Second    Sport
3     -2 Baseball
5      0    Rugby
```

To access an entire row, specify that row while not specifying any column. Likewise, to access an entire column, specify that column while not specifying any row.

```
> # all of column 3
> # since it is only one column a vector is returned
> theDF[, 3]

 [1] Hockey     Football   Baseball   Curling    Rugby      Lacrosse
 [7] Basketball Tennis     Cricket    Soccer
10 Levels: Baseball Basketball Cricket Curling Football ... Tennis

> # all of columns 2 through 3
> theDF[, 2:3]

   Second      Sport
1      -4     Hockey
2      -3   Football
3      -2   Baseball
4      -1    Curling
5       0      Rugby
6       1   Lacrosse
7       2 Basketball
8       3     Tennis
9       4    Cricket
10      5     Soccer

> # all of row 2
> theDF[2, ]

  First Second    Sport
2     9     -3 Football

> # all of rows 2 through 4
> theDF[2:4, ]

  First Second    Sport
2     9     -3 Football
3     8     -2 Baseball
4     7     -1  Curling
```

To access multiple columns by name, make the column argument a `character vector` of the names.

```
> theDF[, c("First", "Sport")]

   First      Sport
1     10     Hockey
2      9   Football
3      8   Baseball
4      7    Curling
5      6      Rugby
```

```
6      5   Lacrosse
7      4 Basketball
8      3     Tennis
9      2    Cricket
10     1     Soccer
```

Yet another way to access a specific column is to use its column name (or its number) either as second argument to the square brackets or as the only argument to either single or double square brackets.

```
> # just the "Sport" column
> # since it is one column it returns as a (factor) vector
> theDF[, "Sport"]

 [1] Hockey     Football   Baseball   Curling    Rugby      Lacrosse
 [7] Basketball Tennis     Cricket    Soccer
10 Levels: Baseball Basketball Cricket Curling Football ... Tennis

> class(theDF[, "Sport"])

[1] "factor"

> # just the "Sport" column
> # this returns a one column data.frame
> theDF["Sport"]

        Sport
1      Hockey
2    Football
3    Baseball
4     Curling
5       Rugby
6    Lacrosse
7  Basketball
8      Tennis
9     Cricket
10     Soccer

> class(theDF["Sport"])

[1] "data.frame"

> # just the "Sport" column
> # this also returns a (factor) vector
> theDF[["Sport"]]

 [1] Hockey     Football   Baseball   Curling    Rugby      Lacrosse
 [7] Basketball Tennis     Cricket    Soccer
10 Levels: Baseball Basketball Cricket Curling Football ... Tennis

> class(theDF[["Sport"]])

[1] "factor"
```

All of these methods have differing outputs. Some return a vector; some return a single-column data.frame. To ensure a single-column data.frame while using single square brackets, there is a third argument: drop=FALSE. This also works when specifying a single column by number.

```
> theDF[, "Sport", drop=FALSE]

        Sport
1      Hockey
2    Football
3    Baseball
4     Curling
5       Rugby
6    Lacrosse
7  Basketball
8      Tennis
9     Cricket
10     Soccer

> class(theDF[, "Sport", drop=FALSE])

[1] "data.frame"

> theDF[, 3, drop=FALSE]

        Sport
1      Hockey
2    Football
3    Baseball
4     Curling
5       Rugby
6    Lacrosse
7  Basketball
8      Tennis
9     Cricket
10     Soccer

> class(theDF[, 3, drop=FALSE])

[1] "data.frame"
```

In Section 4.4.2 we see that factors are stored specially. To see how they would be represented in data.frame, form use **model.matrix** to create a set of indicator (or dummy) variables. That is one column for each level of a factor, with a 1 if a row contains that level or a 0 otherwise.

```
> newFactor <- factor(c("Pennsylvania", "New York", "New Jersey",
+                       "New York", "Tennessee", "Massachusetts",
+                       "Pennsylvania", "New York"))
> model.matrix(~ newFactor - 1)
```

```
  newFactorMassachusetts newFactorNew Jersey newFactorNew York
1                      0                    0                0
2                      0                    0                1
3                      0                    1                0
4                      0                    0                1
5                      0                    0                0
6                      1                    0                0
7                      0                    0                0
8                      0                    0                1
  newFactorPennsylvania newFactorTennessee
1                     1                  0
2                     0                  0
3                     0                  0
4                     0                  0
5                     0                  1
6                     0                  0
7                     1                  0
8                     0                  0
attr(,"assign")
[1] 1 1 1 1 1
attr(,"contrasts")
attr(,"contrasts")$newFactor
[1] "contr.treatment"
```

We learn more about formulas (the argument to **model.matrix**) in Sections 11.2 and 14.3.2 and Chapters 18 and 19.

5.2 Lists

Often a container is needed to hold arbitrary objects of either the same type or varying types. R accomplishes this through `lists`. They store any number of items of any type. A `list` can contain all `numerics` or `characters` or a mix of the two or `data.frames` or, recursively, other `lists`.

`Lists` are created with the **list** function where each argument to the function becomes an element of the `list`.

```
> # creates a three element list
> list(1, 2, 3)

[[1]]
[1] 1

[[2]]
[1] 2

[[3]]
[1] 3
```

```
> # creates a single element list
> # the only element is a vector that has three elements
> list(c(1, 2, 3))

[[1]]
[1] 1 2 3

> # creates a two element list
> # the first is a three element vector
> # the second element is a five element vector
> (list3 <- list(c(1, 2, 3), 3:7))

[[1]]
[1] 1 2 3

[[2]]
[1] 3 4 5 6 7

> # two element list
> # first element is a data.frame
> # second element is a 10 element vector
> list(theDF, 1:10)

[[1]]
   First Second      Sport
1     10     -4     Hockey
2      9     -3   Football
3      8     -2   Baseball
4      7     -1    Curling
5      6      0      Rugby
6      5      1   Lacrosse
7      4      2 Basketball
8      3      3     Tennis
9      2      4    Cricket
10     1      5     Soccer

[[2]]
 [1]  1  2  3  4  5  6  7  8  9 10

> # three element list
> # first is a data.frame
> # second is a vector
> # third is list3 which holds two vectors
> list5 <- list(theDF, 1:10, list3)
> list5
```

```
[[1]]
   First Second      Sport
1     10     -4     Hockey
2      9     -3   Football
3      8     -2   Baseball
4      7     -1    Curling
5      6      0      Rugby
6      5      1   Lacrosse
7      4      2 Basketball
8      3      3     Tennis
9      2      4    Cricket
10     1      5     Soccer

[[2]]
 [1]  1  2  3  4  5  6  7  8  9 10

[[3]]
[[3]][[1]]
[1] 1 2 3

[[3]][[2]]
[1] 3 4 5 6 7
```

Notice in the previous block of code (where `list3` was created) that enclosing an expression in parenthesis displays the results after execution.

Like `data.frames`, `lists` can have names. Each element has a unique name that can be either viewed or assigned using **names**.

```
> names(list5)

NULL

> names(list5) <- c("data.frame", "vector", "list")
> names(list5)

[1] "data.frame" "vector"     "list"

> list5

$data.frame
   First Second      Sport
1     10     -4     Hockey
2      9     -3   Football
3      8     -2   Baseball
4      7     -1    Curling
5      6      0      Rugby
6      5      1   Lacrosse
```

```
7       4       2 Basketball
8       3       3     Tennis
9       2       4    Cricket
10      1       5     Soccer

$vector
 [1]  1  2  3  4  5  6  7  8  9 10

$list
$list[[1]]
[1] 1 2 3

$list[[2]]
[1] 3 4 5 6 7
```

Names can also be assigned to list elements during creation using name-value pairs.

```
> list6 <- list(TheDataFrame=theDF, TheVector=1:10, TheList=list3)
> names(list6)

[1] "TheDataFrame" "TheVector"    "TheList"

> list6

$TheDataFrame
   First Second      Sport
1     10     -4     Hockey
2      9     -3   Football
3      8     -2   Baseball
4      7     -1    Curling
5      6      0      Rugby
6      5      1   Lacrosse
7      4      2 Basketball
8      3      3     Tennis
9      2      4    Cricket
10     1      5     Soccer

$TheVector
 [1]  1  2  3  4  5  6  7  8  9 10

$TheList
$TheList[[1]]
[1] 1 2 3

$TheList[[2]]
[1] 3 4 5 6 7
```

Creating an empty list of a certain size is, perhaps confusingly, done with **vector**.

```
> (emptyList <- vector(mode="list", length=4))

[[1]]
NULL

[[2]]
NULL

[[3]]
NULL

[[4]]
NULL
```

To access an individual element of a `list`, use double square brackets, specifying either the element number or name. Note that this allows access to only one element at a time.

```
> list5[[1]]

   First Second      Sport
1     10     -4     Hockey
2      9     -3   Football
3      8     -2   Baseball
4      7     -1    Curling
5      6      0      Rugby
6      5      1   Lacrosse
7      4      2 Basketball
8      3      3     Tennis
9      2      4    Cricket
10     1      5     Soccer

> list5[["data.frame"]]

   First Second      Sport
1     10     -4     Hockey
2      9     -3   Football
3      8     -2   Baseball
4      7     -1    Curling
5      6      0      Rugby
6      5      1   Lacrosse
7      4      2 Basketball
8      3      3     Tennis
9      2      4    Cricket
10     1      5     Soccer
```

Once an element is accessed it can be treated as if that actual element is being used, allowing nested indexing of elements.

```
> list5[[1]]$Sport

 [1] Hockey     Football   Baseball   Curling    Rugby      Lacrosse
 [7] Basketball Tennis     Cricket    Soccer
10 Levels: Baseball Basketball Cricket Curling Football ... Tennis

> list5[[1]][, "Second"]

 [1] -4 -3 -2 -1  0  1  2  3  4  5

> list5[[1]][, "Second", drop=FALSE]

   Second
1      -4
2      -3
3      -2
4      -1
5       0
6       1
7       2
8       3
9       4
10      5
```

It is possible to append elements to a `list` simply by using an index (either numeric or named) that does not exist.

```
> # see how long it currently is
> length(list5)

[1] 3

> # add a fourth element, unnamed
> list5[[4]] <- 2
> length(list5)

[1] 4

> # add a fifth element, name
> list5[["NewElement"]] <- 3:6
> length(list5)

[1] 5

> names(list5)

[1] "data.frame" "vector"     "list"       ""           "NewElement"

> list5

$data.frame
   First Second      Sport
1     10     -4     Hockey
2      9     -3   Football
```

```
3       8     -2   Baseball
4       7     -1    Curling
5       6      0      Rugby
6       5      1   Lacrosse
7       4      2 Basketball
8       3      3     Tennis
9       2      4    Cricket
10      1      5     Soccer

$vector
 [1]  1  2  3  4  5  6  7  8  9 10

$list
$list[[1]]
[1] 1 2 3

$list[[2]]
[1] 3 4 5 6 7


[[4]]
[1] 2

$NewElement
[1] 3 4 5 6
```

Occasionally appending to a `list`—or `vector` or `data.frame` for that matter—is fine, but doing so repeatedly is computationally expensive. So it is best to create a `list` as long as its final desired size and then fill it in using the appropriate indices.

5.3 Matrices

A very common mathematical structure that is essential to statistics is a `matrix`. This is similar to a `data.frame` in that it is rectangular with rows and columns except that every single element, regardless of column, must be the same type, most commonly all `numerics`. They also act similarly to `vectors` with element-by-element addition, multiplication, subtraction, division and equality. The **nrow**, **ncol** and **dim** functions work just like they do for `data.frames`.

```
> # create a 5x2 matrix
> A <- matrix(1:10, nrow=5)
> # create another 5x2 matrix
> B <- matrix(21:30, nrow=5)
> # create another 5x2 matrix
> C <- matrix(21:40, nrow=2)
> A
```

```
     [,1] [,2]
[1,]    1    6
[2,]    2    7
[3,]    3    8
[4,]    4    9
[5,]    5   10

> B

     [,1] [,2]
[1,]   21   26
[2,]   22   27
[3,]   23   28
[4,]   24   29
[5,]   25   30

> C

     [,1] [,2] [,3] [,4] [,5] [,6] [,7] [,8] [,9] [,10]
[1,]   21   23   25   27   29   31   33   35   37    39
[2,]   22   24   26   28   30   32   34   36   38    40

> nrow(A)

[1] 5

> ncol(A)

[1] 2

> dim(A)

[1] 5 2

> # add them
> A + B

     [,1] [,2]
[1,]   22   32
[2,]   24   34
[3,]   26   36
[4,]   28   38
[5,]   30   40

> # multiply them
> A * B

     [,1] [,2]
[1,]   21  156
[2,]   44  189
```

```
[3,]   69  224
[4,]   96  261
[5,]  125  300

> # see if the elements are equal
> A == B

      [,1]  [,2]
[1,] FALSE FALSE
[2,] FALSE FALSE
[3,] FALSE FALSE
[4,] FALSE FALSE
[5,] FALSE FALSE
```

`Matrix` multiplication is a commonly used operation in mathematics, requiring the number of columns of the left-hand `matrix` to be the same as the number of rows of the right-hand `matrix`. Both `A` and `B` are $5X2$ so we will transpose `B` so it can be used on the right-hand side.

```
> A %*% t(B)

     [,1] [,2] [,3] [,4] [,5]
[1,]  177  184  191  198  205
[2,]  224  233  242  251  260
[3,]  271  282  293  304  315
[4,]  318  331  344  357  370
[5,]  365  380  395  410  425
```

Another similarity with `data.frames` is that `matrices` can also have row and column names.

```
> colnames(A)

NULL

> rownames(A)

NULL

> colnames(A) <- c("Left", "Right")
> rownames(A) <- c("1st", "2nd", "3rd", "4th", "5th")
>
> colnames(B)

NULL

> rownames(B)

NULL

> colnames(B) <- c("First", "Second")
```

```
> rownames(B) <- c("One", "Two", "Three", "Four", "Five")
>
> colnames(C)

NULL

> rownames(C)

NULL

> colnames(C) <- LETTERS[1:10]
> rownames(C) <- c("Top", "Bottom")
```

There are two special vectors, letters and LETTERS, that contain the lower case and upper case letters, respectively.

Notice the effect when transposing a matrix and multiplying matrices. Transposing naturally flips the row and column names. Matrix multiplication keeps the row names from the left matrix and the column names from the right matrix.

```
> t(A)

      1st 2nd 3rd 4th 5th
Left    1   2   3   4   5
Right   6   7   8   9  10

> A %*% C

      A   B   C   D   E   F   G   H   I   J
1st 153 167 181 195 209 223 237 251 265 279
2nd 196 214 232 250 268 286 304 322 340 358
3rd 239 261 283 305 327 349 371 393 415 437
4th 282 308 334 360 386 412 438 464 490 516
5th 325 355 385 415 445 475 505 535 565 595
```

5.4 Arrays

An array is essentially a multidimensional vector. It must all be of the same type, and individual elements are accessed in a similar fashion using square brackets. The first element is the row index, the second is the column index and the remaining elements are for outer dimensions.

```
> theArray <- array(1:12, dim=c(2, 3, 2))
> theArray

, , 1

     [,1] [,2] [,3]
[1,]    1    3    5
[2,]    2    4    6
```

```
, , 2

     [,1] [,2] [,3]
[1,]    7    9   11
[2,]    8   10   12

> theArray[1, , ]

     [,1] [,2]
[1,]    1    7
[2,]    3    9
[3,]    5   11

> theArray[1, , 1]

[1] 1 3 5

> theArray[, , 1]

     [,1] [,2] [,3]
[1,]    1    3    5
[2,]    2    4    6
```

The main difference between an `array` and a `matrix` is that `matrices` are restricted to two dimensions, while `arrays` can have an arbitrary number.

5.5 Conclusion

Data come in many types and structures, which can pose a problem for some analysis environments, but R handles them with aplomb. The most common data structure is the one-dimensional `vector`, which forms the basis of everything in R. The most powerful structure is the `data.frame`—something special in R that most other languages do not have—which handles mixed data types in a spreadsheet-like format. `Lists` are useful for storing collections of items, like a hash in Perl.

6 Reading Data into R

Now that we have seen some of R's basic functionality it is time to load in data. As with everything in R, there are numerous ways to get data; the most common is probably reading comma separated values (CSV) files. Of course there are many other options that we cover as well.

6.1 Reading CSVs

The best way to read data from a CSV file[1] is to use **read.table**. Many people also like to use **read.csv**, which is a wrapper around **read.table** with the `sep` argument preset to a comma (`,`). The result of using **read.table** is a `data.frame`.

The first argument to **read.table** is the full path of the file to be loaded. The file can be sitting on disk or even the Web. For purposes of this book we will read from the Web.

Any CSV will work, but we have posted an incredibly simple file at `http://www.jaredlander.com/data/TomatoFirst.csv`. Let's read that into R using **read.table**.

```
> theUrl <- "http://www.jaredlander.com/data/TomatoFirst.csv"
> tomato <- read.table(file=theUrl, header=TRUE, sep=",")
```

This can now be seen using **head**.

```
> head(tomato)

  Round               Tomato Price      Source Sweet Acid Color Texture
1     1           Simpson SM  3.99 Whole Foods   2.8  2.8   3.7     3.4
2     1  Tuttorosso (blue)  2.99     Pioneer   3.3  2.8   3.4     3.0
3     1 Tuttorosso (green)  0.99     Pioneer   2.8  2.6   3.3     2.8
4     1     La Fede SM DOP  3.99   Shop Rite   2.6  2.8   3.0     2.3
5     2       Cento SM DOP  5.49  D Agostino   3.3  3.1   2.9     2.8
6     2     Cento Organic  4.99  D Agostino   3.2  2.9   2.9     3.1
```

1. Even though CSVs can hold numeric, text, date and other types of data, it is actually stored as text and can be opened in any text editor.

```
  Overall Avg.of.Totals Total.of.Avg
1     3.4          16.1         16.1
2     2.9          15.3         15.3
3     2.9          14.3         14.3
4     2.8          13.4         13.4
5     3.1          14.4         15.2
6     2.9          15.5         15.1
```

As mentioned before, the first argument is the file name in quotes (or as a `character` variable). Notice how we explicitly used the argument names `file`, `header` and `sep`. As discussed in Section 4.5, function arguments can be specified without the name of the argument (positionally indicated), but specifying the arguments is good practice. The second argument, `header`, indicates that the first row of data holds the column names. The third argument, `sep`, gives the delimiter separating data cells. Changing this to other values such as "`\t`" (tab delimited) or "`;`" (semicolon delimited) enables it to read other types of files.

One often unknown argument that is helpful to use is `stringsAsFactors`. Setting this to `FALSE` (the default is `TRUE`) prevents `character` columns from being converted to `factor` columns. This both saves computation time—this can be dramatic if it is a large dataset with many `character` columns with many unique values—and keeps the columns as `characters`, which are easier to work with.

Although we do not mention this argument in Section 5.1, `stringsAsFactors` can be used in **data.frame**. Re-creating that first bit of code results in an easier-to-use "Sport" column.

```
> x <- 10:1
> y <- -4:5
> q <- c("Hockey", "Football", "Baseball", "Curling", "Rugby",
+        "Lacrosse", "Basketball", "Tennis", "Cricket", "Soccer")
> theDF <- data.frame(First=x, Second=y, Sport=q, stringsAsFactors=FALSE)
> theDF$Sport

 [1] "Hockey"     "Football"   "Baseball"   "Curling"    "Rugby"
 [6] "Lacrosse"   "Basketball" "Tennis"     "Cricket"    "Soccer"
```

There are numerous other arguments to **read.table**, the most useful being `quote` and `colClasses`, respectively, specifying the character used for enclosing cells and the data type for each column.

Like **read.csv**, there are other wrapper functions for **read.table** with preset arguments. The main differences are the `sep` and `dec` arguments. These are detailed in Table 6.1.

Large files can be slow to read into memory using **read.table**, and fortunately there are alternatives available. The two most prominent functions for reading large CSVs—and other text files—are **read_delim** from the **readr** package by Hadley Wickham and **fread** from the **data.table** package by Matt Dowle, covered in Sections 6.1.1 and 6.1.2, respectively. Both are very fast, and neither converts `character` data to `factors` automatically.

Table 6.1 **Functions, and their default arguments, for reading plain text data**

Function	sep	dec
read.table	<empty>	.
read.csv	,	.
read.csv2	;	,
read.delim	\t	.
read.delim2	\t	,

6.1.1 read_delim

The **readr** package provides a family of functions for reading text files. The most commonly used will be **read_delim** for reading delimited files such as CSVs. Its first argument is the full filename or URL of the file to be read. The col_names argument is set to TRUE by default to specify that the first row of the file holds the column names.

```
> library(readr)
> theUrl <- "http://www.jaredlander.com/data/TomatoFirst.csv"
> tomato2 <- read_delim(file=theUrl, delim=',')

Parsed with column specification:
  cols(
    Round = col_integer(),
    Tomato = col_character(),
    Price = col_double(),
    Source = col_character(),
    Sweet = col_double(),
    Acid = col_double(),
    Color = col_double(),
    Texture = col_double(),
    Overall = col_double(),
    `Avg of Totals` = col_double(),
    `Total of Avg` = col_double()
  )
```

When **read_delim** is executed, a message is displayed that shows the column names and the type of data they store. The data can be displayed using **head**. **read_delim**, and all the data-reading functions in **readr**, return a tibble, which is an extension of data.frame and is further explained in Section 12.2. The most obvious visual change is that metadata is displayed such as the number of rows and columns and the data types of each column. tibbles also intelligently only print as many rows and columns as will fit on the screen.

```
> tomato2

# A tibble: 16 × 11
   Round                       Tomato Price            Source Sweet  Acid
   <int>                        <chr> <dbl>             <chr> <dbl> <dbl>
1      1                   Simpson SM  3.99       Whole Foods   2.8   2.8
2      1            Tuttorosso (blue)  2.99           Pioneer   3.3   2.8
3      1           Tuttorosso (green)  0.99           Pioneer   2.8   2.6
4      1              La Fede SM DOP  3.99         Shop Rite   2.6   2.8
5      2                Cento SM DOP  5.49        D Agostino   3.3   3.1
6      2              Cento Organic  4.99        D Agostino   3.2   2.9
7      2                 La Valle SM  3.99         Shop Rite   2.6   2.8
8      2             La Valle SM DOP  3.99            Faicos   2.1   2.7
9      3   Stanislaus Alta Cucina  4.53 Restaurant Depot   3.4   3.3
10     3                        Ciao    NA             Other   2.6   2.9
11     3          Scotts Backyard SM  0.00       Home Grown   1.6   2.9
12     3 Di Casa Barone (organic) 12.80            Eataly   1.7   3.6
13     4           Trader Joes Plum  1.49       Trader Joes   3.4   3.3
14     4            365 Whole Foods  1.49       Whole Foods   2.8   2.7
15     4          Muir Glen Organic  3.19       Whole Foods   2.9   2.8
16     4          Bionature Organic  3.39       Whole Foods   2.4   3.3
# ... with 5 more variables: Color <dbl>, Texture <dbl>,
#   Overall <dbl>, `Avg of Totals` <dbl>, `Total of Avg` <dbl>
```

Not only is **read_delim** faster than **read.table**; it also removes the need to set `stringsAsFactors` to `FALSE` since that argument does not even exist. The functions **read_csv**, **read_csv2** and **read_tsv** are special cases for when the delimiters are commas (`,`), semicolons (`;`) and tabs (`\t`), respectively. Note that the data is read into a `tbl_df` object, which is an extension of `tbl`, which is itself an extension of `data.frame`. `tbl` is a special type of `data.frame` that is defined in the **dplyr** package and explained in Section 12.2. A nice feature is that the data type of each column is displayed under the column names.

There are helper functions in **readr** that are wrappers around **read_delim** with specific delimiters preset, such as **read_csv** and **read_tsv**.

6.1.2 fread

Another options for reading large data quickly is **fread** from the **data.table** package. The first argument is the full filename or URL of the file to be read. The `header` argument indicates that the first row of the file holds the column names and `sep` specifies the field delimiter. This function has a `stringsAsFactors` argument that is set to `FALSE` by default.

```
> library(data.table)
> theUrl <- "http://www.jaredlander.com/data/TomatoFirst.csv"
> tomato3 <- fread(input=theUrl, sep=',', header=TRUE)
```

Here, also, **head** can be used to see the first few rows of data.

```
> head(tomato3)

   Round                 Tomato Price      Source Sweet Acid Color
1:     1             Simpson SM  3.99 Whole Foods   2.8  2.8   3.7
2:     1    Tuttorosso (blue)  2.99     Pioneer   3.3  2.8   3.4
3:     1 Tuttorosso (green)  0.99     Pioneer   2.8  2.6   3.3
4:     1        La Fede SM DOP  3.99   Shop Rite   2.6  2.8   3.0
5:     2          Cento SM DOP  5.49  D Agostino   3.3  3.1   2.9
6:     2        Cento Organic  4.99  D Agostino   3.2  2.9   2.9
   Texture Overall Avg of Totals Total of Avg
1:     3.4     3.4          16.1         16.1
2:     3.0     2.9          15.3         15.3
3:     2.8     2.9          14.3         14.3
4:     2.3     2.8          13.4         13.4
5:     2.8     3.1          14.4         15.2
6:     3.1     2.9          15.5         15.1
```

This is also faster than **read.table** and results in a `data.table` object, which is an extension of `data.frame`. This is another special object that improves upon `data.frames` and is explained in Section 11.4.

Both **read_delim** or **fread** are fast, capable functions, so the decision of which to use depends upon whether **dplyr** or **data.table** is preferred for data manipulation.

6.2 Excel Data

Excel may be the world's most popular data analysis tool, and while that has benefits and disadvantages, it means that R users will sometimes be required to read Excel files. When the first edition of this book was published, it was difficult to read Excel data into R. At the time, the simplest method would be to use Excel (or another spreadsheet program) to convert the Excel file to a CSV file and then use **read.csv**. While that seems like an unnecessary step, it was actually the recommended method in the R manual.

Fortunately, for anyone tasked with using Excel data, the package **readxl**, by Hadley Wickham, makes reading Excel files, both `.xls` and `.xlsx`, easy. The main function is **read_excel**, which reads the data from a single Excel sheet. Unlike **read.table**, **read_delim** and **fread**, **read_excel** cannot read data directly from the Internet, and thus the files must be downloaded first. We could do this by visiting a browser or we can stay within R and use **download.file**.

```
> download.file(url='http://www.jaredlander.com/data/ExcelExample.xlsx',
+               destfile='data/ExcelExample.xlsx', method='curl')
```

After we download the file we check the sheets in the Excel file.

```
> library(readxl)
> excel_sheets('data/ExcelExample.xlsx')

[1] "Tomato" "Wine"   "ACS"
```

By default **read_excel** reads the first sheet, in this case the one holding the `tomato` data. The result is a `tibble` rather than a traditional `data.frame`.

```
> tomatoXL <- read_excel('data/ExcelExample.xlsx')
> tomatoXL

# A tibble: 16 × 11
   Round                   Tomato Price           Source Sweet  Acid
   <dbl>                    <chr> <dbl>            <chr> <dbl> <dbl>
1      1               Simpson SM  3.99      Whole Foods   2.8   2.8
2      1        Tuttorosso (blue)  2.99          Pioneer   3.3   2.8
3      1       Tuttorosso (green)  0.99          Pioneer   2.8   2.6
4      1          La Fede SM DOP  3.99        Shop Rite   2.6   2.8
5      2             Cento SM DOP  5.49       D Agostino   3.3   3.1
6      2            Cento Organic  4.99       D Agostino   3.2   2.9
7      2              La Valle SM  3.99        Shop Rite   2.6   2.8
8      2          La Valle SM DOP  3.99           Faicos   2.1   2.7
9      3   Stanislaus Alta Cucina  4.53 Restaurant Depot   3.4   3.3
10     3                     Ciao    NA            Other   2.6   2.9
11     3       Scotts Backyard SM  0.00      Home Grown   1.6   2.9
12     3 Di Casa Barone (organic) 12.80           Eataly   1.7   3.6
13     4        Trader Joes Plum  1.49      Trader Joes   3.4   3.3
14     4          365 Whole Foods  1.49      Whole Foods   2.8   2.7
15     4       Muir Glen Organic  3.19      Whole Foods   2.9   2.8
16     4       Bionature Organic  3.39      Whole Foods   2.4   3.3
# ... with 5 more variables: Color <dbl>, Texture <dbl>,
#   Overall <dbl>, `Avg of Totals` <dbl>, `Total of Avg` <dbl>
```

Since `tomatoXL` is a `tibble` only the columns that fit on the screen (or in this case the page) are printed. This will vary depending on how wide the terminal is set.

Specifying which sheet to read can be done by supplying either the sheet position as a number or the sheet name as a `character`.

```
> # using position
> wineXL1 <- read_excel('data/ExcelExample.xlsx', sheet=2)
> head(wineXL1)

# A tibble: 6 × 14
  Cultivar Alcohol `Malic acid`   Ash `Alcalinity of ash  ` Magnesium
     <dbl>   <dbl>        <dbl> <dbl>                 <dbl>     <dbl>
1        1   14.23         1.71  2.43                  15.6       127
2        1   13.20         1.78  2.14                  11.2       100
3        1   13.16         2.36  2.67                  18.6       101
4        1   14.37         1.95  2.50                  16.8       113
5        1   13.24         2.59  2.87                  21.0       118
6        1   14.20         1.76  2.45                  15.2       112
# ... with 8 more variables: `Total phenols` <dbl>, Flavanoids <dbl>,
#   `Nonflavanoid phenols` <dbl>, Proanthocyanins <dbl>, `Color
#   intensity` <dbl>, Hue <dbl>, `OD280/OD315 of diluted
#   wines` <dbl>, `Proline ` <dbl>
```

```
> # using name
> wineXL2 <- read_excel('data/ExcelExample.xlsx', sheet='Wine')
> head(wineXL2)

# A tibble: 6 × 14
  Cultivar Alcohol `Malic acid`   Ash `Alcalinity of ash  ` Magnesium
     <dbl>   <dbl>        <dbl> <dbl>                 <dbl>     <dbl>
1        1   14.23         1.71  2.43                  15.6       127
2        1   13.20         1.78  2.14                  11.2       100
3        1   13.16         2.36  2.67                  18.6       101
4        1   14.37         1.95  2.50                  16.8       113
5        1   13.24         2.59  2.87                  21.0       118
6        1   14.20         1.76  2.45                  15.2       112
# ... with 8 more variables: `Total phenols` <dbl>, Flavanoids <dbl>,
#   `Nonflavanoid phenols` <dbl>, Proanthocyanins <dbl>, `Color
#   intensity` <dbl>, Hue <dbl>, `OD280/OD315 of diluted
#   wines` <dbl>, `Proline ` <dbl>
```

Reading Excel data used to be a burdensome experience but is now as simple as reading a CSV thanks to Hadley Wickham's **readxl** package.

6.3 Reading from Databases

Databases arguably store the vast majority of the world's data. Most of these, whether they be PostgreSQL, MySQL, Microsoft SQL Server or Microsoft Access, can be accessed either through various drivers, typically via an ODBC connection. The most popular open-source databases have packages such as **RPostgreSQL** and **RMySQL**. Other databases without a specific package can make use of the more generic, and aptly named, **RODBC** package. Database connectivity can be difficult, so the **DBI** package was written to create a uniform experience while working with different databases.

Setting up a database is beyond the scope of this book so we use a simple SQLite database though these steps will be similar for most databases. First, we download the database file[2] using **download.file**.

```
> download.file("http://www.jaredlander.com/data/diamonds.db",
+               destfile = "data/diamonds.db", mode='wb')
```

Since SQLite has its own R package, **RSQLite**, we use that to connect to our database, otherwise we would use **RODBC**.

```
> library(RSQLite)
```

To connect to the database we first specify the driver using **dbDriver**. The function's main argument is the type of driver, such as "SQLite" or "ODBC."

```
> drv <- dbDriver('SQLite')
> class(drv)
```

2. SQLite is special in that the entire database is stored in a single file on disk making it easy for lightweight applications and sharing.

```
[1] "SQLiteDriver"
attr(,"package")
[1] "RSQLite"
```

We then establish a connection to the specific database with **dbConnect**. The first argument is the driver. The most common second argument is the DSN[3] connection string for the database, or the path to the file for SQLite databases. Additional arguments are typically the database username, password, host and port.

```
> con <- dbConnect(drv, 'data/diamonds.db')
> class(con)

[1] "SQLiteConnection"
attr(,"package")
[1] "RSQLite"
```

Now that we are connected to the database we can learn more about the database, such as the table names and the fields within tables, using functions from the **DBI** package.

```
> dbListTables(con)

[1] "DiamondColors" "diamonds"      "sqlite_stat1"

> dbListFields(con, name='diamonds')

 [1] "carat"   "cut"     "color"   "clarity" "depth" "table"
 [7] "price"   "x"       "y"       "z"

> dbListFields(con, name='DiamondColors')

[1] "Color"       "Description" "Details"
```

At this point we are ready to run a query on that database using **dbGetQuery**. This can be any valid SQL query of arbitrary complexity. **dbGetQuery** returns an ordinary `data.frame`, just like any other. Fortunately, **dbGetQuery** has the `stringsAsFactors` argument seen in Section 6.1. Again, setting this to `FALSE` is usually a good idea, as it will save processing time and keep `character` data as `character`.

```
> # simple SELECT * query from one table
> diamondsTable <- dbGetQuery(con,
+                             "SELECT * FROM diamonds",
+                             stringsAsFactors=FALSE)
> # simple SELECT * query from one table
> colorTable <- dbGetQuery(con,
+                          "SELECT * FROM DiamondColors",
+                          stringsAsFactors=FALSE)
```

---

3. This differs by operating system but should result in a string name for that connection.

```
> # do a join between the two tables
> longQuery <- "SELECT * FROM diamonds, DiamondColors
                WHERE
                diamonds.color = DiamondColors.Color"
> diamondsJoin <- dbGetQuery(con, longQuery,
                     stringsAsFactors=FALSE)
```

We can easily check the results of these queries by viewing the resulting `data.frames`.

```
> head(diamondsTable)

  carat       cut color clarity depth table price    x    y    z
1  0.23     Ideal     E     SI2  61.5    55   326 3.95 3.98 2.43
2  0.21   Premium     E     SI1  59.8    61   326 3.89 3.84 2.31
3  0.23      Good     E     VS1  56.9    65   327 4.05 4.07 2.31
4  0.29   Premium     I     VS2  62.4    58   334 4.20 4.23 2.63
5  0.31      Good     J     SI2  63.3    58   335 4.34 4.35 2.75
6  0.24 Very Good     J    VVS2  62.8    57   336 3.94 3.96 2.48

> head(colorTable)

  Color          Description                    Details
1     D Absolutely Colorless                   No color
2     E            Colorless      Minute traces of color
3     F            Colorless      Minute traces of color
4     G       Near Colorless Color is dificult to detect
5     H       Near Colorless Color is dificult to detect
6     I       Near Colorless   Slightly detectable color

> head(diamondsJoin)

  carat       cut color clarity depth table price    x    y    z
1  0.23     Ideal     E     SI2  61.5    55   326 3.95 3.98 2.43
2  0.21   Premium     E     SI1  59.8    61   326 3.89 3.84 2.31
3  0.23      Good     E     VS1  56.9    65   327 4.05 4.07 2.31
4  0.29   Premium     I     VS2  62.4    58   334 4.20 4.23 2.63
5  0.31      Good     J     SI2  63.3    58   335 4.34 4.35 2.75
6  0.24 Very Good     J    VVS2  62.8    57   336 3.94 3.96 2.48
  Color    Description                   Details
1     E      Colorless    Minute traces of color
2     E      Colorless    Minute traces of color
3     E      Colorless    Minute traces of color
4     I Near Colorless Slightly detectable color
5     J Near Colorless Slightly detectable color
6     J Near Colorless Slightly detectable color
```

While it is not necessary, it is good practice to close the ODBC connection using **dbDisconnect**, although it will close automatically when either R closes or we open another connection using **dbConnect**. Only one connection may be open at a time.

6.4 Data from Other Statistical Tools

In an ideal world another tool besides R would never be needed, but in reality data are sometimes locked in a proprietary format such as those from SAS, SPSS or Octave. The **foreign** package provides a number of functions similar to **read.table** to read in data from other tools.

A partial list of functions to read data from commonly used statistical tools is in Table 6.2. The arguments for these functions are generally similar to **read.table**. These functions usually return the data as a `data.frame` but do not always succeed.

Table 6.2 Functions for reading data from some commonly used statistical tools

Function	Format
read.spss	SPSS
read.dta	Stata
read.ssd	SAS
read.octave	Octave
read.mtp	Minitab
read.systat	Systat

While **read.ssd** can read SAS data it requires a valid SAS license. This can be sidestepped by using Microsoft R Server from Microsoft, with their special **RxSasData** function in their **RevoScaleR** package.

Like many concepts in R, Hadley Wickham has written a new package, **haven**, that is very similar to **foreign** but follows his conventions, is optimized for speed and convenience and results in a `tibble` rather than a `data.frame`. The most common **haven** functions for reading data from other statistical tools are listed in Table 6.3.

Table 6.3 Functions for reading data from some commonly used statistical tools

Function	Format
read_spss	SPSS
read_sas	Stata
read_stata	Systat

6.5 R Binary Files

When working with other R programmers, a good way to pass around data—or any R objects such as variables and functions—is to use RData files. These are binary files that represent R objects of any kind. They can store a single object or multiple objects and can be passed among Windows, Mac and Linux without a problem.

First, let's create an RData file, remove the object that created it and then read it back into R.

```
> # save the tomato data.frame to disk
> save(tomato, file="data/tomato.rdata")
> # remove tomato from memory
> rm(tomato)
> # check if it still exists
> head(tomato)

Error in head(tomato): object 'tomato' not found

> # read it from the rdata file
> load("data/tomato.rdata")
> # check if it exists now
> head(tomato)

  Round                Tomato Price      Source Sweet Acid Color Texture
1     1          Simpson SM  3.99 Whole Foods   2.8  2.8   3.7     3.4
2     1   Tuttorosso (blue)  2.99     Pioneer   3.3  2.8   3.4     3.0
3     1 Tuttorosso (green)  0.99     Pioneer   2.8  2.6   3.3     2.8
4     1      La Fede SM DOP  3.99   Shop Rite   2.6  2.8   3.0     2.3
5     2       Cento SM DOP  5.49  D Agostino   3.3  3.1   2.9     2.8
6     2      Cento Organic  4.99  D Agostino   3.2  2.9   2.9     3.1
  Overall Avg.of.Totals Total.of.Avg
1     3.4          16.1         16.1
2     2.9          15.3         15.3
3     2.9          14.3         14.3
4     2.8          13.4         13.4
5     3.1          14.4         15.2
6     2.9          15.5         15.1
```

Now let's create a few objects to store in a single RData file, remove them and then load them again.

```
> # create some objects
> n <- 20
> r <- 1:10
> w <- data.frame(n, r)
```

```
> # check them out
> n

[1] 20

> r

 [1]  1  2  3  4  5  6  7  8  9 10

> w

    n  r
1  20  1
2  20  2
3  20  3
4  20  4
5  20  5
6  20  6
7  20  7
8  20  8
9  20  9
10 20 10

> # save them
> save(n, r, w, file="data/multiple.rdata")
> # delete them
> rm(n, r, w)
> # are they gone?
> n

Error in eval(expr, envir, enclos): object 'n' not found

> r

Error in eval(expr, envir, enclos): object 'r' not found

> w

Error in eval(expr, envir, enclos): object 'w' not found

> # load them back
> load("data/multiple.rdata")
> # check them out again
> n

[1] 20

> r

 [1]  1  2  3  4  5  6  7  8  9 10
```

```
> w

    n  r
1  20  1
2  20  2
3  20  3
4  20  4
5  20  5
6  20  6
7  20  7
8  20  8
9  20  9
10 20 10
```

These objects are restored into the working environment, with the same names they had when they were saved to the `RData` file. That is why we do not assign the result of the **load** function to an object.

The **saveRDS** saves one object in a binary `RDS` file. The object is not saved with a name, so when we use **readRDS** to load the file into the working environment, we assign it to an object.

```
> # create an object
> smallVector <- c(1, 5, 4)
> # view it
> smallVector

[1] 1 5 4

> # save to rds file
> saveRDS(smallVector, file='thisObject.rds')
>
> # read the file and save to a different object
> thatVect <- readRDS('thisObject.rds')
> # display it
> thatVect

[1] 1 5 4

> # check they are the same
> identical(smallVector, thatVect)

[1] TRUE
```

6.6 Data Included with R

R and some packages come with data included, so we can easily have data to use. Accessing this data is simple as long as we know what to look for. **ggplot2**, for instance, comes with a dataset about diamonds. It can be loaded using the **data** function.

```
> data(diamonds, package='ggplot2')
> head(diamonds)

# A tibble: 6 × 10
  carat       cut color clarity depth table price     x     y     z
  <dbl>     <ord> <ord>   <ord> <dbl> <dbl> <int> <dbl> <dbl> <dbl>
1  0.23     Ideal     E     SI2  61.5    55   326  3.95  3.98  2.43
2  0.21   Premium     E     SI1  59.8    61   326  3.89  3.84  2.31
3  0.23      Good     E     VS1  56.9    65   327  4.05  4.07  2.31
4  0.29   Premium     I     VS2  62.4    58   334  4.20  4.23  2.63
5  0.31      Good     J     SI2  63.3    58   335  4.34  4.35  2.75
6  0.24 Very Good     J    VVS2  62.8    57   336  3.94  3.96  2.48
```

To find a list of available data, simply type `data()` into the console.

6.7 Extract Data from Web Sites

These days a lot of data are displayed on Web pages. If we are lucky, then it is stored neatly in an HTML table. If we are not so lucky, we might need to parse the text of the page.

6.7.1 Simple HTML Tables

If the data are stored neatly in an `HTML table`, we can use **readHTMLTable** in the **XML** package to easily extract it. On my site there is a post about a Super Bowl pool I was asked to analyze at `http://www.jaredlander.com/2012/02/another-kind-of-super-bowl-pool`. In that post there is a table with three columns that we wish to extract. It is fairly simple with the following code.

```
> library(XML)
> theURL <- "http://www.jaredlander.com/2012/02/another-kind-of-
+                 super-bowl-pool/"
> bowlPool <- readHTMLTable(theURL, which=1, header=FALSE,
+                           stringsAsFactors=FALSE)
> bowlPool

                V1      V2        V3
1   Participant 1 Giant A Patriot Q
2   Participant 2 Giant B Patriot R
3   Participant 3 Giant C Patriot S
4   Participant 4 Giant D Patriot T
5   Participant 5 Giant E Patriot U
6   Participant 6 Giant F Patriot V
7   Participant 7 Giant G Patriot W
8   Participant 8 Giant H Patriot X
9   Participant 9 Giant I Patriot Y
10 Participant 10 Giant J Patriot Z
```

Here the first argument was the URL, but it could have also been a file on disk. The `which` argument allows us to choose which table to read if there are multiple tables. For this example, there was only one table, but it could have easily been the second or third or fourth.... We set `header` to `FALSE` to indicate that no header was in the table. Lastly, we used `stringsAsFactors=FALSE` so that the `character` columns would not be converted to `factors`.

6.7.2 Scraping Web Data

Information is often scattered about in `tables`, `divs`, `spans` or other `HTML` elements. As an example we put the menu and restaurant details for Ribalta, a beloved New York pizzeria, into an `HTML` file. The address and phone number are stored in an `ordered list`, section identifiers are in `spans` and the items and prices are in `tables`. We use Hadley Wickham's **rvest** package to extract the data into a usable format.

The file can be read directly from the URL, or from disc, using **read_html**. This creates an `xml_document` object that holds all of the `HTML`.

```
> library(rvest)
> ribalta <- read_html('http://www.jaredlander.com/data/ribalta.html')
> class(ribalta)
> ribalta

[1] "xml_document" "xml_node"
{xml_document}
<html xmlns="http://www.w3.org/1999/xhtml">
[1] <head>\n<meta http-equiv="Content-Type" content="text/html; cha ...
[2] <body>\r\n<ul>\n<li class="address">\r\n    <span class="street ...
```

By exploring the `HTML` we see that the address is stored in a `span`, which is an element of an `ordered list`. First we use **html_nodes** to select all `span` elements within `ul` elements.

```
> ribalta %>% html_nodes('ul') %>% html_nodes('span')

{xml_nodeset (6)}
[1] <span class="street">48 E 12th St</span>
[2] <span class="city">New York</span>
[3] <span class="zip">10003</span>
[4] <span>\r\n    \t<span id="latitude" value="40.733384"></span>\r ...
[5] <span id="latitude" value="40.733384"></span>
[6] <span id="longitude" value="-73.9915618"></span>
```

This returns a `list` of all `span` nodes inside `ul` elements. Following a highly nested hierarchy of `HTML` elements can be bothersome and brittle, so we instead identify the element of our interest by specifying its class, in this case `'street'`. In `HTML`, class is denoted with a period (`.`) and ID is denoted with a hash (#). Instead of having **html_nodes** search for a `span`, we have it search for any element with class `'street'`.

```
> ribalta %>% html_nodes('.street')

{xml_nodeset (1)}
[1] <span class="street">48 E 12th St</span>
```

We properly extracted the HTML element but not the information stored in the element. For that we need to call **html_text** to extract the text stored inside the span element.

```
> ribalta %>% html_nodes('.street') %>% html_text()

[1] "48 E 12th St"
```

When information is stored as an attribute of an HTML element, we use **html_attr** rather than **html_text**. In this case we extract the longitude value that is an attribute of the span element with ID 'longitude'.

```
> ribalta %>% html_nodes('#longitude') %>% html_attr('value')

[1] "-73.9915618"
```

In this particular file a lot of information is stored in tables with class 'food-items', so we specify that **html_nodes** should search for all tables with class 'food-items'. Since multiple tables exist, we specify that we want the sixth table using the **extract2** function from the **magrittr** package. The data is finally extracted and stored in a data.frame using **html_table**. In this case the tables do not have headers, so the columns of the data.frame have generic names.

```
> ribalta %>%
+     html_nodes('table.food-items') %>%
+     magrittr::extract2(5) %>%
+     html_table()

                     X1
1    Marinara Pizza Rosse
2         Doc Pizza Rosse
3 Vegetariana Pizza Rosse
4    Brigante Pizza Rosse
5     Calzone Pizza Rosse
6   Americana Pizza Rosse
                                                            X2 X3
1                                  basil, garlic and oregano.  9
2                               buffalo mozzarella and basil. 15
3                mozzarella cheese, basil and baked vegetables. 15
4                      mozzarella cheese, salami and spicy oil. 15
5 ricotta, mozzarella cheese, prosciutto cotto and black pepper. 16
6                         mozzarella cheese, wurstel and fries. 16
```

6.8 Reading JSON Data

A popular format for data, especially for APIs and document databases, is JSON, which stands for JavaScript Object Notation. It is a data format, stored in plain text, which is well suited for nested data. The two main R packages for reading JSON data are **rjson** and **jsonlite**.

The following is a sample from a JSON file listing some of our favorite pizza places in New York. There is an entry for each pizzeria. Within that is a `Name` element and an array, named `Details`, that holds elements for `Address`, `City`, `State`, `Zip` and `Phone`.

```
[
    {
        "Name": "Di Fara Pizza",
        "Details": [
            {
                "Address": "1424 Avenue J",
                "City": "Brooklyn",
                "State": "NY",
                "Zip": "11230"
            }
        ]
    },
    {
        "Name": "Fiore's Pizza",
        "Details": [
            {
                "Address": "165 Bleecker St",
                "City": "New York",
                "State": "NY",
                "Zip": "10012"
            }
        ]
    },
    {
        "Name": "Juliana's",
        "Details": [
            {
                "Address": "19 Old Fulton St",
                "City": "Brooklyn",
                "State": "NY",
                "Zip": "11201"
            }
        ]
    }
]
```

The **fromJSON** function reads the file into R and parses the JSON text. By default, it attempts to simplify the data into a `data.frame`.

```
> library(jsonlite)
> pizza <- fromJSON('http://www.jaredlander.com/data/
                    PizzaFavorites.json')
```

```
> pizza
                      Name                                  Details
1            Di Fara Pizza     1424 Avenue J, Brooklyn, NY, 11230
2           Fiore's Pizza    165 Bleecker St, New York, NY, 10012
3               Juliana's  19 Old Fulton St, Brooklyn, NY, 11201
4      Keste Pizza & Vino    271 Bleecker St, New York, NY, 10014
5   L & B Spumoni Gardens       2725 86th St, Brooklyn, NY, 11223
6 New York Pizza Suprema         413 8th Ave, New York, NY, 10001
7            Paulie Gee's 60 Greenpoint Ave, Brooklyn, NY, 11222
8                 Ribalta       48 E 12th St, New York, NY, 10003
9               Totonno's  1524 Neptune Ave, Brooklyn, NY, 11224
```

The result is a two-column `data.frame` where the first column is the `Name` and the second column, named `Details`, is actually a one-row `data.frame` for each row of the outer `data.frame`. This may seem odd, but storing objects in cells of `data.frames` has long been possible and has recently become more and more the norm. We can see that `Details` is a `list`-column where each element is a `data.frame`.

```
> class(pizza)

[1] "data.frame"

> class(pizza$Name)

[1] "character"

> class(pizza$Details)

[1] "list"

> class(pizza$Details[[1]])

[1] "data.frame"
```

This nested structure of a `data.frame` within a `data.frame` is best unraveled using the tools available in **dplyr**, **tidyr** and **purrr** described in Chapters 12, 15 and 13.

6.9 Conclusion

Reading data is the first step to any analysis; without the data there is nothing to do. The most common way to read data into R is from a CSV using **read.table** or Excel using **read_excel**. The various database packages, and generically **RODBC**, provide an excellent method for reading from databases. Reading from data trapped in `HTML` is made easy using the **XML** and **rvest** packages. R also has special binary file formats, `RData` and `RDS`, for the quick storage, loading and transfer of R objects.

7 Statistical Graphics

One of the hardest parts of an analysis is producing quality supporting graphics. Conversely, a good graph is one of the best ways to present findings. Fortunately, R provides excellent graphing capabilities, both in the base installation and with add-on packages such as **lattice** and **ggplot2**. We will briefly present some simple graphs using base graphics and then show their counterparts in **ggplot2**. This will be supplemented throughout the book where supporting graphics—with code—will be made using **ggplot2** and occasionally base graphics.

Graphics are used in statistics primarily for two reasons: exploratory data analysis (EDA) and presenting results. Both are incredibly important but must be targeted to different audiences.

7.1 Base Graphics

When graphing for the first time with R, most people use base graphics and then move on to **ggplot2** when their needs become more complex. While base graphs can be beautiful creations, we recommend spending the most time learning about **ggplot2** in Section 7.2. This section is here for completeness and because base graphics are just needed, especially for modifying the plots generated by other functions.

Before we can go any further we need some data. Most of the datasets built into R are tiny, even by standards from ten years ago. A good dataset for example graphs is, ironically, included with **ggplot2**. In order to access it, **ggplot2** must first be installed and loaded. Then the `diamonds` data can be loaded and inspected.

```
> library(ggplot2)
> data(diamonds)
> head(diamonds)

# A tibble: 6 × 10
  carat       cut color clarity depth table price     x     y     z
  <dbl>     <ord> <ord>   <ord> <dbl> <dbl> <int> <dbl> <dbl> <dbl>
1  0.23     Ideal     E     SI2  61.5    55   326  3.95  3.98  2.43
2  0.21   Premium     E     SI1  59.8    61   326  3.89  3.84  2.31
3  0.23      Good     E     VS1  56.9    65   327  4.05  4.07  2.31
4  0.29   Premium     I     VS2  62.4    58   334  4.20  4.23  2.63
5  0.31      Good     J     SI2  63.3    58   335  4.34  4.35  2.75
6  0.24 Very Good     J    VVS2  62.8    57   336  3.94  3.96  2.48
```

7.1.1 Base Histograms

The most common graph of data in a single variable is a histogram. This shows the distribution of values for that variable. Creating a histogram is very simple and illustrated below for the `carat` column in `diamonds`.

```
> hist(diamonds$carat, main="Carat Histogram", xlab="Carat")
```

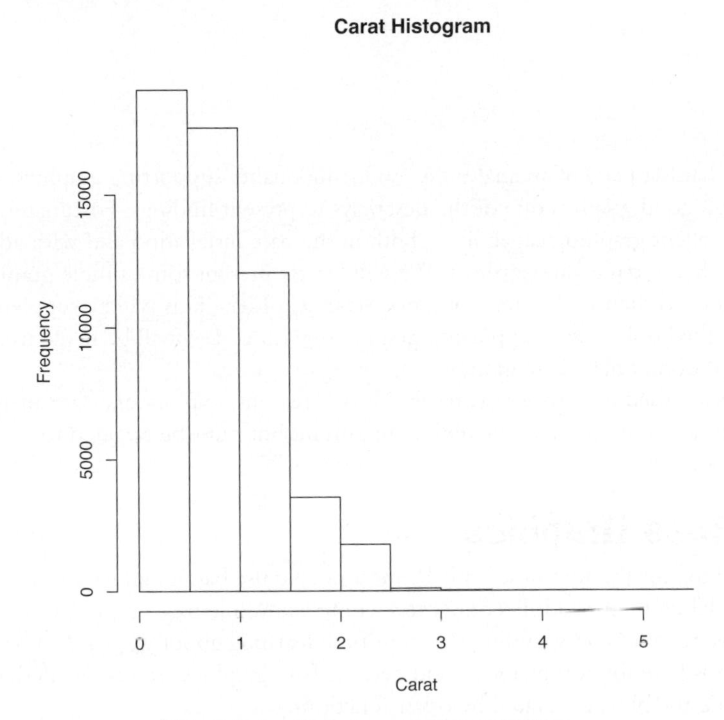

Figure 7.1 Histogram of diamond carats.

This shows the distribution of the carat size. Notice that the title was set using the `main` argument and the x-axis label with the `xlab` argument. More complicated histograms are easier to create with **ggplot2**. These extra capabilities are presented in Section 7.2.1.

Histograms break the data into buckets and the heights of the bars represent the number of observations that fall into each bucket. This can be sensitive to the number and size of buckets, so making a good histogram can require some experimentation.

7.1.2 Base Scatterplot

It is frequently good to see two variables in comparison with each other; this is where the scatterplot is used. Every point represents an observation in two variables where the x-axis represents one variable and the y-axis another. We will plot the price of diamonds against the carat using `formula` notation.

```
> plot(price ~ carat, data=diamonds)
```

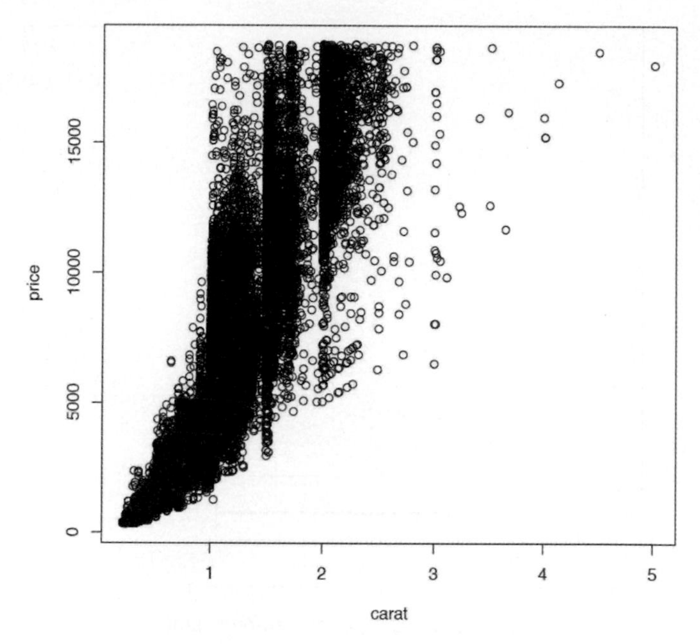

Figure 7.2 Scatterplot of diamond price versus carat.

The ~ separating `price` and `carat` indicates that we are viewing `price` against `carat`, where `price` is the y value and `carat` is the x value. Formulas are explained in more detail in Chapters 18 and 19.

It is also possible to build a scatterplot by simply specifying the x and y variables without the `formula` interface. This allows plotting of variables that are not necessarily in a `data.frame`.

```
> plot(diamonds$carat, diamonds$price)
```

Scatterplots are one of the most frequently used statistical graphs and will be detailed further using **ggplot2** in Section 7.2.2.

7.1.3 Boxplots

Although boxplots are often among the first graphs taught to statistics students, they are a matter of great debate in the statistics community. Andrew Gelman from Columbia University has been very vocal in his displeasure with boxplots.[1] However, other people such as Hadley Wickham[2] and John Tukey are strong proponents of the boxplot. Given their ubiquity (deserved or not) it is important to learn them. Thankfully, R has the **boxplot** function.

```
> boxplot(diamonds$carat)
```

1. `http://andrewgelman.com/2009/02/boxplot_challen/` and `http://andrewgelman.com/2009/10/better_than_a_b/`

2. `http://vita.had.co.nz/papers/boxplots.pdf`

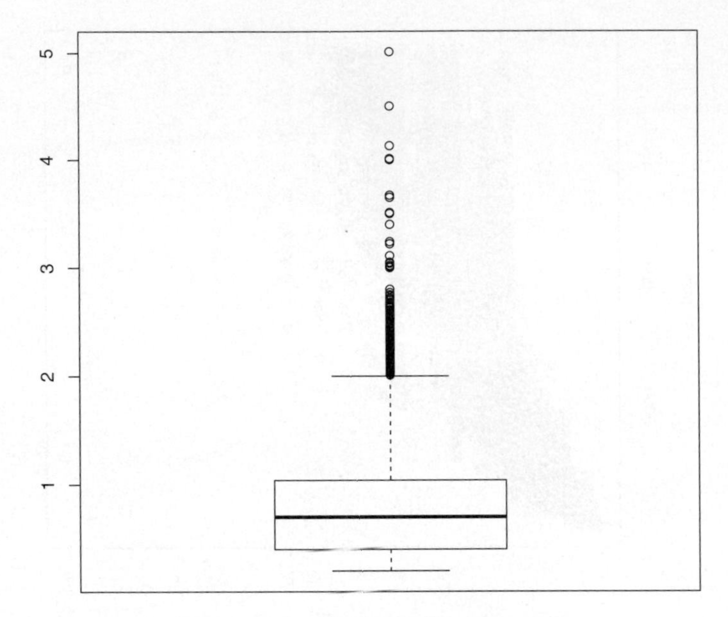

Figure 7.3 Boxplot of diamond carat.

The idea behind the boxplot is that the thick middle line represents the median and the box is bounded by the first and third quartiles. That is, the middle 50 percent of data (the Interquartile Range or IQR) is held in the box. The lines extend out to 1.5*IQR in both directions. Outlier points are then plotted beyond that. It is important to note, that while 50 percent of the data are very visible in the box, that means 50 percent of the data is not really displayed. That is a lot of information to not see.

As with other graphs previously discussed, more details will be provided using **ggplot2** in Section 7.2.3.

Many objects, such as linear models and contingency tables, have built-in plot functions, which we will see later in the book.

7.2 ggplot2

While R's base graphics are extremely powerful and flexible and can be customized to a great extent, using them can be labor intensive. Two packages—**ggplot2** and **lattice**—were built to make graphing easier. Over the past few years **ggplot2** has far exceeded **lattice** in popularity and features. We re-create all the previous graphs in Section 7.1 and expand the examples with more advanced features. Neither this chapter nor this book is an exhaustive review of **ggplot2**. But throughout this book, where there is a plot the accompanying code (mostly with **ggplot2**, although some use base graphics) is included.

Initially, the **ggplot2** syntax is harder to grasp, but the effort is more than worthwhile. It is much easier to delineate data by color, shape or size and add legends with **ggplot2**.

Graphs are quicker to build. Graphs that could take 30 lines of code with base graphics are possible with just one line in **ggplot2**.

The basic structure for **ggplot2** starts with the **ggplot** function,[3] which at its most basic should take the data as its first argument. It can take more arguments, or fewer, but we will stick with that for now. After initializing the object, we add layers using the + symbol. To start, we will just discuss geometric layers such as points, lines and histograms. They are included using functions like **geom_point**, **geom_line** and **geom_histogram**. These functions take multiple arguments, the most important being which variable in the data gets mapped to which axis or other aesthetic using **aes**. Furthermore, each layer can have different aesthetic mappings and even different data.

7.2.1 ggplot2 Histograms and Densities

Returning to the histogram seen in Figure 7.1, we plot the distribution of diamond carats using **ggplot2**. This is built using **ggplot** and **geom_histogram**. Because histograms are one-dimensional displays of data, we need to specify only one aesthetic mapping, the x-axis. Figure 7.4 shows the plot.

```
> ggplot(data=diamonds) + geom_histogram(aes(x=carat))
```

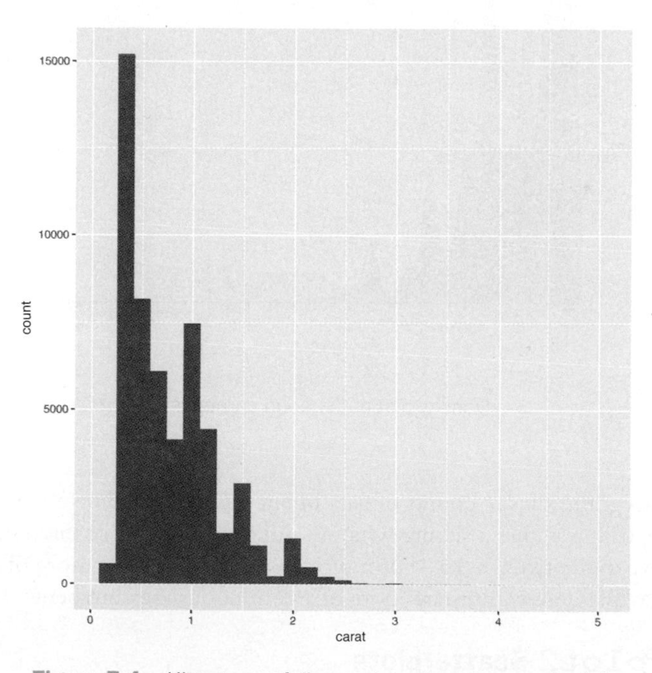

Figure 7.4 Histogram of diamond carats using ggplot2.

3. The package was previously called ggplot, but early on Hadley made massive changes, so he upgraded the name to ggplot2.

A similar display is the density plot, which is done by changing **geom_histogram** to **geom_density**. We also specify the color to fill in the graph using the `fill` argument. This differs from the `color` argument that we will see later. Also notice that the `fill` argument was entered outside the **aes** function. This is because we want the whole graph to be that color. We will see how it can be used inside **aes** later. This results in the graph shown in Figure 7.5.

```
> ggplot(data=diamonds) + geom_density(aes(x=carat), fill="grey50")
```

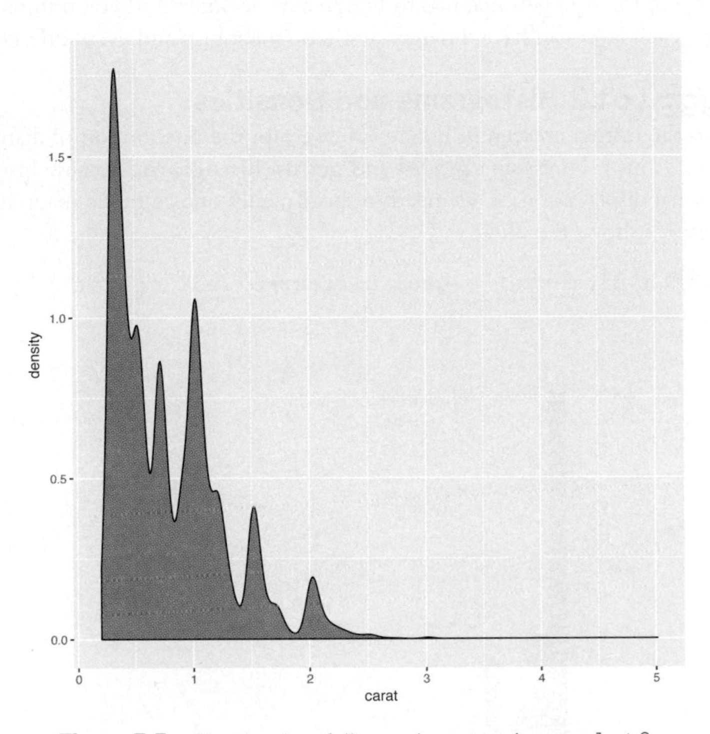

Figure 7.5 Density plot of diamond carats using `ggplot2`.

Whereas histograms display counts of data in buckets, density plots show the probability of observations falling within a sliding window along the variable of interest. The difference between the two is subtle but important. Histograms are more of a discrete measurement, while density plots are more of a continuous measurement.

7.2.2 `ggplot2` Scatterplots

Here we not only show the **ggplot2** way of making scatterplots but also show off some of the power of **ggplot2**. We start by re-creating the simple scatterplot in Figure 7.2. Like before, we use **ggplot** to initialize the object, but this time we include **aes** inside the **ggplot** call instead of using it in the geom. The **ggplot2** version is shown in Figure 7.6.

```
> ggplot(diamonds, aes(x=carat, y=price)) + geom_point()
```

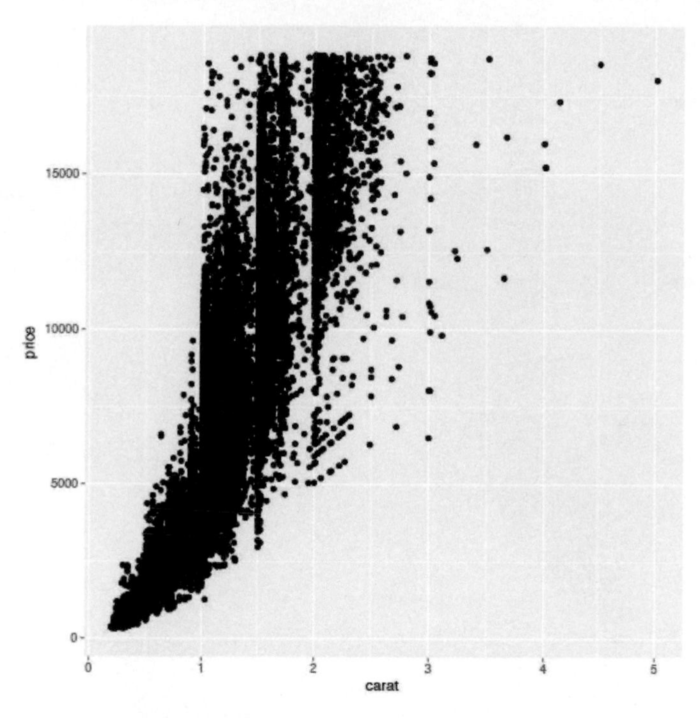

Figure 7.6 Simple ggplot2 scatterplot.

In the next few examples we will be using `ggplot(diamonds, aes(x=carat, y=price))` repeatedly, which ordinarily would require a lot of redundant typing. Fortunately we can save `ggplot` objects to variables and add layers later. We will save it to `g`. Notice that nothing is plotted.

```
> # save basics of ggplot object to a variable
> g <- ggplot(diamonds, aes(x=carat, y=price))
```

Going forward we can add any layer to `g`. Running `g + geom_point()` would re-create the graph shown in Figure 7.6.

The `diamonds` data has many interesting variables we can examine. Let's first look at color, which we will map to the color[4] aesthetic in Figure 7.7.

4. **ggplot** will accept both the American (color) and British (colour) spellings.

```
> g + geom_point(aes(color=color))
```

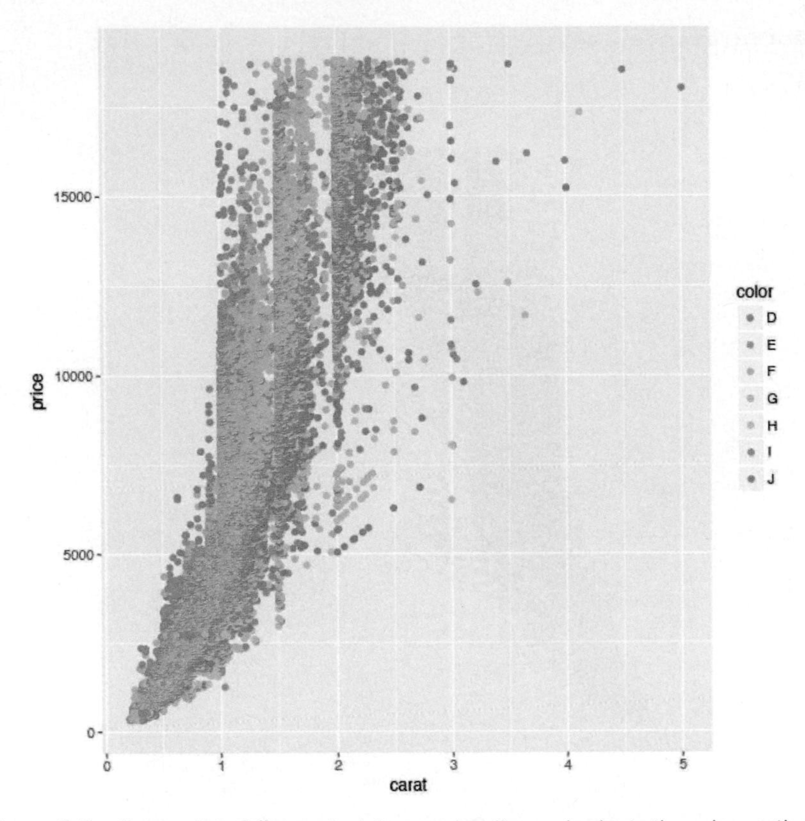

Figure 7.7 Scatterplot of diamonds data mapping diamond color to the color aesthetic.

Notice that we set color=color inside **aes**. This is because the designated color will be determined by the data. Also see that a legend was automatically generated. Recent versions of **ggplot2** have added flexibility with the legend, which we will discuss later.

ggplot2 also has the ability to make faceted plots, or small multiples as Edward Tufte would say. This is done using **facet_wrap** or **facet_grid**. **facet_wrap** takes the levels of one variable, cuts up the underlying data according to them, makes a separate pane for each set and arranges them to fit in the plot, as seen in Figure 7.8. Here the row and column placement have no real meaning. **facet_grid** acts similarly but assigns all levels of a variable to either a row or column as shown in Figure 7.9. In this case the upper left pane displays a scatterplot where the data are only for diamonds with `Fair cut` and `I1 clarity`. The pane to the right is a scatterplot where the data are only for diamonds with `Fair cut` and `SI2 clarity`. The pane in the second row, first column, is a scatterplot where the data are only for diamonds with `Good cut` and `I1 clarity`. After understanding how to read one pane in this plot we can easily understand all the panes and make quick comparisons.

```
> g + geom_point(aes(color=color)) + facet_wrap(~color)
```

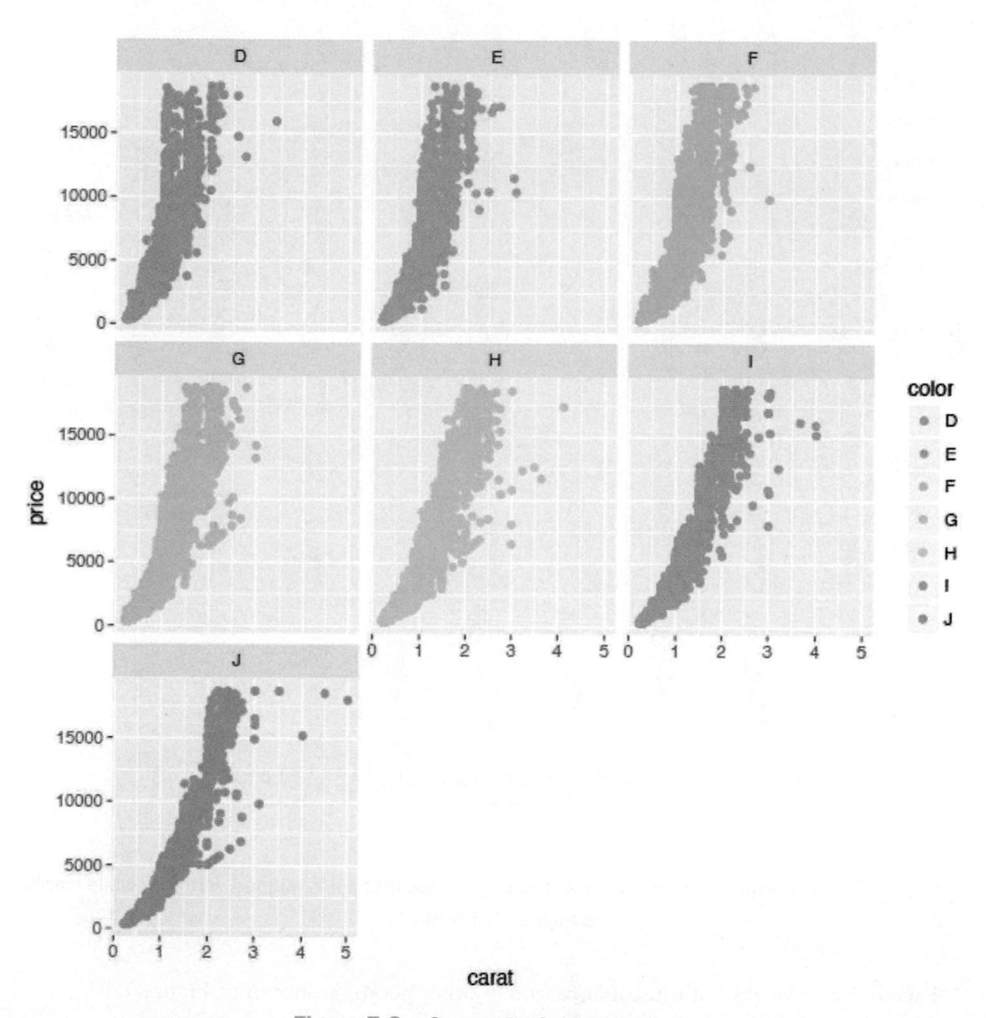

Figure 7.8 Scatterplot faceted by color.

```
> g + geom_point(aes(color=color)) + facet_grid(cut~clarity)
```

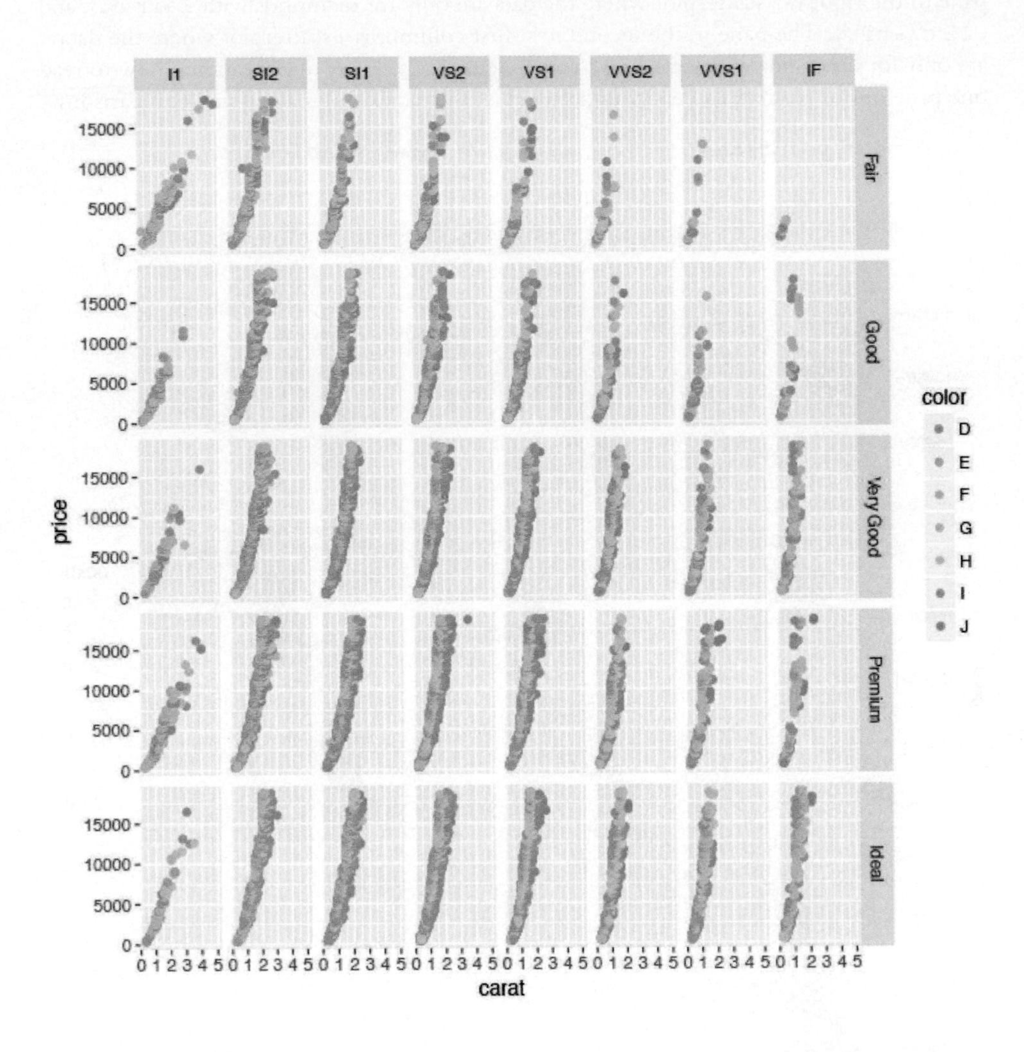

Figure 7.9 Scatterplot faceted by cut and clarity. Notice that cut is aligned vertically while clarity is aligned horizontally.

Faceting also works with histograms or any other geom, as shown in Figure 7.10.

```
> ggplot(diamonds, aes(x=carat)) + geom_histogram() + facet_wrap(~color)
```

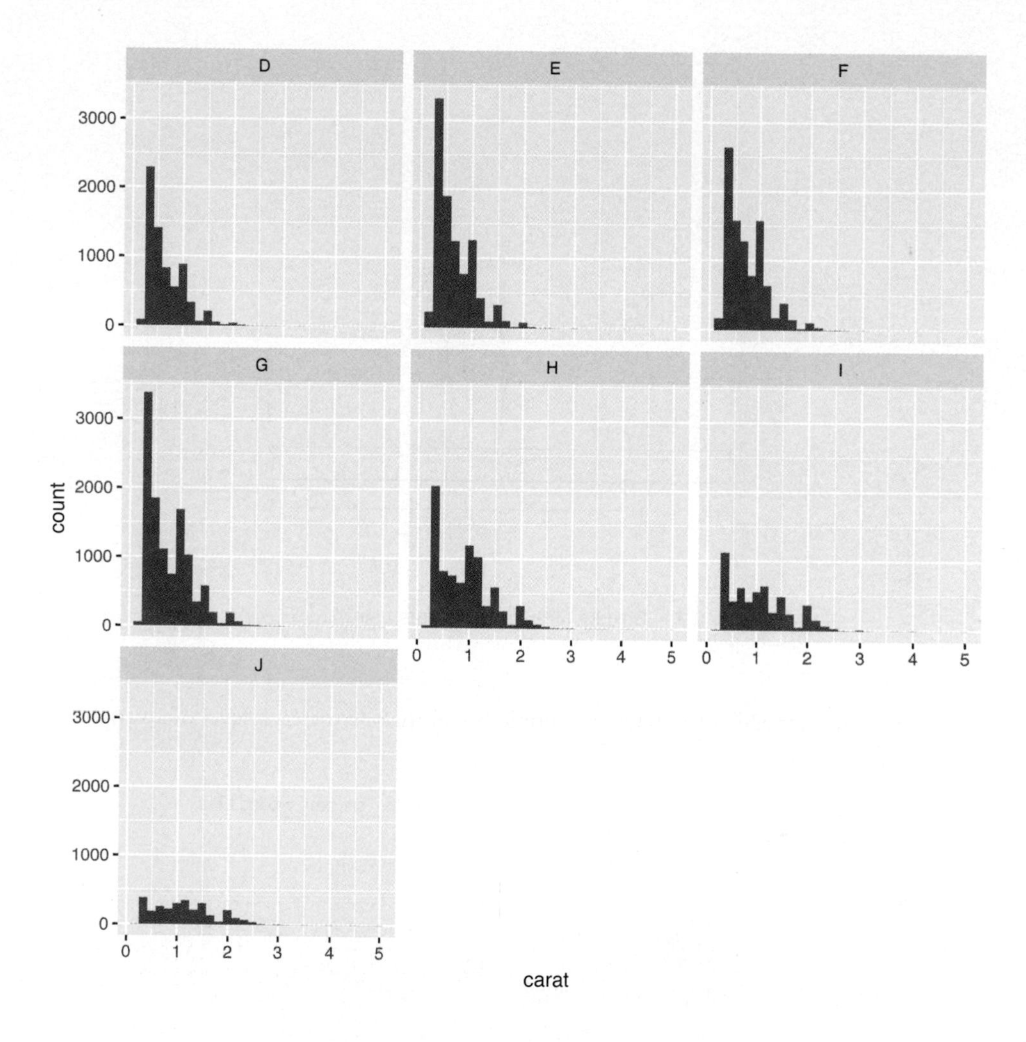

Figure 7.10 Histogram faceted by color.

7.2.3 ggplot2 Boxplots and Violins Plots

Being a complete graphics package, **ggplot2** offers a boxplot geom through **geom_boxplot**. Even though it is one-dimensional, using only a y aesthetic, there needs to be some x aesthetic, so we will use 1. The result is shown in Figure 7.11.

```
> ggplot(diamonds, aes(y=carat, x=1)) + geom_boxplot()
```

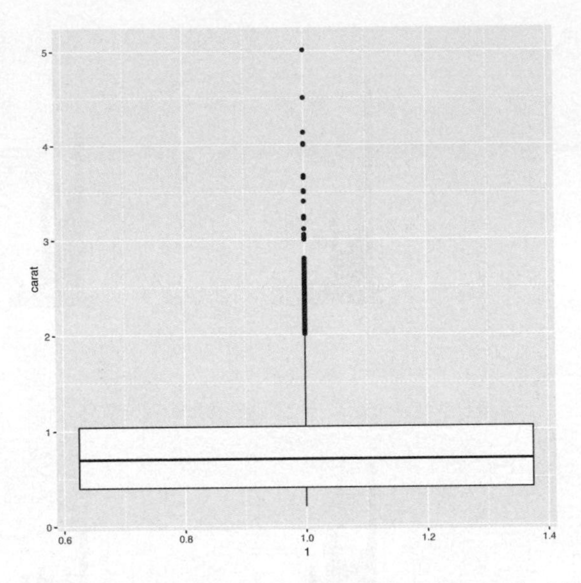

Figure 7.11 Boxplot of diamond carats using `ggplot2`.

This is neatly extended to drawing multiple boxplots, one for each level of a variable, as seen in Figure 7.12.

```
> ggplot(diamonds, aes(y=carat, x=cut)) + geom_boxplot()
```

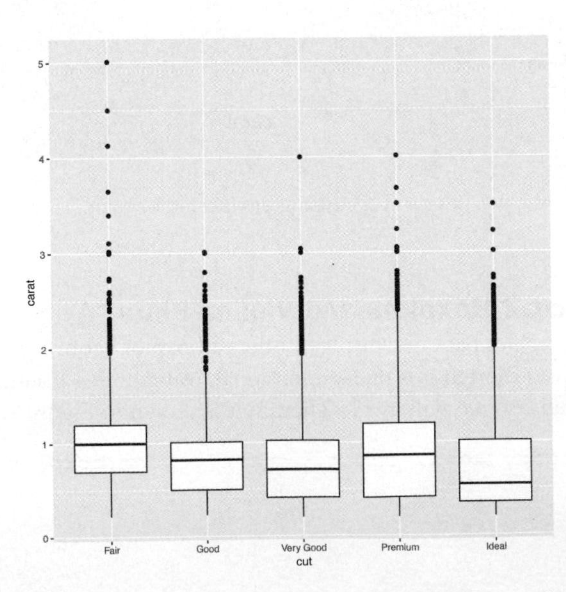

Figure 7.12 Boxplot of diamond carats by cut using `ggplot2`.

Getting fancy, we can swap out the boxplot for violin plots using **geom_violin** as shown in Figure 7.13.

```
> ggplot(diamonds, aes(y=carat, x=cut)) + geom_violin()
```

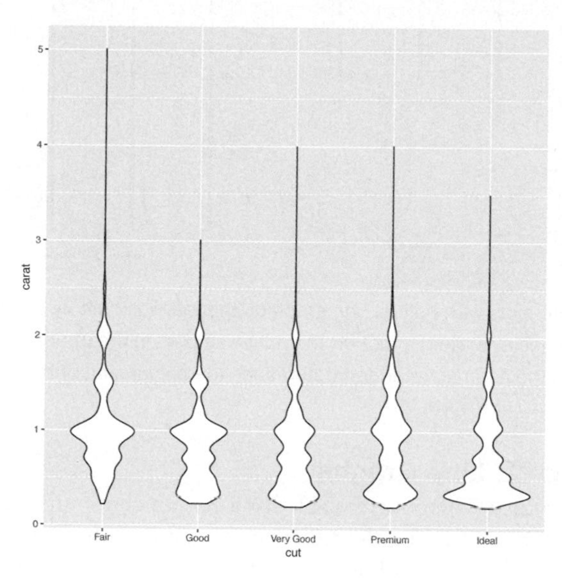

Figure 7.13 Violin plot of diamond carats by cut using `ggplot2`.

Violin plots are similar to boxplots except that the boxes are curved, giving a sense of the density of the data. This provides more information than the straight sides of ordinary boxplots.

We can use multiple layers (geoms) on the same plot, as seen in Figure 7.14. Notice that the order of the layers matters. In the graph on the left, the points are underneath the violins, while in the graph on the right, the points are on top of the violins.

```
> ggplot(diamonds, aes(y=carat, x=cut)) + geom_point() + geom_violin()
> ggplot(diamonds, aes(y=carat, x=cut)) + geom_violin() + geom_point()
```

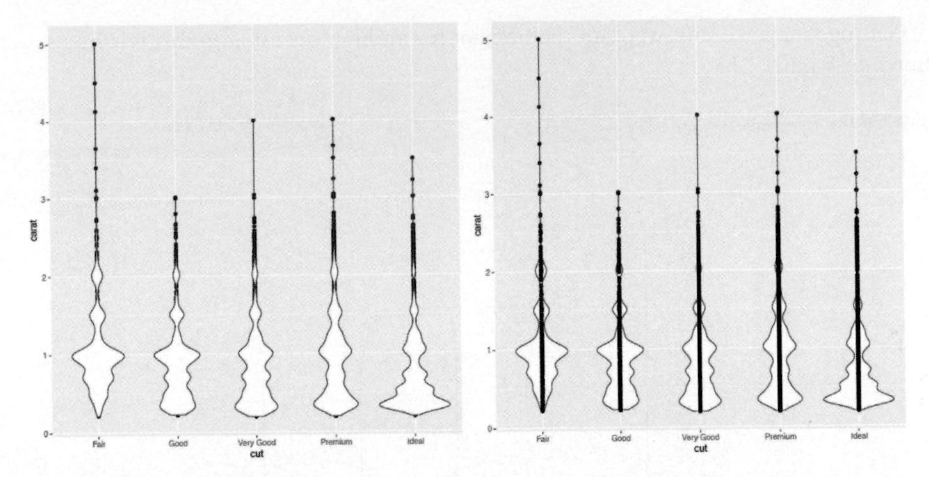

Figure 7.14 Violin plots with points. The graph on the left was built by adding the points geom and then the violin geom, while the plot on the right was built in the opposite order. The order in which the geoms are added determines the positioning of the layers.

7.2.4 `ggplot2` Line Graphs

Line charts are often used when one variable has a certain continuity, but that is not always necessary because there is often a good reason to use a line with categorical data. Figure 7.15 shows an example of a line plot using the `economics` data from **ggplot2**. **ggplot2** intelligently handles `Dates` and plots them on a logical scale.

```
> ggplot(economics, aes(x=date, y=pop)) + geom_line()
```

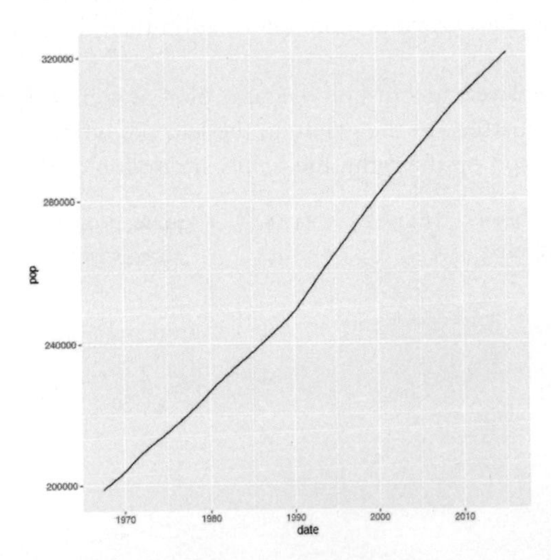

Figure 7.15 Line plot using ggplot2.

While this worked just fine, it is sometimes necessary to use aes(group=1) with **geom_line**. Yes, it is hacky, but it gets the job done, just like when plotting a single boxplot as in Section 7.2.3. It is a quirk of **ggplot2** that sometimes lines cannot be plotted without a group aesthetic.

A common task for line plots is displaying a metric over the course of a year for many years. To prepare the economics data we will use Wickham's **lubridate** package, which has convenient functions for manipulating dates. We need to create two new variables, year and month. To simplify things we will subset the data to include only years starting with 2000.

```
> # load the lubridate package
> library(lubridate)
>
> ## create year and month variables
> economics$year <- year(economics$date)
> # the label argument to month means that the result should be the
> # names of the month instead of the number
> economics$month <- month(economics$date, label=TRUE)
>
> # subset the data
> # the which function returns the indices of observations where the
> # tested condition was TRUE
> econ2000 <- economics[which(economics$year >= 2000), ]
>
> # load the scales package for better axis formatting
> library(scales)
>
> # build the foundation of the plot
> g <- ggplot(econ2000, aes(x=month, y=pop))
> # add lines color coded and grouped by year
> # the group aesthetic breaks the data into separate groups
> g <- g + geom_line(aes(color=factor(year), group=year))
> #  name the legend "Year"
> g <- g + scale_color_discrete(name="Year")
> # format the y axis
> g <- g + scale_y_continuous(labels=comma)
> # add a title and axis labels
> g <- g + labs(title="Population Growth", x="Month", y="Population")
> # plot the graph
> g
```

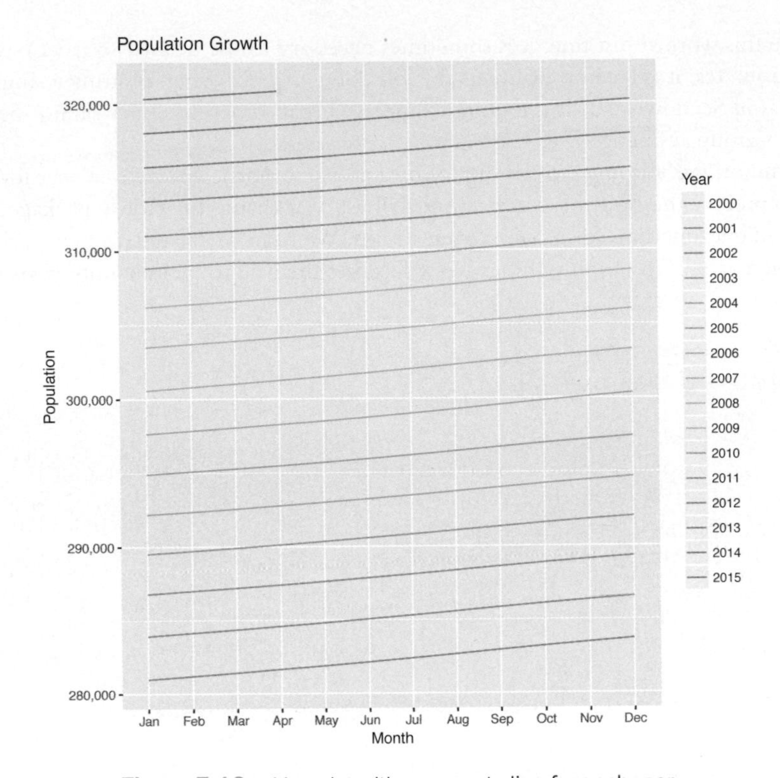

Figure 7.16 Line plot with a seperate line for each year.

Figure 7.16 contains many new concepts. The first part, `ggplot(econ2000, aes(x=month, y=pop)) + geom_line(aes(color=factor(year), group=year))`, is code we have seen before; it creates the line graph with a separate line and color for each year. Notice that we converted `year` to a `factor` so that it would get a discrete color scale. That scale was named by using `scale_color_discrete( name="Year")`. The y-axis was formatted to have commas using `scale_y_continuous(labels=comma)`. Lastly, the title, x-label and y-label were set with `labs(title="Population Growth", x="Month", y="Population")`. All of these pieces put together built a professional looking, publication-quality graph.

Also note the use of **which** to subset the data. This is similar to a `where` clause in SQL.

7.2.5 Themes

A great part of **ggplot2** is the ability to use themes to easily change the way plots look. While building a theme from scratch can be daunting, Jeffrey Arnold from the University of Rochester has put together **ggthemes**, a package of themes to re-create commonly used styles of graphs. Just a few styles—*The Economist*, Excel, Edward Tufte and the *Wall Street Journal*—are exhibited in Figure 7.17.

```
> library(ggthemes)
> # build a plot and store it in g2
```

```
> g2 <- ggplot(diamonds, aes(x=carat, y=price)) +
+     geom_point(aes(color=color))
>
> # apply a few themes
> g2 + theme_economist() + scale_colour_economist()
> g2 + theme_excel() + scale_colour_excel()
> g2 + theme_tufte()
> g2 + theme_wsj()
```

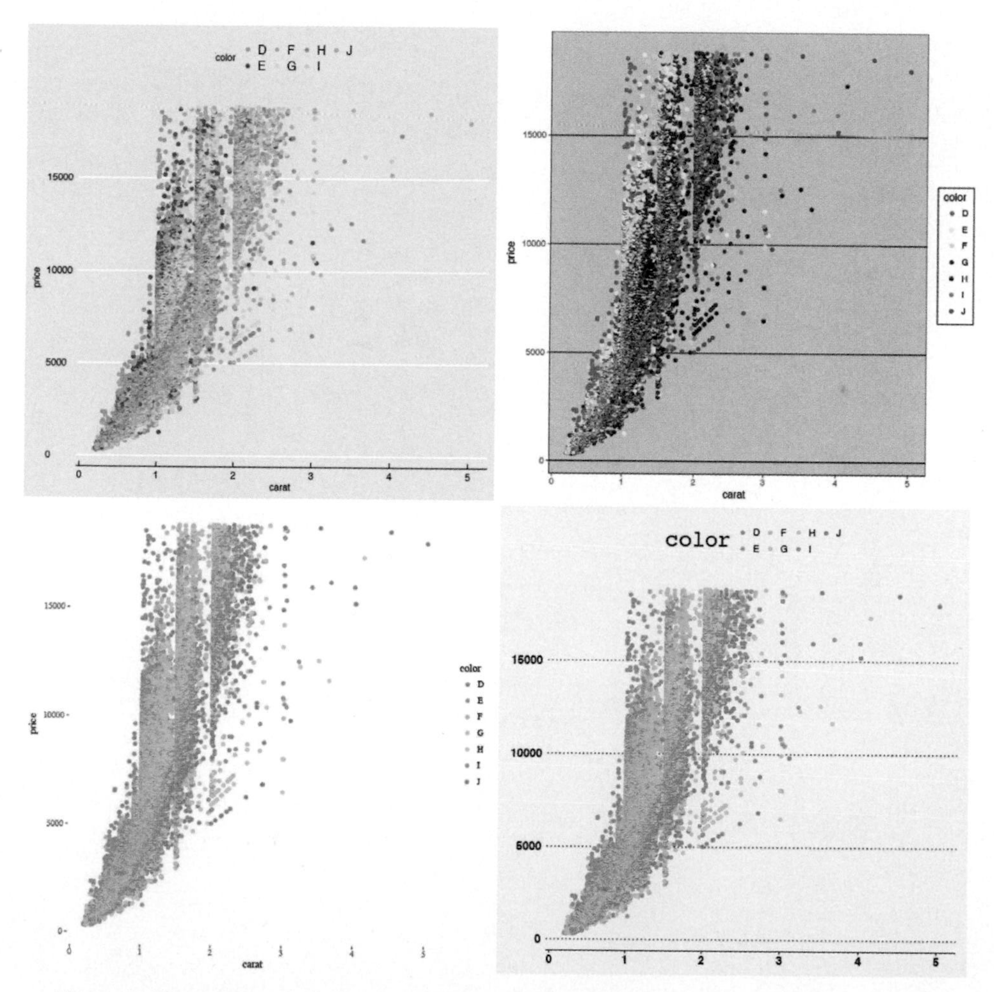

Figure 7.17 Various themes from the ggthemes package. Starting from top left and going clockwise: *The Economist*, Excel (for those with bosses who demand Excel output), Edward Tufte and the *Wall Street Journal*.

7.3 Conclusion

We have seen both basic graphs and **ggplot2** graphs that are both nicer and easier to create. We have covered histograms, scatterplots, boxplots, line plots and density graphs. We have also looked at using colors and small multiples for distinguishing data. There are many other features in **ggplot2** such as jittering, stacking, dodging and alpha, which we will demonstrate in context throughout the book.

8

Writing R functions

If we find ourselves running the same code repeatedly, it is probably a good idea to turn it into a function. In programming it is best to reduce redundancy whenever possible. There are several reasons for doing so, including maintainability and ease of reuse. R has a convenient way to make functions, but it is very different from other languages, so some expectation adjustment might be necessary.

8.1 Hello, World!

This would not be a serious book about a programming language if we did not include a "Hello, World!" example, so we will start with that. Let's build a function that simply prints "Hello, World!" to the console.

```
> say.hello <- function()
+ {
+     print("Hello, World!")
+ }
```

First, note that in R the period (.) is just another character and has no special meaning,[1] unlike in other languages. This allows us to call this function `say.hello`.

Next, we see that functions are assigned to objects just like any other variable, using the <- operator. This is the strangest part of writing functions for people coming from other languages.

Following `function` are a set of parentheses that can either be empty—not have any arguments—or contain any number of arguments. We will cover those in Section 8.2.

The body of the function is enclosed in curly braces ({ and }). This is not necessary if the function contains only one line, but that is rare. Notice the indenting for the commands inside the function. While not required, it is good practice to properly indent code to ensure readability. It is here in the body that we put the lines of code we want the function to perform. A semicolon (`;`) can be used to indicate the end of the line but is not necessary, and its use is actually frowned upon.

Calling `say.hello()` prints as desired.

1. One exception is that objects with names starting with a period are accessible but invisible, so they will not be found by **ls**. A second exception is when dealing with generic functions that dispatch to type-specific methods, though this will not make a significant impact on our usage.

8.2 Function Arguments

More often than not we want to pass arguments to our function. These are easily added inside the parentheses of the function declaration. We will use an argument to print "Hello Jared."

Before we do that, however, we need to briefly learn about the **sprintf** function. Its first argument is a string with special input characters and subsequent arguments that will be substituted into the special input characters.

```
> # one substitution
> sprintf("Hello %s", "Jared")

[1] "Hello Jared"

> # two substitutions
> sprintf("Hello %s, today is %s", "Jared", "Sunday")

[1] "Hello Jared, today is Sunday"
```

We now use **sprintf** to build a string to print based on a function's arguments.

```
> hello.person <- function(name)
+ {
+     print(sprintf("Hello %s", name))
+ }
> hello.person("Jared")

[1] "Hello Jared"

> hello.person("Bob")

[1] "Hello Bob"

> hello.person("Sarah")

[1] "Hello Sarah"
```

The argument `name` can be used as a variable inside the function (it does not exist outside the function). It can also be used like any other variable and as an argument to further function calls.

We can add a second argument to be printed as well. When calling functions with more than one argument, there are two ways to specify which argument goes with which value, either positionally or by name.

```
> hello.person <- function(first, last)
+ {
+     print(sprintf("Hello %s %s", first, last))
+ }
> # by position
> hello.person("Jared", "Lander")
```

```
[1] "Hello Jared Lander"

> # by name
> hello.person(first="Jared", last="Lander")

[1] "Hello Jared Lander"

> # the other order
> hello.person(last="Lander", first="Jared")

[1] "Hello Jared Lander"

> # just specify one name
> hello.person("Jared", last="Lander")

[1] "Hello Jared Lander"

> # specify the other
> hello.person(first="Jared", "Lander")

[1] "Hello Jared Lander"

> # specify the second argument first
> # then provide the first argument with no name
> hello.person(last="Lander", "Jared")

[1] "Hello Jared Lander"
```

Being able to specify the arguments by name adds a lot of flexibility to calling functions. Even partial argument names can be supplied, but this should be done with care.

```
> hello.person(fir="Jared", l="Lander")

[1] "Hello Jared Lander"
```

8.2.1 Default Arguments

When using multiple arguments it is sometimes desirable to not have to enter a value for each. In other languages functions can be overloaded by defining the function mutliple times, each with a differering number of arguments. R instead provides the capability to specify default arguments. These can be `NULL`, characters, numbers or any valid R object.

Let's rewrite `hello.person` to provide "Doe" as the default last name.

```
> hello.person <- function(first, last="Doe")
+ {
+     print(sprintf("Hello %s %s", first, last))
+ }
>
> # call without specifying last
> hello.person("Jared")

[1] "Hello Jared Doe"
```

```
> # call with a different last
> hello.person("Jared", "Lander")

[1] "Hello Jared Lander"
```

8.2.2 Extra Arguments

R offers a special operator that enables functions to take an arbitrary number of arguments that do not need to be specified in the function definition. This is the `dot-dot-dot` argument (...). This should be used very carefully, although it can provide great flexibility. For now we will just see how it can absorb extra arguments; later we will find a use for it when passing arguments between functions.

```
> # call hello.person with an extra argument
> hello.person("Jared", extra="Goodbye")

Error in hello.person("Jared", extra = "Goodbye"): unused argument (extra =
"Goodbye")

> # call it with two valid arguments and a third
> hello.person("Jared", "Lander", "Goodbye")

Error in hello.person("Jared", "Lander", "Goodbye"): unused argument ("Goodbye")

> # now build hello.person with ... so that it absorbs extra arguments
> hello.person <- function(first, last="Doe", ...)
+ {
+     print(sprintf("Hello %s %s", first, last))
+ }
> # call hello.person with an extra argument
> hello.person("Jared", extra="Goodbye")

[1] "Hello Jared Doe"

> # call it with two valid arguments and a third
> hello.person("Jared", "Lander", "Goodbye")

[1] "Hello Jared Lander"
```

8.3 Return Values

Functions are generally used for computing some value, so they need a mechanism to supply that value back to the caller. This is called returning and is done quite easily. There are two ways to accomplish this with R. The value of the last line of code in a function is automatically returned, although this can be bad practice. The **return** command more explicitly specifies that a value should be returned and the function should be exited.

To illustrate, we will build a function that doubles its only argument and returns that value.

```
> # first build it without an explicit return
> double.num <- function(x)
+ {
```

```
+     x * 2
+ }
>
> double.num(5)

[1] 10

> # now build it with an explicit return
> double.num <- function(x)
+ {
+     return(x * 2)
+ }
>
> double.num(5)

[1] 10

> # build it again, this time with another argument after the explicit return
> double.num <- function(x)
+ {
+     return(x * 2)

+     # below here is not executed because the function already exited
+     print("Hello!")
+     return(17)
+ }
>
> double.num(5)

[1] 10
```

8.4 do.call

A particularly underused trick is the **do.call** function. This allows us to specify the name of a function either as a `character` or as an object, and provide arguments as a list.

```
> do.call("hello.person", args=list(first="Jared", last="Lander"))

[1] "Hello Jared Lander"

> do.call(hello.person, args=list(first="Jared", last="Lander"))

[1] "Hello Jared Lander"
```

This is particularly useful when building a function that allows the user to specify an action. In the following example the user supplies a `vector` and a function to be run.

```
> run.this <- function(x, func=mean)
+ {
+     do.call(func, args=list(x))
+ }
>
> # finds the mean by default
```

```
> run.this(1:10)

[1] 5.5

> # specify to calculate the mean
> run.this(1:10, mean)

[1] 5.5

> # calculate the sum
> run.this(1:10, sum)

[1] 55

> # calculate the standard deviation
> run.this(1:10, sd)

[1] 3.02765
```

8.5 Conclusion

Functions allow us to create reusable code that avoids repetition and allows easy modification. Important points to remember are function arguments, default values and returned values. Later in this book we will see functions that get far more complicated than the ones we have seen so far.

9

Control Statements

Control statements allow us to control the flow of our programming and cause different things to happen, depending on the values of tests. Tests result in a `logical`, `TRUE` or `FALSE`, which is used in if-like statements. The main control statements are **if**, **else**, **ifelse** and **switch**.

9.1 `if` and `else`

The most common test is the **if** command. It essentially says. If something is `TRUE`, then perform some action; otherwise, do not perform that action. The thing we are testing goes inside parentheses following the **if** command. The most basic checks are: equal to (==), less than (<), less than or equal to (<=), greater than (>), greater than or equal to (>=) and not equal (!=).

If these tests pass they result in `TRUE`, and if they fail they result in `FALSE`. As noted in Section 4.3.4, `TRUE` is numerically equivalent to 1 and `FALSE` is equivalent to 0.

```
> as.numeric(TRUE)

[1] 1

> as.numeric(FALSE)

[1] 0
```

These tests do not need to be used inside **if** statements. The following are some simple examples.

```
> 1 == 1   # TRUE

[1] TRUE

> 1 < 1    # FALSE

[1] FALSE

> 1 <= 1   # TRUE

[1] TRUE
```

```
> 1 > 1   # FALSE

[1] FALSE

> 1 >= 1  # TRUE

[1] TRUE

> 1 != 1  # FALSE

[1] FALSE
```

We can now show that using this test inside an **if** statement controls actions that follow.

```
> # set up a variable to hold 1
> toCheck <- 1
>
> # if toCheck is equal to 1, print hello
> if(toCheck == 1)
+ {
+     print("hello")
+ }

[1] "hello"

> # now if toCheck is equal to 0, print hello
> if(toCheck == 0)
+ {
+     print("hello")
+ }
> # notice nothing was printed
```

Notice that **if** statements are similar to functions, in that all statements (there can be one or multiple) go inside curly braces.

Life is not always so simple that we want an action only if some relationship is `TRUE`. We often want a different action if that relationship is `FALSE`. In the following example we put an **if** statement followed by an **else** statement inside a function, so that it can be used repeatedly.

```
> # first create the function
> check.bool <- function(x)
+ {
+     if(x == 1)
+     {
+         # if the input is equal to 1, print hello
+         print("hello")
+     }else
+     {
+         # otherwise print goodbye
```

```
+             print("goodbye")
+        }
+ }
```

Notice that **else** is on the same line as its preceding closing curly brace (}). This is important, as the code will fail otherwise.

Now let's use that function and see if it works.

```
> check.bool(1)

[1] "hello"

> check.bool(0)

[1] "goodbye"

> check.bool("k")

[1] "goodbye"

> check.bool(TRUE)

[1] "hello"
```

Anything other than 1 caused the function to print "goodbye." That is exactly what we wanted. Passing TRUE printed "hello" because TRUE is numerically the same as 1.

Perhaps we want to successively test a few cases. That is where we can use **else if**. We first test a single statement, then make another test, and then perhaps fall over to catch all. We will modify check.bool to test for one condition and then another.

```
> check.bool <- function(x)
+ {
+      if(x == 1)
+      {
+           # if the input is equal to 1, print hello
+           print("hello")
+      }else if(x == 0)
+      {
+           # if the input is equal to 0, print goodbye
+           print("goodbye")
+      }else
+      {
+           # otherwise print confused
+           print("confused")
+      }
+ }
>
> check.bool(1)

[1] "hello"
```

```
> check.bool(0)

[1] "goodbye"

> check.bool(2)

[1] "confused"

> check.bool("k")

[1] "confused"
```

9.2 switch

If we have multiple cases to check, writing **else if** repeatedly can be cumbersome and inefficient. This is where **switch** is most useful. The first argument is the value we are testing. Subsequent arguments are a particular value and what should be the result. The last argument, if not given a value, is the default result.

To illustrate we build a function that takes in a value and returns a corresponding result.

```
> use.switch <- function(x)
+ {
+     switch(x,
+            "a"="first",
+            "b"="second",
+            "z"="last",
+            "c"="third",
+            "other")
+ }
>
> use.switch("a")

[1] "first"

> use.switch("b")

[1] "second"

> use.switch("c")

[1] "third"

> use.switch("d")

[1] "other"

> use.switch("e")

[1] "other"
```

```
> use.switch("z")

[1] "last"
```

If the first argument is numeric, it is matched positionally to the following arguments, regardless of the names of the subsequent arguments. If the numeric argument is greater than the number of subsequent arguments, NULL is returned.

```
> use.switch(1)

[1] "first"

> use.switch(2)

[1] "second"

> use.switch(3)

[1] "last"

> use.switch(4)

[1] "third"

> use.switch(5)

[1] "other"

> use.switch(6)         # nothing is returned
> is.null(use.switch(6))

[1] TRUE
```

Here we introduced a new function, **is.null**, which, as the name implies, tests whether an object is NULL.

9.3 ifelse

While **if** is like the if statement in traditional languages, **ifelse** is more like the **if** function in Excel. The first argument is the condition to be tested (much like in a traditional **if** statement), the second argument is the return value if the test is TRUE and the third argument is the return value if the test if FALSE. The beauty here—unlike with the traditional **if**—is that this works with vectorized arguments. As is often the case in R, using vectorization avoids **for** loops and speeds up our code. The nuances of **ifelse** can be tricky, so we show numerous examples.

We start with a very simple example, testing whether 1 is equal to 1 and printing “Yes” if that is TRUE and “No” if it is FALSE.

```
> # see if 1 == 1
> ifelse(1 == 1, "Yes", "No")
```

```
[1] "Yes"

> # see if 1 == 0
> ifelse(1 == 0, "Yes", "No")

[1] "No"
```

This clearly gives us the results we want. **ifelse** uses all the regular equality tests seen in Section 9.1 and any other `logical` test. It is worth noting, however, that if testing just a single element (a `vector` of length 1 or a simple **is.na**), it is more efficient to use **if** than **ifelse**. This can result in a nontrivial speedup of our code.

Next we will illustrate a vectorized first argument.

```
> toTest <- c(1, 1, 0, 1, 0, 1)
> ifelse(toTest == 1, "Yes", "No")

[1] "Yes" "Yes" "No"  "Yes" "No"  "Yes"
```

This returned "Yes" for each element of `toTest` that equaled 1 and "No" for each element of `toTest` that did not equal 1.

The `TRUE` and `FALSE` arguments can even refer to the testing element.

```
> ifelse(toTest == 1, toTest*3, toTest)

[1] 3 3 0 3 0 3

> # the FALSE argument is repeated as needed
> ifelse(toTest == 1, toTest*3, "Zero")

[1] "3"    "3"    "Zero" "3"    "Zero" "3"
```

Now let's say that `toTest` has `NA` elements. In that case the corresponding result from **ifelse** is `NA`.

```
> toTest[2] <- NA
> ifelse(toTest == 1, "Yes", "No")

[1] "Yes" NA    "No"  "Yes" "No"  "Yes"
```

This would be the same if the `TRUE` and `FALSE` arguments are `vectors`.

```
> ifelse(toTest == 1, toTest*3, toTest)

[1]  3 NA  0  3  0  3

> ifelse(toTest == 1, toTest*3, "Zero")

[1] "3"    NA     "Zero" "3"    "Zero" "3"
```

9.4 Compound Tests

The statement being tested with **if**, **ifelse** and **switch** can be any argument that results in a logical TRUE or FALSE. This can be an equality check or even the result of **is.numeric** or **is.na**. Sometimes we want to test more than one relationship at a time. This is done using logical and and or operators. These are & and && for and and | and || for or. The differences are subtle but can impact our code's speed.

The double form (&& or ||) is best used in **if** and the single form (& or |) is necessary for **ifelse**. The double form compares only one element from each side, while the single form compares each element of each side.

```
> a <- c(1, 1, 0, 1)
> b <- c(2, 1, 0, 1)
>
> # this checks each element of a and each element of b
> ifelse(a == 1 & b == 1, "Yes", "No")

[1] "No"  "Yes" "No"  "Yes"

> # this only checks the first element of a and the first element of b
> # it only returns one result
> ifelse(a == 1 && b == 1, "Yes", "No")

[1] "No"
```

Another difference between the double and single forms is how they are processed. When using the single form, both sides of the operator are always checked. With the double form, sometimes only the left side needs to be checked. For instance, if testing 1 == 0 && 2 == 2, the left side fails, so there is no reason to check the right hand side. Similarly, when testing 3 == 3 || 0 == 0, the left side passes, so there is no need to check the right side. This can be particularly helpful when the right side would throw an error if the left side had failed.

There can be more than just two conditions tested. Many conditions can be strung together using multiple and or or operators. The different clauses can be grouped by parentheses just like mathematical operations. Without parentheses, the order of operations is similar to PEMDAS, seen in Section 4.1, where and is equivalent to multiplication and or is equivalent to addition, so and takes precedence over or.

9.5 Conclusion

Controlling the flow of our program, both at the command line and in functions, plays an important role when processing and analyzing our data. If statements, along with **else**, are the most common—and efficient—for testing single element objects, although **ifelse** is far more common in R programming because of its vectorized nature. Switch statements are often forgotten but can come in very handy. The and (& and &&) and or (| and ||) operators allow us to combine multiple tests into one.

10

Loops, the Un-R Way to Iterate

When starting to use R, most people use loops whenever they need to iterate over elements of a vector, list or data.frame. While it is natural to do this in other languages, with R we generally want to use vectorization. That said, sometimes loops are unavoidable, so R offers both **for** and **while** loops.

10.1 for Loops

The most commonly used loop is the **for** loop. It iterates over an index—provided as a vector—and performs some operations. For a first simple example, we print out the first ten numbers.

The loop is declared using **for**, which takes one English-seeming argument in three parts. The third part is any vector of values of any kind, most commonly numeric or character. The first part is the variable that is iteratively assigned the values in the vector from the third part. The middle part is simply the word in indicating that the variable (the first part) is in the vector (the third part).

```
> for(i in 1:10)
+ {
+     print(i)
+ }

[1] 1
[1] 2
[1] 3
[1] 4
[1] 5
[1] 6
[1] 7
[1] 8
[1] 9
[1] 10
```

Here we generated a vector holding the numbers 1 through 10, and then printed each. Notice that this could have been performed simply by using the built-in vectorization of the **print** function.

```
> print(1:10)

 [1]  1  2  3  4  5  6  7  8  9 10
```

Sure, it does not look exactly the same, but that is just cosmetic.

The `vector` in **for** loops does not have to be sequential, it can be any `vector`.

```
> # build a vector holding fruit names
> fruit <- c("apple", "banana", "pomegranate")
> # make a variable to hold their lengths, with all NA to start
> fruitLength <- rep(NA, length(fruit))
> # show it, all NAs
> fruitLength

[1] NA NA NA

> # give it names
> names(fruitLength) <- fruit
> # show it again, still NAs
> fruitLength

      apple      banana pomegranate
         NA          NA          NA

> # loop through the fruit assigning their lengths to the result vector
> for(a in fruit)
+ {
+     fruitLength[a] <- nchar(a)
+ }
> # show the lengths
> fruitLength

      apple      banana pomegranate
          5           6          11
```

Again, R's built-in vectorization could have made all of this much easier.

```
> # simply call nchar
> fruitLength2 <- nchar(fruit)
> # give it names
> names(fruitLength2) <- fruit
> # show it
> fruitLength2

      apple      banana pomegranate
          5           6          11
```

This, as expected, provides identical results, as seen next.

```
> identical(fruitLength, fruitLength2)

[1] TRUE
```

10.2 while Loops

Although used far less frequently in R than the **for** loop, the **while** loop is just as simple to implement. It simply runs the code inside the braces repeatedly as long as the tested condition proves true. In the following example, we print the value of x and iterate it until it reaches 5. This is a highly trivial example but shows the functionality nonetheless.

```
> x <- 1
> while(x <= 5)
+ {
+     print(x)
+     x <- x + 1
+ }

[1] 1
[1] 2
[1] 3
[1] 4
[1] 5
```

10.3 Controlling Loops

Sometimes we have to skip to the next iteration of the loop or completely break out of it. This is accomplished with **next** and **break**. We use a **for** loop to demonstrate.

```
> for(i in 1:10)
+ {
+     if(i == 3)
+     {
+         next
+     }
+     print(i)
+ }

[1] 1
[1] 2
[1] 4
[1] 5
[1] 6
[1] 7
[1] 8
[1] 9
[1] 10
```

Notice that the number 3 did not get printed.

```
> for(i in 1:10)
+ {
```

```
+       if(i == 4)
+       {
+           break
+       }
+       print(i)
+ }

[1] 1
[1] 2
[1] 3
```

Here, even though we told R to iterate over the first ten integers, it stopped after 3 because we broke the loop at 4.

10.4 Conclusion

The two primary loops are **for**, which iterates over a fixed sequence of elements, and **while** which continues a loop as long as some condition holds true. As stated earlier, if a solution can be done without loops, via vectorization or matrix algebra, then avoid the loop. It is particularly important to avoid nested loops. Loops inside other loops are extremely slow in R.

11

Group Manipulation

A general rule of thumb for data analysis is that manipulating the data (or "data munging," a term popularized by Simple founder Josh Reich) consumes about 80 percent of the effort. This often requires repeated operations on different sections of the data, something Hadley Wickham coined "split-apply-combine." That is, we split the data into discrete sections based on some metric, apply a transformation of some kind to each section, and then combine all the sections together. This is somewhat like the Map Reduce[1] paradigm of Hadoop.[2] There are many different ways to iterate over data in R, and we look at some of the more convenient functions. Much of the functionality seen in this chapter is constantly being improved, with the latest tools covered in Chapters 12 and 13.

11.1 Apply Family

Built into R is the **apply** function and all of its relatives such as **tapply**, **lapply** and **mapply**. Each has its quirks and necessities and is best used in different situations.

11.1.1 apply

apply is the first member of this family that users usually learn, and it is also the most restrictive. It must be used on a `matrix`, meaning all of the elements must be of the same type whether they are `character`, `numeric` or `logical`. If used on some other object, such as a `data.frame`, it will be converted to a `matrix` first.

The first argument to **apply** is the object we are working with. The second argument is the margin to apply the function over, 1 meaning to operate over the rows and 2 meaning to operate over the columns. The third argument is the function we want to apply. Any following arguments will be passed on to that function. **apply** will iterate over each row (or column) of the `matrix`, treating them as individual inputs to the first argument of the specified function.

To illustrate its use we start with a trivial example, summing the rows or columns of a `matrix`.

```
> # build the matrix
> theMatrix <- matrix(1:9, nrow=3)
```

1. Map Reduce is where data are split into discrete sets, computed on and then recombined in some fashion.
2. Hadoop is a framework for distributing data and computations across a grid of computers.

```
> # sum the rows
> apply(theMatrix, 1, sum)

[1] 12 15 18

> # sum the columns
> apply(theMatrix, 2, sum)

[1]  6 15 24
```

Notice that this could alternatively be accomplished using the built-in **rowSums** and **colSums** functions, yielding the same results.

```
> rowSums(theMatrix)

[1] 12 15 18

> colSums(theMatrix)

[1]  6 15 24
```

For a moment, let's set an element of `theMatrix` to `NA` to see how we handle missing data using the `na.rm` argument and the use of additional arguments. As explained in Sections 4.7.1 and 18.1, if even a single element of a `vector` is `NA`, then the result of **sum** will be `NA`. This can avoided by setting `na.rm=TRUE`, in which case the `NA`s will be removed and the sum computed on the remaining elements. When using **sum**—or any other function—with **apply**, additional arguments (such as `na.rm`) are specified after the function itself. Any argument from the applied function can be specified. Unlike when calling the function directly, the arguments must be named.

```
> theMatrix[2, 1] <- NA
> apply(theMatrix, 1, sum)

[1] 12 NA 18

> apply(theMatrix, 1, sum, na.rm=TRUE)

[1] 12 13 18

> rowSums(theMatrix)

[1] 12 NA 18

> rowSums(theMatrix, na.rm=TRUE)

[1] 12 13 18
```

11.1.2 `lapply` and `sapply`

lapply works by applying a function to each element of a `list` and returning the results as a `list`.

```
> theList <- list(A=matrix(1:9, 3), B=1:5, C=matrix(1:4, 2), D=2)
> lapply(theList, sum)

$A
[1] 45

$B
[1] 15

$C
[1] 10

$D
[1] 2
```

Dealing with `lists` can be cumbersome, so to return the result of **lapply** as a `vector` instead, use **sapply**. It is exactly the same as **lapply** in every other way.

```
> sapply(theList, sum)

 A  B  C  D
45 15 10  2
```

Because a `vector` is technically a form of a `list`, **lapply** and **sapply** can also take a `vector` as their input.

```
> theNames <- c("Jared", "Deb", "Paul")
> lapply(theNames, nchar)

[[1]]
[1] 5

[[2]]
[1] 3

[[3]]
[1] 4
```

11.1.3 mapply

Perhaps the most-overlooked-when-so-useful member of the **apply** family is **mapply**, which applies a function to each element of multiple `lists`. Often, when confronted with this scenario, people will resort to using a loop, which is certainly not necessary.

```
> ## build two lists
> firstList <- list(A=matrix(1:16, 4), B=matrix(1:16, 2), C=1:5)
> secondList <- list(A=matrix(1:16, 4), B=matrix(1:16, 8), C=15:1)
> # test element-by-element if they are identical
> mapply(identical, firstList, secondList)
```

```
   A     B     C
 TRUE FALSE FALSE

> ## build a simple function
> ## it adds the number of rows (or length) of each corresponding element
> simpleFunc <- function(x, y)
+ {
+     NROW(x) + NROW(y)
+ }
> # apply the function to the two lists
> mapply(simpleFunc, firstList, secondList)

 A  B  C
 8 10 20
```

11.1.4 Other apply Functions

There are many other members of the **apply** family that either do not get used much or have been superseded by functions in the **plyr**, **dplyr** and **purrr** packages. (Some would argue that **lapply** and **sapply** have been superseded, but they do have their advantages over their corresponding **dplyr** and **purrr** functions.)

These include

- **tapply**
- **rapply**
- **eapply**
- **vapply**
- **by**

11.2 aggregate

People experienced with SQL generally want to run an aggregation and group by as their first R task. The way to do this is to use the aptly named **aggregate** function. There are a number of different ways to call **aggregate**, so we will look at perhaps its most convenient method, using a `formula`.

We will see `formulas` used to great extent with linear models in Chapter 19, and they play a useful role in R. `Formulas` consist of a left side and a right side separated by a tilde (~). The left side represents a variable that we want to make a calculation on, and the right side represents a variable (or more) that we want to group the calculation by.[3]

To demonstrate **aggregate** we once again turn to the `diamonds` data in **ggplot2**.

```
> data(diamonds, package='ggplot2')
> head(diamonds)
```

3. As we show in Chapter 19, the right side can be numeric, although for the **aggregate** function we will just use categorical variables.

```
# A tibble: 6 × 10
  carat       cut color clarity depth table price     x     y     z
  <dbl>     <ord> <ord>   <ord> <dbl> <dbl> <int> <dbl> <dbl> <dbl>
1  0.23     Ideal     E     SI2  61.5    55   326  3.95  3.98  2.43
2  0.21   Premium     E     SI1  59.8    61   326  3.89  3.84  2.31
3  0.23      Good     E     VS1  56.9    65   327  4.05  4.07  2.31
4  0.29   Premium     I     VS2  62.4    58   334  4.20  4.23  2.63
5  0.31      Good     J     SI2  63.3    58   335  4.34  4.35  2.75
6  0.24 Very Good     J    VVS2  62.8    57   336  3.94  3.96  2.48
```

We calculate the average `price` for each type of `cut`: Fair, Good, Very Good, Premium and Ideal. The first argument to **aggregate** is the `formula` specifying that `price` should be broken up (or `group by` in SQL terms) by `cut`. The second argument is the data to use, in this case `diamonds`. The third argument is the function to apply to each subset of the data; for us this will be the **mean**.

```
> aggregate(price ~ cut, diamonds, mean)

        cut    price
1      Fair 4358.758
2      Good 3928.864
3 Very Good 3981.760
4   Premium 4584.258
5     Ideal 3457.542
```

For the first argument we specified that `price` should be aggregated by `cut`. Notice that we only specified the column name and did not have to identify the data because that is given in the second argument. After the third argument specifying the function, additional named arguments to that function can be passed, such as `aggregate(price ~ cut, diamonds, mean, na.rm=TRUE)`.

To group the data by more than one variable, add the additional variable to the right side of the `formula` separating it with a plus sign (+).

```
> aggregate(price ~ cut + color, diamonds, mean)

         cut color    price
1       Fair     D 4291.061
2       Good     D 3405.382
3  Very Good     D 3470.467
4    Premium     D 3631.293
5      Ideal     D 2629.095
6       Fair     E 3682.312
7       Good     E 3423.644
8  Very Good     E 3214.652
9    Premium     E 3538.914
10     Ideal     E 2597.550
```

```
11      Fair      F 3827.003
12      Good      F 3495.750
13 Very Good      F 3778.820
14   Premium      F 4324.890
15     Ideal      F 3374.939
16      Fair      G 4239.255
17      Good      G 4123.482
18 Very Good      G 3872.754
19   Premium      G 4500.742
20     Ideal      G 3720.706
21      Fair      H 5135.683
22      Good      H 4276.255
23 Very Good      H 4535.390
24   Premium      H 5216.707
25     Ideal      H 3889.335
26      Fair      I 4685.446
27      Good      I 5078.533
28 Very Good      I 5255.880
29   Premium      I 5946.181
30     Ideal      I 4451.970
31      Fair      J 4975.655
32      Good      J 4574.173
33 Very Good      J 5103.513
34   Premium      J 6294.592
35     Ideal      J 4918.186
```

To aggregate two variables (for now we still just group by `cut`), they must be combined using **cbind** on the left side of the `formula`.

```
> aggregate(cbind(price, carat) ~ cut, diamonds, mean)

        cut    price     carat
1      Fair 4358.758 1.0461366
2      Good 3928.864 0.8491847
3 Very Good 3981.760 0.8063814
4   Premium 4584.258 0.8919549
5     Ideal 3457.542 0.7028370
```

This finds the mean of both `price` and `carat` for each value of `cut`. It is important to note that only one function can be supplied, and hence applied to the variables. To apply more than one function, it is easier to use the **plyr** and **dplyr** packages, which are explained in Section 11.3 and Chapter 12.

Of course, multiple variables can be supplied to both the left and right sides at the same time.

```
> aggregate(cbind(price, carat) ~ cut + color, diamonds, mean)

         cut color    price     carat
1       Fair     D 4291.061 0.9201227
2       Good     D 3405.382 0.7445166
3  Very Good     D 3470.467 0.6964243
4    Premium     D 3631.293 0.7215471
5      Ideal     D 2629.095 0.5657657
6       Fair     E 3682.312 0.8566071
7       Good     E 3423.644 0.7451340
8  Very Good     E 3214.652 0.6763167
9    Premium     E 3538.914 0.7177450
10     Ideal     E 2597.550 0.5784012
11      Fair     F 3827.003 0.9047115
12      Good     F 3495.750 0.7759296
13 Very Good     F 3778.820 0.7409612
14   Premium     F 4324.890 0.8270356
15     Ideal     F 3374.939 0.6558285
16      Fair     G 4239.255 1.0238217
17      Good     G 4123.482 0.8508955
18 Very Good     G 3872.754 0.7667986
19   Premium     G 4500.742 0.8414877
20     Ideal     G 3720.706 0.7007146
21      Fair     H 5135.683 1.2191749
22      Good     H 4276.255 0.9147293
23 Very Good     H 4535.390 0.9159485
24   Premium     H 5216.707 1.0164492
25     Ideal     H 3889.335 0.7995249
26      Fair     I 4685.446 1.1980571
27      Good     I 5078.533 1.0572222
28 Very Good     I 5255.880 1.0469518
29   Premium     I 5946.181 1.1449370
30     Ideal     I 4451.970 0.9130291
31      Fair     J 4975.655 1.3411765
32      Good     J 4574.173 1.0995440
33 Very Good     J 5103.513 1.1332153
34   Premium     J 6294.592 1.2930941
35     Ideal     J 4918.186 1.0635937
```

Unfortunately, **aggregate** can be quite slow. Fortunately, there are other options, such as **plyr**, **dplyr** and **data.table**, that are considerably faster.

11.3 plyr

One of the best things to ever happen to R was the development of the **plyr**[4] package by Hadley Wickham. It epitomizes the "split-apply-combine" method of data manipulation. The core of **plyr** consists of functions such as **ddply**, **llply** and **ldply**. All of the manipulation functions consist of five letters, with the last three always being `ply`. The first letter indicates the type of input and the second letter indicates the type of output. For instance, **ddply** takes in a `data.frame` and outputs a `data.frame`, **llply** takes in a `list` and outputs a `list` and **ldply** takes in a `list` and outputs a `data.frame`. A full enumeration is listed in Table 11.1.

Table 11.1 **`plyr` functions and their corresponding inputs and outputs**

Function	Input Type	Output Type
ddply	`data.frame`	`data.frame`
llply	`list`	`list`
aaply	`array/vector/matrix`	`array/vector/matrix`
dlply	`data.frame`	`list`
daply	`data.frame`	`array/vector/matrix`
d_ply	`data.frame`	none (used for side effects)
ldply	`list`	`data.frame`
laply	`list`	`array/vector/matrix`
l_ply	`list`	none (used for side effects)
adply	`array/vector/matrix`	`data.frame`
alply	`array/vector/matrix`	`list`
a_ply	`array/vector/matrix`	none (used for side effects)

11.3.1 ddply

ddply takes a `data.frame`, splits it according to some variable(s), performs a desired action on it and returns a `data.frame`. To learn about **ddply** we look at the `baseball` data that come with **plyr**.

```
> library(plyr)
> head(baseball)

           id year stint team lg  g  ab  r  h X2b X3b hr rbi sb cs bb
4   ansonca01 1871     1  RC1     25 120 29 39  11   3  0  16  6  2  2
44  forceda01 1871     1  WS3     32 162 45 45   9   4  0  29  8  0  4
68  mathebo01 1871     1  FW1     19  89 15 24   3   1  0  10  2  1  2
99  startjo01 1871     1  NY2     33 161 35 58   5   1  1  34  4  2  3
102 suttoez01 1871     1  CL1     29 128 35 45   3   7  3  23  3  1  1
106 whitede01 1871     1  CL1     29 146 40 47   6   5  1  21  2  2  4
```

4. A play on the word *pliers* because it is one of the most versatile and essential tools.

```
    so ibb hbp sh sf gidp
4    1  NA  NA NA NA   NA
44   0  NA  NA NA NA   NA
68   0  NA  NA NA NA   NA
99   0  NA  NA NA NA   NA
102  0  NA  NA NA NA   NA
106  1  NA  NA NA NA   NA
```

A common statistic in baseball is On Base Percentage (OBP), which is calculated as

$$OBP = \frac{H + BB + HBP}{AB + BB + HBP + SF} \tag{11.1}$$

where

H = Hits
BB = Bases on Balls (Walks)
HBP = Times Hit by Pitch
AB = At Bats
SF = Sacrifice Flies

Before 1954 sacrifice flies were counted as part of sacrifice hits, which includes bunts, so for players before 1954 sacrifice flies should be assumed to be 0. That will be the first change we make to the data. There are many instances of `hbp` (hit by pitch) that are `NA`, so we set those to 0 as well. We also exclude players with less than 50 at bats in a season.

```
> # subsetting with [ is faster than using ifelse
> baseball$sf[baseball$year < 1954] <- 0
> # check that it worked
> any(is.na(baseball$sf))

[1] FALSE

> # set NA hbp's to 0
> baseball$hbp[is.na(baseball$hbp)] <- 0
> # check that it worked
> any(is.na(baseball$hbp))

[1] FALSE

> # only keep players with at least 50 at bats in a season
> baseball <- baseball[baseball$ab >= 50, ]
```

Calculating the OBP for a given player in a given year is easy enough with just vector operations.

```
> # calculate OBP
> baseball$OBP <- with(baseball, (h + bb + hbp) / (ab + bb + hbp + sf))
> tail(baseball)
```

```
             id year stint team lg   g  ab  r   h X2b X3b hr rbi sb
89499 claytro01 2007     1  TOR AL  69 189 23  48  14   0  1  12  2
89502 cirilje01 2007     1  MIN AL  50 153 18  40   9   2  2  21  2
89521 bondsba01 2007     1  SFN NL 126 340 75  94  14   0 28  66  5
89523 biggicr01 2007     1  HOU NL 141 517 68 130  31   3 10  50  4
89530 ausmubr01 2007     1  HOU NL 117 349 38  82  16   3  3  25  6
89533  aloumo01 2007     1  NYN NL  87 328 51 112  19   1 13  49  3
      cs  bb  so ibb hbp sh sf gidp       OBP
89499  1  14  50   0   1  3  3    8 0.3043478
89502  0  15  13   0   1  3  2    9 0.3274854
89521  0 132  54  43   3  0  2   13 0.4800839
89523  3  23 112   0   3  7  5    5 0.2846715
89530  1  37  74   3   6  4  1   11 0.3180662
89533  0  27  30   5   2  0  3   13 0.3916667
```

Here we used a new function, **with**. This allows us to specify the columns of a `data.frame` without having to specify the `data.frame` name each time.

To calculate the OBP for a player's entire career we cannot just average his individual season OBPs; we need to calculate and sum the numerator, and then divide by the sum of the denominator. This requires the use of **ddply**.

First we make a function to do that calculation; then we will use **ddply** to run that calculation for each player.

```
> # this function assumes that the column names for the data are as below
> obp <- function(data)
+ {
+     c(OBP=with(data, sum(h + bb + hbp) / sum(ab + bb + hbp + sf)))
+ }
>
> # use ddply to calculate career OBP for each player
> careerOBP <- ddply(baseball, .variables="id", .fun=obp)
> # sort the results by OBP
> careerOBP <- careerOBP[order(careerOBP$OBP, decreasing=TRUE), ]
> # see the results
> head(careerOBP, 10)
```

```
            id       OBP
1089 willite01 0.4816861
875   ruthba01 0.4742209
658  mcgrajo01 0.4657478
356  gehrilo01 0.4477848
85   bondsba01 0.4444622
476  hornsro01 0.4339068
184   cobbty01 0.4329655
327   foxxji01 0.4290509
953  speaktr01 0.4283386
191  collied01 0.4251246
```

This nicely returns the top ten players by career on base percentage. Notice that Billy Hamilton and Bill Joyce are absent from our results because they are mysteriously missing from the `baseball` data.

11.3.2 llply

In Section 11.1.2 we use **lapply** to sum each element of a list.

```
> theList <- list(A=matrix(1:9, 3), B=1:5, C=matrix(1:4, 2), D=2)
> lapply(theList, sum)

$A
[1] 45

$B
[1] 15

$C
[1] 10

$D
[1] 2
```

This can be done with **llply**, yielding identical results.

```
> llply(theList, sum)

$A
[1] 45

$B
[1] 15

$C
[1] 10

$D
[1] 2

> identical(lapply(theList, sum), llply(theList, sum))

[1] TRUE
```

To get the result as a vector, **laply** can be used similarly to **sapply**.

```
> sapply(theList, sum)

 A  B  C  D
45 15 10  2

> laply(theList, sum)

[1] 45 15 10  2
```

Notice, however, that while the results are the same, **laply** did not include names for the `vector`. These little nuances can be maddening but help dictate when to use which function.

11.3.3 `plyr` Helper Functions

plyr has a great number of useful helper functions such as **each**, which lets us supply multiple functions to a function like **aggregate**. One drawback, however, is that when using **each**, additional arguments can no longer be supplied to the functions.

```
> aggregate(price ~ cut, diamonds, each(mean, median))

        cut price.mean price.median
1      Fair   4358.758     3282.000
2      Good   3928.864     3050.500
3 Very Good   3981.760     2648.000
4   Premium   4584.258     3185.000
5     Ideal   3457.542     1810.000
```

Another great function is **idata.frame**, which creates a reference to a `data.frame` so that subsetting is much faster and more memory efficient. To illustrate this, we do a simple operation on the `baseball` data with the regular `data.frame` and an `idata.frame`.

```
> system.time(dlply(baseball, "id", nrow))

   user  system elapsed
   0.27    0.00    0.31

> iBaseball <- idata.frame(baseball)
> system.time(dlply(iBaseball, "id", nrow))

   user  system elapsed
   0.18    0.00    0.19
```

The speed gain depends on the size of the data and the complexity of the calculation. In this case, there was actually a decrease in performance. It is a bit of a moot point, as in most situations where speed is a concern **dplyr** should be used over **plyr** and **idata.frame**.

11.3.4 Speed versus Convenience

A criticism often leveled at **plyr** is that it can run slowly. The typical response to this is that using **plyr** is a question of speed versus convenience. Most of the functionality in **plyr** can be accomplished using base functions or other packages, but few of those offer the ease of use of **plyr**. Over the years, Hadley Wickham has taken great steps to speed up **plyr**, including optimized R code, C++ code and parallelization. The next evolution of this process was the introduction of the **dplyr** package, detailed in Chapter 14.

11.4 `data.table`

For speed junkies there is a package called **data.table**, written by Matt Dowle that extends and enhances the functionality of `data.frames`. The syntax is a little different from regular data.frames, so it will take getting used to, which is probably the primary reason it has not seen near-universal adoption.

The secret to the speed is that data.tables have an index like databases. This enables faster value accessing, group by operations and joins.

Creating data.tables is just like creating data.frames, and the two are very similar.

```
> library(data.table)
> # create a regular data.frame
> theDF <- data.frame(A=1:10,
+                     B=letters[1:10],
+                     C=LETTERS[11:20],
+                     D=rep(c("One", "Two", "Three"), length.out=10))
> # create a data.table
> theDT <- data.table(A=1:10,
+                     B=letters[1:10],
+                     C=LETTERS[11:20],
+                     D=rep(c("One", "Two", "Three"), length.out=10))
> # print them and compare
> theDF

    A B C     D
1   1 a K   One
2   2 b L   Two
3   3 c M Three
4   4 d N   One
5   5 e O   Two
6   6 f P Three
7   7 g Q   One
8   8 h R   Two
9   9 i S Three
10 10 j T   One

> theDT

     A B C     D
 1:  1 a K   One
 2:  2 b L   Two
 3:  3 c M Three
 4:  4 d N   One
 5:  5 e O   Two
 6:  6 f P Three
 7:  7 g Q   One
 8:  8 h R   Two
 9:  9 i S Three
10: 10 j T   One

> # notice by default data.frame turns character data into factors
> # while data.table does not
> class(theDF$B)
```

```
[1] "factor"

> class(theDT$B)

[1] "character"
```

The data is identical—except that **data.frame** turned B into a factor while **data.table** did not—and only the way it was printed looks different.

It is also possible to create a data.table out of an existing data.frame.

```
> diamondsDT <- data.table(diamonds)
> diamondsDT

        carat       cut color clarity depth table price    x    y    z
    1:   0.23     Ideal     E     SI2  61.5    55   326 3.95 3.98 2.43
    2:   0.21   Premium     E     SI1  59.8    61   326 3.89 3.84 2.31
    3:   0.23      Good     E     VS1  56.9    65   327 4.05 4.07 2.31
    4:   0.29   Premium     I     VS2  62.4    58   334 4.20 4.23 2.63
    5:   0.31      Good     J     SI2  63.3    58   335 4.34 4.35 2.75
   ---
53936:   0.72     Ideal     D     SI1  60.8    57  2757 5.75 5.76 3.50
53937:   0.72      Good     D     SI1  63.1    55  2757 5.69 5.75 3.61
53938:   0.70 Very Good     D     SI1  62.8    60  2757 5.66 5.68 3.56
53939:   0.86   Premium     H     SI2  61.0    58  2757 6.15 6.12 3.74
53940:   0.75     Ideal     D     SI2  62.2    55  2757 5.83 5.87 3.64
```

Notice that printing the diamonds data would try to print out all the data but data.table intelligently just prints the first five and last five rows.

Accessing rows can be done similarly to accessing rows in a data.frame.

```
> theDT[1:2, ]

   A B C   D
1: 1 a K One
2: 2 b L Two

> theDT[theDT$A >= 7, ]

    A B C     D
1:  7 g Q   One
2:  8 h R   Two
3:  9 i S Three
4: 10 j T   One

> theDT[A >= 7, ]

    A B C     D
1:  7 g Q   One
2:  8 h R   Two
3:  9 i S Three
4: 10 j T   One
```

While the second line in the preceding code is valid syntax, it is not necessarily efficient syntax. That line creates a vector of length nrow(theDT) = 10 consisting of TRUE or FALSE entries, which is a vector scan. After we create a key for the data.table we can use different syntax to pick rows through a binary search, which will be much faster and is covered in Section 11.4.1. The third line computes the same result as the second and is nicely without the dollar sign notation. This is because the **data.table** function knows to find the column A within the theDT data.table.

Accessing individual columns must be done a little differently than accessing columns in data.frames. In Section 5.1 we show that multiple columns in a data.frame should be specified as a character vector. With data.tables the columns should be specified as a list of the actual names, not as characters.

```
> theDT[, list(A, C)]

     A C
 1:  1 K
 2:  2 L
 3:  3 M
 4:  4 N
 5:  5 O
 6:  6 P
 7:  7 Q
 8:  8 R
 9:  9 S
10: 10 T

> # just one column
> theDT[, B]

 [1] "a" "b" "c" "d" "e" "f" "g" "h" "i" "j"

> # one column while maintaining data.table structure
> theDT[, list(B)]

    B
 1: a
 2: b
 3: c
 4: d
 5: e
 6: f
 7: g
 8: h
 9: i
10: j
```

If we must specify the column names as characters (perhaps because they were passed as arguments to a function), the with argument should be set to FALSE.

```
> theDT[, "B", with=FALSE]

    B
 1: a
 2: b
 3: c
 4: d
 5: e
 6: f
 7: g
 8: h
 9: i
10: j

> theDT[, c("A", "C"), with=FALSE]

     A C
 1:  1 K
 2:  2 L
 3:  3 M
 4:  4 N
 5:  5 O
 6:  6 P
 7:  7 Q
 8:  8 R
 9:  9 S
10: 10 T

> theCols <- c("A", "C")
> theDT[, theCols, with=FALSE]

     A C
 1:  1 K
 2:  2 L
 3:  3 M
 4:  4 N
 5:  5 O
 6:  6 P
 7:  7 Q
 8:  8 R
 9:  9 S
10: 10 T
```

This time we used a `vector` to hold the column names instead of a `list`. These nuances are important to proper functions of `data.tables` but can lead to a great deal of frustration.

11.4.1 Keys

Now that we have a few data.tables in memory, we might be interested in seeing some information about them.

```
> # show tables
> tables()

     NAME         NROW NCOL MB
[1,] diamondsDT 53,940   10  4
[2,] theDT          10    4  1
[3,] tomato3        16   11  1
     COLS
[1,] carat,cut,color,clarity,depth,table,price,x,y,z
[2,] A,B,C,D
[3,] Round,Tomato,Price,Source,Sweet,Acid,Color,Texture,Overall,Avg of Totals,Total o
     KEY
[1,]
[2,]
[3,]
Total: 6MB
```

This shows, for each data.table in memory, the name, the number of rows, the size in megabytes, the column names and the key. We have not assigned keys for any of the tables so that column is blank. The key is used to index the data.table and will provide the extra speed.

We start by adding a key to theDT. We will use the D column to index the data.table. This is done using **setkey**, which takes the name of the data.table as its first argument and the name of the desired column (without quotes, as is consistent with column selection) as the second argument.

```
> # set the key
> setkey(theDT, D)
> # show the data.table again
> theDT

     A B C     D
 1:  1 a K   One
 2:  4 d N   One
 3:  7 g Q   One
 4: 10 j T   One
 5:  3 c M Three
 6:  6 f P Three
 7:  9 i S Three
 8:  2 b L   Two
 9:  5 e O   Two
10:  8 h R   Two
```

The data have been reordered according to column D, which is sorted alphabetically. We can confirm the key was set with **key**.

```
> key(theDT)

[1] "D"
```

Or **tables**.

```
> tables()

     NAME          NROW NCOL MB
[1,] diamondsDT 53,940   10  4
[2,] theDT          10    4  1
[3,] tomato3        16   11  1
     COLS
[1,] carat,cut,color,clarity,depth,table,price,x,y,z
[2,] A,B,C,D
[3,] Round,Tomato,Price,Source,Sweet,Acid,Color,Texture,Overall,Avg of Totals,Total o
     KEY
[1,]
[2,] D
[3,]
Total: 6MB
```

This adds some new functionality to selecting rows from `data.tables`. In addition to selecting rows by the row number or by some expression that evaluates to `TRUE` or `FALSE`, a value of the key column can be specified.

```
> theDT["One", ]

    A B C   D
1:  1 a K One
2:  4 d N One
3:  7 g Q One
4: 10 j T One

> theDT[c("One", "Two"), ]

    A B C   D
1:  1 a K One
2:  4 d N One
3:  7 g Q One
4: 10 j T One
5:  2 b L Two
6:  5 e O Two
7:  8 h R Two
```

More than one column can be set as the key.

```
> # set the key
> setkey(diamondsDT, cut, color)
```

To access rows according to both keys, there is a special function named **J**. It takes multiple arguments, each of which is a `vector` of values to select.

```
> # access some rows
> diamondsDT[J("Ideal", "E"), ]

      carat   cut color clarity depth table price    x    y    z
   1:  0.23 Ideal     E     SI2  61.5    55   326 3.95 3.98 2.43
   2:  0.26 Ideal     E    VVS2  62.9    58   554 4.02 4.06 2.54
   3:  0.70 Ideal     E     SI1  62.5    57  2757 5.70 5.72 3.57
   4:  0.59 Ideal     E    VVS2  62.0    55  2761 5.38 5.43 3.35
   5:  0.74 Ideal     E     SI2  62.2    56  2761 5.80 5.84 3.62
  ---
3899:  0.70 Ideal     E     SI1  61.7    55  2745 5.71 5.74 3.53
3900:  0.51 Ideal     E    VVS1  61.9    54  2745 5.17 5.11 3.18
3901:  0.56 Ideal     E    VVS1  62.1    56  2750 5.28 5.29 3.28
3902:  0.77 Ideal     E     SI2  62.1    56  2753 5.84 5.86 3.63
3903:  0.71 Ideal     E     SI1  61.9    56  2756 5.71 5.73 3.54

> diamondsDT[J("Ideal", c("E", "D")), ]

      carat   cut color clarity depth table price    x    y    z
   1:  0.23 Ideal     E     SI2  61.5    55   326 3.95 3.98 2.43
   2:  0.26 Ideal     E    VVS2  62.9    58   554 4.02 4.06 2.54
   3:  0.70 Ideal     E     SI1  62.5    57  2757 5.70 5.72 3.57
   4:  0.59 Ideal     E    VVS2  62.0    55  2761 5.38 5.43 3.35
   5:  0.74 Ideal     E     SI2  62.2    56  2761 5.80 5.84 3.62
  ---
6733:  0.51 Ideal     D    VVS2  61.7    56  2742 5.16 5.14 3.18
6734:  0.51 Ideal     D    VVS2  61.3    57  2742 5.17 5.14 3.16
6735:  0.81 Ideal     D     SI1  61.5    57  2748 6.00 6.03 3.70
6736:  0.72 Ideal     D     SI1  60.8    57  2757 5.75 5.76 3.50
6737:  0.75 Ideal     D     SI2  62.2    55  2757 5.83 5.87 3.64
```

11.4.2 data.table Aggregation

The primary benefit of indexing is faster aggregation. While **aggregate** and the various **d*ply** functions will work because `data.tables` are just enhanced `data.frames`, they will be slower than using the built-in aggregation functionality of `data.table`.

In Section 11.2 we calculate the mean price of diamonds for each type of cut.

```
> aggregate(price ~ cut, diamonds, mean)

        cut    price
1      Fair 4358.758
2      Good 3928.864
3 Very Good 3981.760
4   Premium 4584.258
5     Ideal 3457.542
```

To get the same result using data.table, we do this:

```
> diamondsDT[, mean(price), by=cut]

         cut       V1
1:      Fair 4358.758
2:      Good 3928.864
3: Very Good 3981.760
4:   Premium 4584.258
5:     Ideal 3457.542
```

The only difference between this and the previous result is that the columns have different names. To specify the name of the resulting column, pass the aggregation function as a named list.

```
> diamondsDT[, list(price=mean(price)), by=cut]

         cut    price
1:      Fair 4358.758
2:      Good 3928.864
3: Very Good 3981.760
4:   Premium 4584.258
5:     Ideal 3457.542
```

To aggregate on multiple columns, specify them as a list().

```
> diamondsDT[, mean(price), by=list(cut, color)]

          cut color       V1
 1:      Fair     D 4291.061
 2:      Fair     E 3682.312
 3:      Fair     F 3827.003
 4:      Fair     G 4239.255
 5:      Fair     H 5135.683
 6:      Fair     I 4685.446
 7:      Fair     J 4975.655
 8:      Good     D 3405.382
 9:      Good     E 3423.644
10:      Good     F 3495.750
11:      Good     G 4123.482
12:      Good     H 4276.255
13:      Good     I 5078.533
14:      Good     J 4574.173
15: Very Good     D 3470.467
16: Very Good     E 3214.652
17: Very Good     F 3778.820
18: Very Good     G 3872.754
```

```
19: Very Good     H 4535.390
20: Very Good     I 5255.880
21: Very Good     J 5103.513
22:   Premium     D 3631.293
23:   Premium     E 3538.914
24:   Premium     F 4324.890
25:   Premium     G 4500.742
26:   Premium     H 5216.707
27:   Premium     I 5946.181
28:   Premium     J 6294.592
29:     Ideal     D 2629.095
30:     Ideal     E 2597.550
31:     Ideal     F 3374.939
32:     Ideal     G 3720.706
33:     Ideal     H 3889.335
34:     Ideal     I 4451.970
35:     Ideal     J 4918.186
          cut color       V1
```

To aggregate multiple arguments, pass them as a `list`. Unlike with **aggregate**, a different metric can be measured for each column.

```
> diamondsDT[, list(price=mean(price), carat=mean(carat)), by=cut]

         cut    price     carat
1:      Fair 4358.758 1.0461366
2:      Good 3928.864 0.8491847
3: Very Good 3981.760 0.8063814
4:   Premium 4584.258 0.8919549
5:     Ideal 3457.542 0.7028370

> diamondsDT[, list(price=mean(price), carat=mean(carat),
+                   caratSum=sum(carat)), by=cut]

         cut    price     carat caratSum
1:      Fair 4358.758 1.0461366  1684.28
2:      Good 3928.864 0.8491847  4166.10
3: Very Good 3981.760 0.8063814  9742.70
4:   Premium 4584.258 0.8919549 12300.95
5:     Ideal 3457.542 0.7028370 15146.84
```

Finally, both multiple metrics can be calculated and multiple grouping variables can be specified at the same time.

```
> diamondsDT[, list(price=mean(price), carat=mean(carat)),
+            by=list(cut, color)]
```

```
          cut color    price     carat
 1:      Fair     D 4291.061 0.9201227
 2:      Fair     E 3682.312 0.8566071
 3:      Fair     F 3827.003 0.9047115
 4:      Fair     G 4239.255 1.0238217
 5:      Fair     H 5135.683 1.2191749
 6:      Fair     I 4685.446 1.1980571
 7:      Fair     J 4975.655 1.3411765
 8:      Good     D 3405.382 0.7445166
 9:      Good     E 3423.644 0.7451340
10:      Good     F 3495.750 0.7759296
11:      Good     G 4123.482 0.8508955
12:      Good     H 4276.255 0.9147293
13:      Good     I 5078.533 1.0572222
14:      Good     J 4574.173 1.0995440
15: Very Good     D 3470.467 0.6964243
16: Very Good     E 3214.652 0.6763167
17: Very Good     F 3778.820 0.7409612
18: Very Good     G 3872.754 0.7667986
19: Very Good     H 4535.390 0.9159485
20: Very Good     I 5255.880 1.0469518
21: Very Good     J 5103.513 1.1332153
22:   Premium     D 3631.293 0.7215471
23:   Premium     E 3538.914 0.7177450
24:   Premium     F 4324.890 0.8270356
25:   Premium     G 4500.742 0.8414877
26:   Premium     H 5216.707 1.0164492
27:   Premium     I 5946.181 1.1449370
28:   Premium     J 6294.592 1.2930941
29:     Ideal     D 2629.095 0.5657657
30:     Ideal     E 2597.550 0.5784012
31:     Ideal     F 3374.939 0.6558285
32:     Ideal     G 3720.706 0.7007146
33:     Ideal     H 3889.335 0.7995249
34:     Ideal     I 4451.970 0.9130291
35:     Ideal     J 4918.186 1.0635938
          cut color    price     carat
```

11.5 Conclusion

Aggregating data is a very important step in the analysis process. Sometimes it is the end goal, and other times it is in preparation for applying more advanced methods. No matter the reason for aggregation, there are plenty of functions to make it possible. These include **aggregate**, **apply** and **lapply** in base; **ddply**, **llply** and the rest in **plyr**; and the group by functionality in **data.table**.

12

Faster Group Manipulation with dplyr

Not to be outdone by Matt Dowle, Hadley Wickham has written a sequel to his famous **plyr** package that focuses on speed called **dplyr**. The d in the name reinforces that the package is meant to work with `data.frames`, while `list` and `vector` functionality has been moved to the **purrr** package, which is detailed in Chapter 13. More and more **dplyr** is becoming the de facto choice for data munging, having nearly replaced **plyr**. Fortunately for R users, there is an arms race between Hadley Wickham and Matt Dowle to write the fastest code, and **dplyr** offers a great mix of speed and ease of use.

Writing code with **dplyr** involves using the "grammar of data" to perform data munging. Each step is done by a single function that represents a verb. These verbs will be somewhat familiar to SQL users. Selecting columns is done with **select**, filtering rows with **filter**, grouping data with **group_by** and changing or adding columns with **mutate**. These are just some of the many functions in **dplyr**.

When working with both **dplyr** and **plyr** it is important to load **plyr** first and then **dplyr**, because they have a number of functions with the same names, and in R, functions from the last package loaded take precedence. This sometimes creates a need to explicitly specify both the package and function using the double colon operator (`::`), such as `plyr::summarize` or `dplyr::summarize`.

12.1 Pipes

Not only is **dplyr** incredibly fast; it popularized the new piping paradigm made possible by the **magrittr** package. That is, rather than nesting functions within each other, storing temporary steps to variables, we pipe the result of one function into another using the pipe (%>%) operator.

With pipes we pipe objects into the first arguments of functions. These operations can be chained together, piping the result of one function into the first argument of the next. As an example, we pipe the `diamonds` data into the **head** function and that into the **dim** function.

```
> library(magrittr)
> data(diamonds, package='ggplot2')
> dim(head(diamonds, n=4))
```

```
[1]  4 10

> diamonds %>% head(4) %>% dim

[1]  4 10
```

12.2 tbl

Just as **data.table** introduced the `data.table` object to extend `data.frames`, **dplyr** brought us `tbl` objects which are also an extension of `data.frames`. While they have beneficial underlying properties, the most outwardly noticeable feature of `tbls` is that when they are printed to the screen, only a subset of the rows are displayed by default and only as many columns as will fit on the screen are printed.[1] Another feature is that the data type stored in each column is displayed under the column names.

The `diamonds` data—with more recent versions of **ggplot2**—are stored in a `tbl`, specifically a `tbl_df`, which itself is an extension of `tbl`. Without **dplyr** or similar `tbl`-based package loaded they will print as a normal `data.frame`.

```
> class(diamonds)

[1] "tbl_df"     "tbl"        "data.frame"

> head(diamonds)

# A tibble: 6 × 10
  carat       cut color clarity depth table price     x     y     z
  <dbl>     <ord> <ord>   <ord> <dbl> <dbl> <int> <dbl> <dbl> <dbl>
1  0.23     Ideal     E     SI2  61.5    55   326  3.95  3.98  2.43
2  0.21   Premium     E     SI1  59.8    61   326  3.89  3.84  2.31
3  0.23      Good     E     VS1  56.9    65   327  4.05  4.07  2.31
4  0.29   Premium     I     VS2  62.4    58   334  4.20  4.23  2.63
5  0.31      Good     J     SI2  63.3    58   335  4.34  4.35  2.75
6  0.24 Very Good     J    VVS2  62.8    57   336  3.94  3.96  2.48
```

After **dplyr** is loaded, they print like `tbls`.

```
> library(dplyr)
> head(diamonds)

# A tibble: 6 × 10
  carat       cut color clarity depth table price     x     y     z
  <dbl>     <ord> <ord>   <ord> <dbl> <dbl> <int> <dbl> <dbl> <dbl>
1  0.23     Ideal     E     SI2  61.5    55   326  3.95  3.98  2.43
2  0.21   Premium     E     SI1  59.8    61   326  3.89  3.84  2.31
3  0.23      Good     E     VS1  56.9    65   327  4.05  4.07  2.31
4  0.29   Premium     I     VS2  62.4    58   334  4.20  4.23  2.63
5  0.31      Good     J     SI2  63.3    58   335  4.34  4.35  2.75
6  0.24 Very Good     J    VVS2  62.8    57   336  3.94  3.96  2.48
```

1. The number of displayed columns varies depending on console width.

Since tbls are printed with only a subset of the rows, we do not need to use **head**.

```
> diamonds

# A tibble: 53,940 × 10
   carat       cut color clarity depth table price     x     y     z
   <dbl>     <ord> <ord>   <ord> <dbl> <dbl> <int> <dbl> <dbl> <dbl>
1   0.23     Ideal     E     SI2  61.5    55   326  3.95  3.98  2.43
2   0.21   Premium     E     SI1  59.8    61   326  3.89  3.84  2.31
3   0.23      Good     E     VS1  56.9    65   327  4.05  4.07  2.31
4   0.29   Premium     I     VS2  62.4    58   334  4.20  4.23  2.63
5   0.31      Good     J     SI2  63.3    58   335  4.34  4.35  2.75
6   0.24 Very Good     J    VVS2  62.8    57   336  3.94  3.96  2.48
7   0.24 Very Good     I    VVS1  62.3    57   336  3.95  3.98  2.47
8   0.26 Very Good     H     SI1  61.9    55   337  4.07  4.11  2.53
9   0.22      Fair     E     VS2  65.1    61   337  3.87  3.78  2.49
10  0.23 Very Good     H     VS1  59.4    61   338  4.00  4.05  2.39
# ... with 53,930 more rows
```

The tbl object was originally introduced in **dplyr** and then further expanded in the **tibble** package. After their inclusion in this new package they began to be refereed to as tibbles, though their class is still tbl.

12.3 select

The **select** function takes a data.frame (or tbl) as its first argument then the desired columns as subsequent arguments. The function, like all **dplyr** functions, can be used in the traditional, nested manner or with pipes.

```
> select(diamonds, carat, price)

# A tibble: 53,940 × 2
   carat price
   <dbl> <int>
1   0.23   326
2   0.21   326
3   0.23   327
4   0.29   334
5   0.31   335
6   0.24   336
7   0.24   336
8   0.26   337
9   0.22   337
10  0.23   338
# ... with 53,930 more rows

> diamonds %>% select(carat, price)

# A tibble: 53,940 × 2
   carat price
   <dbl> <int>
1   0.23   326
```

```
2   0.21   326
3   0.23   327
4   0.29   334
5   0.31   335
6   0.24   336
7   0.24   336
8   0.26   337
9   0.22   337
10  0.23   338
# ... with 53,930 more rows
```

```
> # the columns can be specified as a vector of column names as well
> diamonds %>% select(c(carat, price))
```

```
# A tibble: 53,940 × 2
   carat price
   <dbl> <int>
1   0.23   326
2   0.21   326
3   0.23   327
4   0.29   334
5   0.31   335
6   0.24   336
7   0.24   336
8   0.26   337
9   0.22   337
10  0.23   338
# ... with 53,930 more rows
```

The regular **select** function is designed to take unquoted column names to make interactive use easier. The names can be passed either as individual arguments or as a `vector`. If quoted column names are necessary, they can be used with the standard evaluation version of **select**, which has an underscore (_) at the end of the function.

```
> diamonds %>% select_('carat', 'price')
```

```
# A tibble: 53,940 × 2
   carat price
   <dbl> <int>
1   0.23   326
2   0.21   326
3   0.23   327
4   0.29   334
5   0.31   335
6   0.24   336
7   0.24   336
8   0.26   337
9   0.22   337
10  0.23   338
# ... with 53,930 more rows
```

If the column names are stored in a variable, they should be passed to the .dots argument.

```
> theCols <- c('carat', 'price')
> diamonds %>% select_(.dots=theCols)

# A tibble: 53,940 × 2
   carat price
   <dbl> <int>
1   0.23   326
2   0.21   326
3   0.23   327
4   0.29   334
5   0.31   335
6   0.24   336
7   0.24   336
8   0.26   337
9   0.22   337
10  0.23   338
# ... with 53,930 more rows
```

Starting with **dplyr** version 0.6.0 **select_** is deprecated, though it remains for backward comparability. An alternative method, that uses the regular **select**, is to use the **one_of** function.

```
> diamonds %>% select(one_of('carat', 'price'))

# A tibble: 53,940 × 2
   carat price
   <dbl> <int>
 1  0.23   326
 2  0.21   326
 3  0.23   327
 4  0.29   334
 5  0.31   335
 6  0.24   336
 7  0.24   336
 8  0.26   337
 9  0.22   337
10  0.23   338
# ... with 53,930 more rows

> # as a variable
> theCols <- c('carat', 'price')
> diamonds %>% select(one_of(theCols))

# A tibble: 53,940 × 2
   carat price
```

```
   <dbl> <int>
 1  0.23   326
 2  0.21   326
 3  0.23   327
 4  0.29   334
 5  0.31   335
 6  0.24   336
 7  0.24   336
 8  0.26   337
 9  0.22   337
10  0.23   338
# ... with 53,930 more rows
```

It is possible to use traditional R square bracket syntax, though the **dplyr** printing rules still apply.

```
> diamonds[, c('carat', 'price')]

# A tibble: 53,940 × 2
   carat price
   <dbl> <int>
1   0.23   326
2   0.21   326
3   0.23   327
4   0.29   334
5   0.31   335
6   0.24   336
7   0.24   336
8   0.26   337
9   0.22   337
10  0.23   338
# ... with 53,930 more rows
```

As with the square bracket syntax, column names can be specified by position using their indices.

```
> select(diamonds, 1, 7)

# A tibble: 53,940 × 2
   carat price
   <dbl> <int>
1   0.23   326
2   0.21   326
3   0.23   327
4   0.29   334
5   0.31   335
6   0.24   336
7   0.24   336
```

```
8   0.26   337
9   0.22   337
10  0.23   338
# ... with 53,930 more rows
```

```
> diamonds %>% select(1, 7)
```

```
# A tibble: 53,940 × 2
   carat price
   <dbl> <int>
1   0.23   326
2   0.21   326
3   0.23   327
4   0.29   334
5   0.31   335
6   0.24   336
7   0.24   336
8   0.26   337
9   0.22   337
10  0.23   338
# ... with 53,930 more rows
```

Searching for a partial match is done with **dplyr** functions **starts_with**, **ends_with** and **contains**.

```
> diamonds %>% select(starts_with('c'))
```

```
# A tibble: 53,940 × 4
   carat       cut color clarity
   <dbl>     <ord> <ord>   <ord>
1   0.23     Ideal     E     SI2
2   0.21   Premium     E     SI1
3   0.23      Good     E     VS1
4   0.29   Premium     I     VS2
5   0.31      Good     J     SI2
6   0.24 Very Good     J    VVS2
7   0.24 Very Good     I    VVS1
8   0.26 Very Good     H     SI1
9   0.22      Fair     E     VS2
10  0.23 Very Good     H     VS1
# ... with 53,930 more rows
```

```
> diamonds %>% select(ends_with('e'))
```

```
# A tibble: 53,940 × 2
   table price
   <dbl> <int>
1     55   326
2     61   326
```

```
3      65   327
4      58   334
5      58   335
6      57   336
7      57   336
8      55   337
9      61   337
10     61   338
# ... with 53,930 more rows
```

```
> diamonds %>% select(contains('l'))
```

```
# A tibble: 53,940 × 3
   color clarity table
   <ord>   <ord> <dbl>
1      E     SI2    55
2      E     SI1    61
3      E     VS1    65
4      I     VS2    58
5      J     SI2    58
6      J    VVS2    57
7      I    VVS1    57
8      H     SI1    55
9      E     VS2    61
10     H     VS1    61
# ... with 53,930 more rows
```

Regular expression searches are done with **matches**. The following code searches for columns that contain the letter "r", followed by any number of wildcard matches and then the letter "t". Regular expressions are further explained in Section 16.4.

```
> diamonds %>% select(matches('r.+t'))
```

```
# A tibble: 53,940 × 2
   carat clarity
   <dbl>   <ord>
1   0.23     SI2
2   0.21     SI1
3   0.23     VS1
4   0.29     VS2
5   0.31     SI2
6   0.24    VVS2
7   0.24    VVS1
8   0.26     SI1
9   0.22     VS2
10  0.23     VS1
# ... with 53,930 more rows
```

Columns can be designated not to be selected by preceding the column names or numbers with the minus sign (-).

```
> # by name
> diamonds %>% select(-carat, -price)

# A tibble: 53,940 × 8
         cut color clarity depth table     x     y     z
       <ord> <ord>   <ord> <dbl> <dbl> <dbl> <dbl> <dbl>
1      Ideal     E     SI2  61.5    55  3.95  3.98  2.43
2    Premium     E     SI1  59.8    61  3.89  3.84  2.31
3       Good     E     VS1  56.9    65  4.05  4.07  2.31
4    Premium     I     VS2  62.4    58  4.20  4.23  2.63
5       Good     J     SI2  63.3    58  4.34  4.35  2.75
6  Very Good     J    VVS2  62.8    57  3.94  3.96  2.48
7  Very Good     I    VVS1  62.3    57  3.95  3.98  2.47
8  Very Good     H     SI1  61.9    55  4.07  4.11  2.53
9       Fair     E     VS2  65.1    61  3.87  3.78  2.49
10 Very Good     H     VS1  59.4    61  4.00  4.05  2.39
# ... with 53,930 more rows

> diamonds %>% select(-c(carat, price))

# A tibble: 53,940 × 8
         cut color clarity depth table     x     y     z
       <ord> <ord>   <ord> <dbl> <dbl> <dbl> <dbl> <dbl>
1      Ideal     E     SI2  61.5    55  3.95  3.98  2.43
2    Premium     E     SI1  59.8    61  3.89  3.84  2.31
3       Good     E     VS1  56.9    65  4.05  4.07  2.31
4    Premium     I     VS2  62.4    58  4.20  4.23  2.63
5       Good     J     SI2  63.3    58  4.34  4.35  2.75
6  Very Good     J    VVS2  62.8    57  3.94  3.96  2.48
7  Very Good     I    VVS1  62.3    57  3.95  3.98  2.47
8  Very Good     H     SI1  61.9    55  4.07  4.11  2.53
9       Fair     E     VS2  65.1    61  3.87  3.78  2.49
10 Very Good     H     VS1  59.4    61  4.00  4.05  2.39
# ... with 53,930 more rows

> # by number
> diamonds %>% select(-1, -7)

# A tibble: 53,940 × 8
         cut color clarity depth table     x     y     z
       <ord> <ord>   <ord> <dbl> <dbl> <dbl> <dbl> <dbl>
1      Ideal     E     SI2  61.5    55  3.95  3.98  2.43
2    Premium     E     SI1  59.8    61  3.89  3.84  2.31
3       Good     E     VS1  56.9    65  4.05  4.07  2.31
4    Premium     I     VS2  62.4    58  4.20  4.23  2.63
```

```
5        Good     J      SI2  63.3    58  4.34  4.35  2.75
6   Very Good     J     VVS2  62.8    57  3.94  3.96  2.48
7   Very Good     I     VVS1  62.3    57  3.95  3.98  2.47
8   Very Good     H      SI1  61.9    55  4.07  4.11  2.53
9        Fair     E      VS2  65.1    61  3.87  3.78  2.49
10  Very Good     H      VS1  59.4    61  4.00  4.05  2.39
# ... with 53,930 more rows
```

```
> diamonds %>% select(-c(1, 7))

# A tibble: 53,940 × 8
          cut color clarity depth table     x     y     z
        <ord> <ord>   <ord> <dbl> <dbl> <dbl> <dbl> <dbl>
1       Ideal     E     SI2  61.5    55  3.95  3.98  2.43
2     Premium     E     SI1  59.8    61  3.89  3.84  2.31
3        Good     E     VS1  56.9    65  4.05  4.07  2.31
4     Premium     I     VS2  62.4    58  4.20  4.23  2.63
5        Good     J     SI2  63.3    58  4.34  4.35  2.75
6   Very Good     J    VVS2  62.8    57  3.94  3.96  2.48
7   Very Good     I    VVS1  62.3    57  3.95  3.98  2.47
8   Very Good     H     SI1  61.9    55  4.07  4.11  2.53
9        Fair     E     VS2  65.1    61  3.87  3.78  2.49
10  Very Good     H     VS1  59.4    61  4.00  4.05  2.39
# ... with 53,930 more rows
```

Specifying columns not to select using quoted names requires putting the minus sign inside the quotes surrounding the names of undesired columns that are given to the `.dots` argument.

```
> diamonds %>% select_(.dots=c('-carat', '-price'))

# A tibble: 53,940 × 8
          cut color clarity depth table     x     y     z
        <ord> <ord>   <ord> <dbl> <dbl> <dbl> <dbl> <dbl>
1       Ideal     E     SI2  61.5    55  3.95  3.98  2.43
2     Premium     E     SI1  59.8    61  3.89  3.84  2.31
3        Good     E     VS1  56.9    65  4.05  4.07  2.31
4     Premium     I     VS2  62.4    58  4.20  4.23  2.63
5        Good     J     SI2  63.3    58  4.34  4.35  2.75
6   Very Good     J    VVS2  62.8    57  3.94  3.96  2.48
7   Very Good     I    VVS1  62.3    57  3.95  3.98  2.47
8   Very Good     H     SI1  61.9    55  4.07  4.11  2.53
9        Fair     E     VS2  65.1    61  3.87  3.78  2.49
10  Very Good     H     VS1  59.4    61  4.00  4.05  2.39
# ... with 53,930 more rows
```

When using **one_of** the minus sign goes before the **one_of** function.

```
> diamonds %>% select(-one_of('carat', 'price'))

# A tibble: 53,940 × 8
         cut color clarity depth table     x     y     z
       <ord> <ord>   <ord> <dbl> <dbl> <dbl> <dbl> <dbl>
 1     Ideal     E     SI2  61.5    55  3.95  3.98  2.43
 2   Premium     E     SI1  59.8    61  3.89  3.84  2.31
 3      Good     E     VS1  56.9    65  4.05  4.07  2.31
 4   Premium     I     VS2  62.4    58  4.20  4.23  2.63
 5      Good     J     SI2  63.3    58  4.34  4.35  2.75
 6 Very Good     J    VVS2  62.8    57  3.94  3.96  2.48
 7 Very Good     I    VVS1  62.3    57  3.95  3.98  2.47
 8 Very Good     H     SI1  61.9    55  4.07  4.11  2.53
 9      Fair     E     VS2  65.1    61  3.87  3.78  2.49
10 Very Good     H     VS1  59.4    61  4.00  4.05  2.39
# ... with 53,930 more rows
```

12.4 filter

Specifying rows based on a logical expression is done with **filter**.

```
> diamonds %>% filter(cut == 'Ideal')

# A tibble: 21,551 × 10
   carat   cut color clarity depth table price     x     y     z
   <dbl> <ord> <ord>   <ord> <dbl> <dbl> <int> <dbl> <dbl> <dbl>
1   0.23 Ideal     E     SI2  61.5    55   326  3.95  3.98  2.43
2   0.23 Ideal     J     VS1  62.8    56   340  3.93  3.90  2.46
3   0.31 Ideal     J     SI2  62.2    54   344  4.35  4.37  2.71
4   0.30 Ideal     I     SI2  62.0    54   348  4.31  4.34  2.68
5   0.33 Ideal     I     SI2  61.8    55   403  4.49  4.51  2.78
6   0.33 Ideal     I     SI2  61.2    56   403  4.49  4.50  2.75
7   0.33 Ideal     J     SI1  61.1    56   403  4.49  4.55  2.76
8   0.23 Ideal     G     VS1  61.9    54   404  3.93  3.95  2.44
9   0.32 Ideal     I     SI1  60.9    55   404  4.45  4.48  2.72
10  0.30 Ideal     I     SI2  61.0    59   405  4.30  4.33  2.63
# ... with 21,541 more rows
```

The base R equivalent is more verbose and uses square brackets.

```
> diamonds[diamonds$cut == 'Ideal', ]

# A tibble: 21,551 × 10
   carat   cut color clarity depth table price     x     y     z
   <dbl> <ord> <ord>   <ord> <dbl> <dbl> <int> <dbl> <dbl> <dbl>
1   0.23 Ideal     E     SI2  61.5    55   326  3.95  3.98  2.43
2   0.23 Ideal     J     VS1  62.8    56   340  3.93  3.90  2.46
```

```
3    0.31 Ideal     J      SI2   62.2    54   344  4.35  4.37  2.71
4    0.30 Ideal     I      SI2   62.0    54   348  4.31  4.34  2.68
5    0.33 Ideal     I      SI2   61.8    55   403  4.49  4.51  2.78
6    0.33 Ideal     I      SI2   61.2    56   403  4.49  4.50  2.75
7    0.33 Ideal     J      SI1   61.1    56   403  4.49  4.55  2.76
8    0.23 Ideal     G      VS1   61.9    54   404  3.93  3.95  2.44
9    0.32 Ideal     I      SI1   60.9    55   404  4.45  4.48  2.72
10   0.30 Ideal     I      SI2   61.0    59   405  4.30  4.33  2.63
# ... with 21,541 more rows
```

To filter on a column being equal to one of many possible values the `%in%` operator is used.

```
> diamonds %>% filter(cut %in% c('Ideal', 'Good'))

# A tibble: 26,457 × 10
   carat   cut color clarity depth table price     x     y     z
   <dbl> <ord> <ord>   <ord> <dbl> <dbl> <int> <dbl> <dbl> <dbl>
1   0.23 Ideal     E     SI2  61.5    55   326  3.95  3.98  2.43
2   0.23  Good     E     VS1  56.9    65   327  4.05  4.07  2.31
3   0.31  Good     J     SI2  63.3    58   335  4.34  4.35  2.75
4   0.30  Good     J     SI1  64.0    55   339  4.25  4.28  2.73
5   0.23 Ideal     J     VS1  62.8    56   340  3.93  3.90  2.46
6   0.31 Ideal     J     SI2  62.2    54   344  4.35  4.37  2.71
7   0.30 Ideal     I     SI2  62.0    54   348  4.31  4.34  2.68
8   0.30  Good     J     SI1  63.4    54   351  4.23  4.29  2.70
9   0.30  Good     J     SI1  63.8    56   351  4.23  4.26  2.71
10  0.30  Good     I     SI2  63.3    56   351  4.26  4.30  2.71
# ... with 26,447 more rows
```

All the standard equality operators can all be used with **filter**.

```
> diamonds %>% filter(price >= 1000)

# A tibble: 39,441 × 10
   carat       cut color clarity depth table price     x     y     z
   <dbl>     <ord> <ord>   <ord> <dbl> <dbl> <int> <dbl> <dbl> <dbl>
1   0.70     Ideal     E     SI1  62.5    57  2757  5.70  5.72  3.57
2   0.86      Fair     E     SI2  55.1    69  2757  6.45  6.33  3.52
3   0.70     Ideal     G     VS2  61.6    56  2757  5.70  5.67  3.50
4   0.71 Very Good     E     VS2  62.4    57  2759  5.68  5.73  3.56
5   0.78 Very Good     G     SI2  63.8    56  2759  5.81  5.85  3.72
6   0.70      Good     E     VS2  57.5    58  2759  5.85  5.90  3.38
7   0.70      Good     F     VS1  59.4    62  2759  5.71  5.76  3.40
8   0.96      Fair     F     SI2  66.3    62  2759  6.27  5.95  4.07
9   0.73 Very Good     E     SI1  61.6    59  2760  5.77  5.78  3.56
10  0.80   Premium     H     SI1  61.5    58  2760  5.97  5.93  3.66
# ... with 39,431 more rows
```

```
> diamonds %>% filter(price != 1000)

# A tibble: 53,915 × 10
   carat       cut color clarity depth table price     x     y     z
   <dbl>     <ord> <ord>   <ord> <dbl> <dbl> <int> <dbl> <dbl> <dbl>
1   0.23     Ideal     E     SI2  61.5    55   326  3.95  3.98  2.43
2   0.21   Premium     E     SI1  59.8    61   326  3.89  3.84  2.31
3   0.23      Good     E     VS1  56.9    65   327  4.05  4.07  2.31
4   0.29   Premium     I     VS2  62.4    58   334  4.20  4.23  2.63
5   0.31      Good     J     SI2  63.3    58   335  4.34  4.35  2.75
6   0.24 Very Good     J    VVS2  62.8    57   336  3.94  3.96  2.48
7   0.24 Very Good     I    VVS1  62.3    57   336  3.95  3.98  2.47
8   0.26 Very Good     H     SI1  61.9    55   337  4.07  4.11  2.53
9   0.22      Fair     E     VS2  65.1    61   337  3.87  3.78  2.49
10  0.23 Very Good     H     VS1  59.4    61   338  4.00  4.05  2.39
# ... with 53,905 more rows
```

Compound filtering is accomplished by either separating the expressions with a comma (`,`) or an ampersand (`&`).

```
> diamonds %>% filter(carat > 2, price < 14000)

# A tibble: 644 × 10
   carat     cut color clarity depth table price     x     y     z
   <dbl>   <ord> <ord>   <ord> <dbl> <dbl> <int> <dbl> <dbl> <dbl>
1   2.06 Premium     J      I1  61.2    58  5203  8.10  8.07  4.95
2   2.14    Fair     J      I1  69.4    57  5405  7.74  7.70  5.36
3   2.15    Fair     J      I1  65.5    57  5430  8.01  7.95  5.23
4   2.22    Fair     J      I1  66.7    56  5607  8.04  8.02  5.36
5   2.01    Fair     I      I1  67.4    58  5696  7.71  7.64  5.17
6   2.01    Fair     I      I1  55.9    64  5696  8.48  8.39  4.71
7   2.27    Fair     J      I1  67.6    55  5733  8.05  8.00  5.43
8   2.03    Fair     H      I1  64.4    59  6002  7.91  7.85  5.07
9   2.03    Fair     H      I1  66.6    57  6002  7.81  7.75  5.19
10  2.06    Good     H      I1  64.3    58  6091  8.03  7.99  5.15
# ... with 634 more rows
```

```
> diamonds %>% filter(carat > 2 & price < 14000)

# A tibble: 644 × 10
   carat     cut color clarity depth table price     x     y     z
   <dbl>   <ord> <ord>   <ord> <dbl> <dbl> <int> <dbl> <dbl> <dbl>
1   2.06 Premium     J      I1  61.2    58  5203  8.10  8.07  4.95
2   2.14    Fair     J      I1  69.4    57  5405  7.74  7.70  5.36
3   2.15    Fair     J      I1  65.5    57  5430  8.01  7.95  5.23
4   2.22    Fair     J      I1  66.7    56  5607  8.04  8.02  5.36
5   2.01    Fair     I      I1  67.4    58  5696  7.71  7.64  5.17
6   2.01    Fair     I      I1  55.9    64  5696  8.48  8.39  4.71
7   2.27    Fair     J      I1  67.6    55  5733  8.05  8.00  5.43
8   2.03    Fair     H      I1  64.4    59  6002  7.91  7.85  5.07
```

```
9   2.03    Fair     H       I1  66.6     57  6002  7.81  7.75  5.19
10  2.06    Good     H       I1  64.3     58  6091  8.03  7.99  5.15
# ... with 634 more rows
```

A logical or statement is expressed with a vertical pipe (|).

```
> diamonds %>% filter(carat < 1 | carat > 5)

# A tibble: 34,881 × 10
   carat       cut color clarity depth table price     x     y     z
   <dbl>     <ord> <ord>   <ord> <dbl> <dbl> <int> <dbl> <dbl> <dbl>
1   0.23     Ideal     E     SI2  61.5    55   326  3.95  3.98  2.43
2   0.21   Premium     E     SI1  59.8    61   326  3.89  3.84  2.31
3   0.23      Good     E     VS1  56.9    65   327  4.05  4.07  2.31
4   0.29   Premium     I     VS2  62.4    58   334  4.20  4.23  2.63
5   0.31      Good     J     SI2  63.3    58   335  4.34  4.35  2.75
6   0.24 Very Good     J    VVS2  62.8    57   336  3.94  3.96  2.48
7   0.24 Very Good     I    VVS1  62.3    57   336  3.95  3.98  2.47
8   0.26 Very Good     H     SI1  61.9    55   337  4.07  4.11  2.53
9   0.22      Fair     E     VS2  65.1    61   337  3.87  3.78  2.49
10  0.23 Very Good     H     VS1  59.4    61   338  4.00  4.05  2.39
# ... with 34,871 more rows
```

When filtering based on the value of a variable, **filter_** is used with a quoted expression. Quoted expressions can be text or expressions preceded with a tilde (~). Switching between unquoted expressions (considered non-standard evaluation) and quoted expressions (standard evaluation) can be difficult but is necessary to make **dplyr** easy to use interactively and practical inside of functions.

```
> diamonds %>% filter_("cut == 'Ideal'")

# A tibble: 21,551 × 10
    carat   cut color clarity depth table price     x     y     z
    <dbl> <ord> <ord>   <ord> <dbl> <dbl> <int> <dbl> <dbl> <dbl>
1    0.23 Ideal     E     SI2  61.5    55   326  3.95  3.98  2.43
2    0.23 Ideal     J     VS1  62.8    56   340  3.93  3.90  2.46
3    0.31 Ideal     J     SI2  62.2    54   344  4.35  4.37  2.71
4    0.30 Ideal     I     SI2  62.0    54   348  4.31  4.34  2.68
5    0.33 Ideal     I     SI2  61.8    55   403  4.49  4.51  2.78
6    0.33 Ideal     I     SI2  61.2    56   403  4.49  4.50  2.75
7    0.33 Ideal     J     SI1  61.1    56   403  4.49  4.55  2.76
8    0.23 Ideal     G     VS1  61.9    54   404  3.93  3.95  2.44
9    0.32 Ideal     I     SI1  60.9    55   404  4.45  4.48  2.72
10   0.30 Ideal     I     SI2  61.0    59   405  4.30  4.33  2.63
# ... with 21,541 more rows

> diamonds %>% filter_(~cut == 'Ideal')

# A tibble: 21,551 × 10
```

```
   carat   cut color clarity depth table price     x     y     z
   <dbl> <ord> <ord>   <ord> <dbl> <dbl> <int> <dbl> <dbl> <dbl>
1   0.23 Ideal     E     SI2  61.5    55   326  3.95  3.98  2.43
2   0.23 Ideal     J     VS1  62.8    56   340  3.93  3.90  2.46
3   0.31 Ideal     J     SI2  62.2    54   344  4.35  4.37  2.71
4   0.30 Ideal     I     SI2  62.0    54   348  4.31  4.34  2.68
5   0.33 Ideal     I     SI2  61.8    55   403  4.49  4.51  2.78
6   0.33 Ideal     I     SI2  61.2    56   403  4.49  4.50  2.75
7   0.33 Ideal     J     SI1  61.1    56   403  4.49  4.55  2.76
8   0.23 Ideal     G     VS1  61.9    54   404  3.93  3.95  2.44
9   0.32 Ideal     I     SI1  60.9    55   404  4.45  4.48  2.72
10  0.30 Ideal     I     SI2  61.0    59   405  4.30  4.33  2.63
# ... with 21,541 more rows
```

```
> # store value as a variable first
> theCut <- 'Ideal'
> diamonds %>% filter_(~cut == theCut)
```

```
# A tibble: 21,551 × 10
   carat   cut color clarity depth table price     x     y     z
   <dbl> <ord> <ord>   <ord> <dbl> <dbl> <int> <dbl> <dbl> <dbl>
1   0.23 Ideal     E     SI2  61.5    55   326  3.95  3.98  2.43
2   0.23 Ideal     J     VS1  62.8    56   340  3.93  3.90  2.46
3   0.31 Ideal     J     SI2  62.2    54   344  4.35  4.37  2.71
4   0.30 Ideal     I     SI2  62.0    54   348  4.31  4.34  2.68
5   0.33 Ideal     I     SI2  61.8    55   403  4.49  4.51  2.78
6   0.33 Ideal     I     SI2  61.2    56   403  4.49  4.50  2.75
7   0.33 Ideal     J     SI1  61.1    56   403  4.49  4.55  2.76
8   0.23 Ideal     G     VS1  61.9    54   404  3.93  3.95  2.44
9   0.32 Ideal     I     SI1  60.9    55   404  4.45  4.48  2.72
10  0.30 Ideal     I     SI2  61.0    59   405  4.30  4.33  2.63
# ... with 21,541 more rows
```

The tricky part is specifying both the value and column as variables, something that might be done while using **filter_** inside of a function. The easiest, though perhaps not intended, way to do this is to construct the entire expression as a string using **sprintf**.

```
> theCol <- 'cut'
> theCut <- 'Ideal'
> diamonds %>% filter_(sprintf("%s == '%s'", theCol, theCut))
```

```
# A tibble: 21,551 × 10
   carat   cut color clarity depth table price     x     y     z
   <dbl> <ord> <ord>   <ord> <dbl> <dbl> <int> <dbl> <dbl> <dbl>
1   0.23 Ideal     E     SI2  61.5    55   326  3.95  3.98  2.43
2   0.23 Ideal     J     VS1  62.8    56   340  3.93  3.90  2.46
3   0.31 Ideal     J     SI2  62.2    54   344  4.35  4.37  2.71
4   0.30 Ideal     I     SI2  62.0    54   348  4.31  4.34  2.68
```

```
5    0.33 Ideal     I     SI2   61.8    55   403  4.49  4.51  2.78
6    0.33 Ideal     I     SI2   61.2    56   403  4.49  4.50  2.75
7    0.33 Ideal     J     SI1   61.1    56   403  4.49  4.55  2.76
8    0.23 Ideal     G     VS1   61.9    54   404  3.93  3.95  2.44
9    0.32 Ideal     I     SI1   60.9    55   404  4.45  4.48  2.72
10   0.30 Ideal     I     SI2   61.0    59   405  4.30  4.33  2.63
# ... with 21,541 more rows
```

The intended way of working with standard evaluation, prior to **dplyr** version 0.6.0, is to use **interp** from the **lazyeval** package to construct a formula out of variables. Since part of the expression is a column name, that part must be wrapped in **as.name**.

```
> library(lazyeval)
> # build a formula expression using variables
> interp(~ a == b, a=as.name(theCol), b=theCut)

~cut == "Ideal"

> # use that as an argument to filter_
> diamonds %>% filter_(interp(~ a == b, a=as.name(theCol), b=theCut))

# A tibble: 21,551 × 10
   carat   cut color clarity depth table price     x     y     z
   <dbl> <ord> <ord>   <ord> <dbl> <dbl> <int> <dbl> <dbl> <dbl>
1   0.23 Ideal     E     SI2  61.5    55   326  3.95  3.98  2.43
2   0.23 Ideal     J     VS1  62.8    56   340  3.93  3.90  2.46
3   0.31 Ideal     J     SI2  62.2    54   344  4.35  4.37  2.71
4   0.30 Ideal     I     SI2  62.0    54   348  4.31  4.34  2.68
5   0.33 Ideal     I     SI2  61.8    55   403  4.49  4.51  2.78
6   0.33 Ideal     I     SI2  61.2    56   403  4.49  4.50  2.75
7   0.33 Ideal     J     SI1  61.1    56   403  4.49  4.55  2.76
8   0.23 Ideal     G     VS1  61.9    54   404  3.93  3.95  2.44
9   0.32 Ideal     I     SI1  60.9    55   404  4.45  4.48  2.72
10  0.30 Ideal     I     SI2  61.0    59   405  4.30  4.33  2.63
# ... with 21,541 more rows
```

After **dplyr** version 0.6.0 the regular **filter** function can be used in conjunction with **UQE** from the **rlang** package to specify rows using variables for the column of interest and the values. The trick is to store the column name as a `character` and convert it to a name object with **as.name**. Then this is unquoted with **UQE**.

```
> diamonds %>% filter(UQE(as.name(theCol)) == theCut)

# A tibble: 21,551 × 10
   carat   cut color clarity depth table price     x     y     z
   <dbl> <ord> <ord>   <ord> <dbl> <dbl> <int> <dbl> <dbl> <dbl>
1   0.23 Ideal     E     SI2  61.5    55   326  3.95  3.98  2.43
2   0.23 Ideal     J     VS1  62.8    56   340  3.93  3.90  2.46
3   0.31 Ideal     J     SI2  62.2    54   344  4.35  4.37  2.71
```

```
4  0.30 Ideal     I      SI2  62.0    54   348  4.31  4.34  2.68
5  0.33 Ideal     I      SI2  61.8    55   403  4.49  4.51  2.78
6  0.33 Ideal     I      SI2  61.2    56   403  4.49  4.50  2.75
7  0.33 Ideal     J      SI1  61.1    56   403  4.49  4.55  2.76
8  0.23 Ideal     G      VS1  61.9    54   404  3.93  3.95  2.44
9  0.32 Ideal     I      SI1  60.9    55   404  4.45  4.48  2.72
10 0.30 Ideal     I      SI2  61.0    59   405  4.30  4.33  2.63
# ... with 21,541 more rows
```

12.5 slice

While **filter** is used for specifying rows based on a logical expression, **slice** is used for specifying rows by row number. The desired indices are passed as a `vector` to **slice**.

```
> diamonds %>% slice(1:5)

# A tibble: 5 × 10
  carat     cut color clarity depth table price     x     y     z
  <dbl>   <ord> <ord>   <ord> <dbl> <dbl> <int> <dbl> <dbl> <dbl>
1  0.23   Ideal     E     SI2  61.5    55   326  3.95  3.98  2.43
2  0.21 Premium     E     SI1  59.8    61   326  3.89  3.84  2.31
3  0.23    Good     E     VS1  56.9    65   327  4.05  4.07  2.31
4  0.29 Premium     I     VS2  62.4    58   334  4.20  4.23  2.63
5  0.31    Good     J     SI2  63.3    58   335  4.34  4.35  2.75

> diamonds %>% slice(c(1:5, 8, 15:20))

# A tibble: 12 × 10
   carat       cut color clarity depth table price     x     y     z
   <dbl>     <ord> <ord>   <ord> <dbl> <dbl> <int> <dbl> <dbl> <dbl>
1   0.23     Ideal     E     SI2  61.5    55   326  3.95  3.98  2.43
2   0.21   Premium     E     SI1  59.8    61   326  3.89  3.84  2.31
3   0.23      Good     E     VS1  56.9    65   327  4.05  4.07  2.31
4   0.29   Premium     I     VS2  62.4    58   334  4.20  4.23  2.63
5   0.31      Good     J     SI2  63.3    58   335  4.34  4.35  2.75
6   0.26 Very Good     H     SI1  61.9    55   337  4.07  4.11  2.53
7   0.20   Premium     E     SI2  60.2    62   345  3.79  3.75  2.27
8   0.32   Premium     E      I1  60.9    58   345  4.38  4.42  2.68
9   0.30     Ideal     I     SI2  62.0    54   348  4.31  4.34  2.68
10  0.30      Good     J     SI1  63.4    54   351  4.23  4.29  2.70
11  0.30      Good     J     SI1  63.8    56   351  4.23  4.26  2.71
12  0.30 Very Good     J     SI1  62.7    59   351  4.21  4.27  2.66
```

Note that the row numbers displayed on the left of the results are not the row numbers indicated by **slice** but rather the row numbers of the returned results.

Negative indices are used to indicate rows that should not be returned.

```
> diamonds %>% slice(-1)
```

```
# A tibble: 53,939 × 10
   carat       cut color clarity depth table price     x     y     z
   <dbl>     <ord> <ord>   <ord> <dbl> <dbl> <int> <dbl> <dbl> <dbl>
1   0.21   Premium     E     SI1  59.8    61   326  3.89  3.84  2.31
2   0.23      Good     E     VS1  56.9    65   327  4.05  4.07  2.31
3   0.29   Premium     I     VS2  62.4    58   334  4.20  4.23  2.63
4   0.31      Good     J     SI2  63.3    58   335  4.34  4.35  2.75
5   0.24 Very Good     J    VVS2  62.8    57   336  3.94  3.96  2.48
6   0.24 Very Good     I    VVS1  62.3    57   336  3.95  3.98  2.47
7   0.26 Very Good     H     SI1  61.9    55   337  4.07  4.11  2.53
8   0.22      Fair     E     VS2  65.1    61   337  3.87  3.78  2.49
9   0.23 Very Good     H     VS1  59.4    61   338  4.00  4.05  2.39
10  0.30      Good     J     SI1  64.0    55   339  4.25  4.28  2.73
# ... with 53,929 more rows
```

12.6 mutate

Creating new columns or modifying existing columns is done with the **mutate** function. Creating a new column that is the ratio of `price` and `carat` is as simple as providing that ratio as an argument to **mutate**.

```
> diamonds %>% mutate(price/carat)
```

```
# A tibble: 53,940 × 11
   carat       cut color clarity depth table price     x     y     z
   <dbl>     <ord> <ord>   <ord> <dbl> <dbl> <int> <dbl> <dbl> <dbl>
1   0.23     Ideal     E     SI2  61.5    55   326  3.95  3.98  2.43
2   0.21   Premium     E     SI1  59.8    61   326  3.89  3.84  2.31
3   0.23      Good     E     VS1  56.9    65   327  4.05  4.07  2.31
4   0.29   Premium     I     VS2  62.4    58   334  4.20  4.23  2.63
5   0.31      Good     J     SI2  63.3    58   335  4.34  4.35  2.75
6   0.24 Very Good     J    VVS2  62.8    57   336  3.94  3.96  2.48
7   0.24 Very Good     I    VVS1  62.3    57   336  3.95  3.98  2.47
8   0.26 Very Good     H     SI1  61.9    55   337  4.07  4.11  2.53
9   0.22      Fair     E     VS2  65.1    61   337  3.87  3.78  2.49
10  0.23 Very Good     H     VS1  59.4    61   338  4.00  4.05  2.39
# ... with 53,930 more rows, and 1 more variables:
#   `price/carat` <dbl>
```

Depending on the size of the terminal, not all columns will be printed to the screen. To ensure this new column can fit on the screen, we select a few columns of interest using **select** and then pipe that result into **mutate**.

```
> diamonds %>% select(carat, price) %>% mutate(price/carat)
```

```
# A tibble: 53,940 × 3
  carat price `price/carat`
  <dbl> <int>         <dbl>
1  0.23   326      1417.391
```

```
2   0.21   326        1552.381
3   0.23   327        1421.739
4   0.29   334        1151.724
5   0.31   335        1080.645
6   0.24   336        1400.000
7   0.24   336        1400.000
8   0.26   337        1296.154
9   0.22   337        1531.818
10  0.23   338        1469.565
# ... with 53,930 more rows
```

The resulting column is unnamed, which is easily remedied by assigning the expression (price/carat) to a name.

```
> diamonds %>% select(carat, price) %>% mutate(Ratio=price/carat)

# A tibble: 53,940 × 3
   carat price    Ratio
   <dbl> <int>    <dbl>
1   0.23   326 1417.391
2   0.21   326 1552.381
3   0.23   327 1421.739
4   0.29   334 1151.724
5   0.31   335 1080.645
6   0.24   336 1400.000
7   0.24   336 1400.000
8   0.26   337 1296.154
9   0.22   337 1531.818
10  0.23   338 1469.565
# ... with 53,930 more rows
```

Columns created with **mutate** can be used immediately in the same **mutate** call.

```
> diamonds %>%
+     select(carat, price) %>%
+     mutate(Ratio=price/carat, Double=Ratio*2)

# A tibble: 53,940 × 4
   carat price    Ratio   Double
   <dbl> <int>    <dbl>    <dbl>
1   0.23   326 1417.391 2834.783
2   0.21   326 1552.381 3104.762
3   0.23   327 1421.739 2843.478
4   0.29   334 1151.724 2303.448
5   0.31   335 1080.645 2161.290
6   0.24   336 1400.000 2800.000
7   0.24   336 1400.000 2800.000
8   0.26   337 1296.154 2592.308
```

```
9    0.22    337 1531.818 3063.636
10   0.23    338 1469.565 2939.130
# ... with 53,930 more rows
```

Notice this did not change the diamonds data. In order to save the changes, the new result needs to be explicitly assigned to the object diamonds.

A nice feature of the **magrittr** package is the assignment pipe (%<>%), which both pipes the left-hand side into the function on the right-hand side and assigns the result back to the object on the left-hand side.

```
> library(magrittr)
> diamonds2 <- diamonds
> diamonds2

# A tibble: 53,940 × 10
   carat       cut color clarity depth table price     x     y     z
   <dbl>     <ord> <ord>   <ord> <dbl> <dbl> <int> <dbl> <dbl> <dbl>
1   0.23     Ideal     E     SI2  61.5    55   326  3.95  3.98  2.43
2   0.21   Premium     E     SI1  59.8    61   326  3.89  3.84  2.31
3   0.23      Good     E     VS1  56.9    65   327  4.05  4.07  2.31
4   0.29   Premium     I     VS2  62.4    58   334  4.20  4.23  2.63
5   0.31      Good     J     SI2  63.3    58   335  4.34  4.35  2.75
6   0.24 Very Good     J    VVS2  62.8    57   336  3.94  3.96  2.48
7   0.24 Very Good     I    VVS1  62.3    57   336  3.95  3.98  2.47
8   0.26 Very Good     H     SI1  61.9    55   337  4.07  4.11  2.53
9   0.22      Fair     E     VS2  65.1    61   337  3.87  3.78  2.49
10  0.23 Very Good     H     VS1  59.4    61   338  4.00  4.05  2.39
# ... with 53,930 more rows

> diamonds2 %<>%
+     select(carat, price) %>%
+     mutate(Ratio=price/carat, Double=Ratio*2)
> diamonds2

# A tibble: 53,940 × 4
   carat price    Ratio   Double
   <dbl> <int>    <dbl>    <dbl>
1   0.23   326 1417.391 2834.783
2   0.21   326 1552.381 3104.762
3   0.23   327 1421.739 2843.478
4   0.29   334 1151.724 2303.448
5   0.31   335 1080.645 2161.290
6   0.24   336 1400.000 2800.000
7   0.24   336 1400.000 2800.000
8   0.26   337 1296.154 2592.308
9   0.22   337 1531.818 3063.636
10  0.23   338 1469.565 2939.130
# ... with 53,930 more rows
```

This new pipe does not preclude the traditional assignment operator.

```
> diamonds2 <- diamonds2 %>%
+     mutate(Quadruple=Double*2)
> diamonds2

# A tibble: 53,940 × 5
   carat price    Ratio   Double Quadruple
   <dbl> <int>    <dbl>    <dbl>     <dbl>
1   0.23   326 1417.391 2834.783  5669.565
2   0.21   326 1552.381 3104.762  6209.524
3   0.23   327 1421.739 2843.478  5686.957
4   0.29   334 1151.724 2303.448  4606.897
5   0.31   335 1080.645 2161.290  4322.581
6   0.24   336 1400.000 2800.000  5600.000
7   0.24   336 1400.000 2800.000  5600.000
8   0.26   337 1296.154 2592.308  5184.615
9   0.22   337 1531.818 3063.636  6127.273
10  0.23   338 1469.565 2939.130  5878.261
# ... with 53,930 more rows
```

12.7 summarize

While **mutate** applies vectorized functions over columns, **summarize** applies functions that return a result of length one such as **mean**, **max**, **median** or other similar functions. The **summarize** function (or spelled the British way, **summarise**), gives named access to columns in a `data.frame` for applying functions. This behavior is similar to the **with** function in base R. For instance, we calculate the mean of a column from the `diamonds` data.

```
> summarize(diamonds, mean(price))

# A tibble: 1 × 1
  `mean(price)`
          <dbl>
1        3932.8

> # with pipe semantics
> diamonds %>% summarize(mean(price))

# A tibble: 1 × 1
  `mean(price)`
          <dbl>
1        3932.8
```

This may seem like more typing than required by base R, but ultimately will result in less typing and easier to understand code when more complicated expressions are involved.

Another nice feature of **summarize** is the capability to name the resulting calculation and to perform multiple calculations in the same call.

```
> diamonds %>%
+     summarize(AvgPrice=mean(price),
+               MedianPrice=median(price),
+               AvgCarat=mean(carat))

# A tibble: 1 × 3
  AvgPrice MedianPrice  AvgCarat
     <dbl>       <dbl>     <dbl>
1   3932.8        2401 0.7979397
```

12.8 group_by

The **summarize** function is moderately useful by itself but really shines when used with **group_by** to first partition the data and then apply a function to each partition independently. To split the data according to a variable and then apply a summary function to each partition, the data is first passed to **group_by** and the resulting grouped `data.frame` or `tbl` is passed to **summarize**, which allows functions to be applied to individual columns. This usage illustrates the power and ease of pipes.

```
> diamonds %>%
+     group_by(cut) %>%
+     summarize(AvgPrice=mean(price))

# A tibble: 5 × 2
        cut AvgPrice
      <ord>    <dbl>
1      Fair 4358.758
2      Good 3928.864
3 Very Good 3981.760
4   Premium 4584.258
5     Ideal 3457.542
```

This is a more eloquent, and faster, way to aggregate data than the **aggregate** function, and it more easily enables multiple calculations and grouping variables.

```
> diamonds %>%
+     group_by(cut) %>%
+     summarize(AvgPrice=mean(price), SumCarat=sum(carat))

# A tibble: 5 × 3
        cut AvgPrice SumCarat
      <ord>    <dbl>    <dbl>
1      Fair 4358.758  1684.28
2      Good 3928.864  4166.10
3 Very Good 3981.760  9742.70
4   Premium 4584.258 12300.95
5     Ideal 3457.542 15146.84
```

```
> diamonds %>%
+     group_by(cut, color) %>%
+     summarize(AvgPrice=mean(price), SumCarat=sum(carat))

Source: local data frame [35 x 4]
Groups: cut [?]

     cut color AvgPrice SumCarat
   <ord> <ord>    <dbl>    <dbl>
1   Fair     D 4291.061   149.98
2   Fair     E 3682.312   191.88
3   Fair     F 3827.003   282.27
4   Fair     G 4239.255   321.48
5   Fair     H 5135.683   369.41
6   Fair     I 4685.446   209.66
7   Fair     J 4975.655   159.60
8   Good     D 3405.382   492.87
9   Good     E 3423.644   695.21
10  Good     F 3495.750   705.32
# ... with 25 more rows
```

When run on a grouped `data.frame`, the **summarize** function drops the innermost level of grouping. That is why the first statement in the previous code returned a `data.frame` with no groups and the second statement returned a `data.frame` with one group.

12.9 arrange

Sorting is performed with the **arrange** function, which is much easier to understand and use than the **order** and **sort** functions from base R.

```
> diamonds %>%
+     group_by(cut) %>%
+     summarize(AvgPrice=mean(price), SumCarat=sum(carat)) %>%
+     arrange(AvgPrice)

# A tibble: 5 × 3
        cut AvgPrice SumCarat
      <ord>    <dbl>    <dbl>
1     Ideal 3457.542 15146.84
2      Good 3928.864  4166.10
3 Very Good 3981.760  9742.70
4      Fair 4358.758  1684.28
5   Premium 4584.258 12300.95

> diamonds %>%
+     group_by(cut) %>%
```

```
+       summarize(AvgPrice=mean(price), SumCarat=sum(carat)) %>%
+       arrange(desc(AvgPrice))

# A tibble: 5 × 3
        cut AvgPrice SumCarat
      <ord>    <dbl>    <dbl>
1   Premium 4584.258 12300.95
2      Fair 4358.758  1684.28
3 Very Good 3981.760  9742.70
4      Good 3928.864  4166.10
5     Ideal 3457.542 15146.84
```

12.10 do

For general purpose calculations not covered by the specialized manipulation functions in **dplyr**, such as **filter**, **mutate** and **summarize**, there is **do**, which enables any arbitrary function on the data. For a simple example we create a function that sorts the `diamonds` data and returns the first N rows.

```
> topN <- function(x, N=5)
+ {
+     x %>% arrange(desc(price)) %>% head(N)
+ }
```

By combining **do** with **group_by** we return the top N rows, sorted by price, for each cut of diamonds. When using pipes, the left-hand side becomes the first argument of the function on the right-hand side. For **do** the first argument is supposed to be a function, not what is on the left-hand side of the pipe, in this case the grouped `diamonds` data. Since the left-hand side is not going to its default location, we specify where it goes by using a period (`.`).

```
> diamonds %>% group_by(cut) %>% do(topN(., N=3))

Source: local data frame [15 x 10]
Groups: cut [5]


   carat       cut color clarity depth table price     x     y     z
   <dbl>     <ord> <ord>   <ord> <dbl> <dbl> <int> <dbl> <dbl> <dbl>
1   2.01      Fair     G     SI1  70.6    64 18574  7.43  6.64  4.69
2   2.02      Fair     H     VS2  64.5    57 18565  8.00  7.95  5.14
3   4.50      Fair     J      I1  65.8    58 18531 10.23 10.16  6.72
4   2.80      Good     G     SI2  63.8    58 18788  8.90  8.85  0.00
5   2.07      Good     I     VS2  61.8    61 18707  8.12  8.16  5.03
6   2.67      Good     F     SI2  63.8    58 18686  8.69  8.64  5.54
7   2.00 Very Good     G     SI1  63.5    56 18818  7.90  7.97  5.04
8   2.00 Very Good     H     SI1  62.8    57 18803  7.95  8.00  5.01
9   2.03 Very Good     H     SI1  63.0    60 18781  8.00  7.93  5.02
```

```
10  2.29   Premium     I     VS2  60.8     60 18823  8.50  8.47  5.16
11  2.29   Premium     I     SI1  61.8     59 18797  8.52  8.45  5.24
12  2.04   Premium     H     SI1  58.1     60 18795  8.37  8.28  4.84
13  1.51     Ideal     G      IF  61.7     55 18806  7.37  7.41  4.56
14  2.07     Ideal     G     SI2  62.5     55 18804  8.20  8.13  5.11
15  2.15     Ideal     G     SI2  62.6     54 18791  8.29  8.35  5.21
```

When using **do** with a single, unnamed argument, such as the previous example, the result is a `data.frame`. If we had named the argument, then the expression would result in a `data.frame` where the calculated column is actually a `list`.

```
> diamonds %>%
+     # group the data according to cut
+     # this essentially creates a separate dataset for each
+     group_by(cut) %>%
+     # apply the topN function, with the second argument set to 3
+     # this is done independently to each group of data
+     do(Top=topN(., 3))

Source: local data frame [5 x 2]
Groups: <by row>

# A tibble: 5 × 2
        cut               Top
*     <ord>            <list>
1      Fair <tibble [3 × 10]>
2      Good <tibble [3 × 10]>
3 Very Good <tibble [3 × 10]>
4   Premium <tibble [3 × 10]>
5     Ideal <tibble [3 × 10]>

> topByCut <- diamonds %>% group_by(cut) %>% do(Top=topN(., 3))
> class(topByCut)

[1] "rowwise_df" "tbl_df"     "tbl"        "data.frame"

> class(topByCut$Top)

[1] "list"

> class(topByCut$Top[[1]])

[1] "tbl_df"     "tbl"        "data.frame"

> topByCut$Top[[1]]

# A tibble: 3 × 10
  carat   cut color clarity depth table price     x     y     z
  <dbl> <ord> <ord>   <ord> <dbl> <dbl> <int> <dbl> <dbl> <dbl>
1  2.01  Fair     G     SI1  70.6    64 18574  7.43  6.64  4.69
2  2.02  Fair     H     VS2  64.5    57 18565  8.00  7.95  5.14
3  4.50  Fair     J      I1  65.8    58 18531 10.23 10.16  6.72
```

In this example, the calculated column was a `list` where each entry was a `data.frame` of the top three rows, by price, for each cut of diamond. It may seem odd to store `lists` in columns of `data.frames`, but that is a built-in use for `data.frames`.

Using **do** with named arguments is equivalent to **ldply** from **plyr**.

12.11 dplyr with Databases

An important feature of **dplyr** is its capability to work with data stored in a database in much the same way it works with data in `data.frames`. As of writing, **dplyr** works with PostgreSQL, MySQL, SQLite, MonetDB, Google Big Query and Spark DataFrames. For more standard computations, the R code is translated into equivalent SQL code. For arbitrary R code that cannot be easily translated into SQL, **dplyr** (experimentally) chunks the data into memory and runs the computations independently. This enables data munging and analysis on data that would otherwise not fit in memory. While database operations will be slower than the equivalent `data.frame` operations, this is of little concern as the data would not have fit into memory anyway.

To illustrate, we look at a SQLite database with two tables holding the `diamonds` data and an additional, related, dataset. We download this database using **download.file**.

```
> download.file("http://www.jaredlander.com/data/diamonds.db",
+               destfile="data/diamonds.db", mode='wb')
```

The first step is to create a connection to the database. Starting with **dplyr** version 0.6.0, in order to work with databases **dbplyr** must also be installed, though not necessarily loaded.

```
> diaDBSource <- src_sqlite("data/diamonds.db")
> diaDBSource

src:  sqlite 3.11.1 [data/diamonds.db]
tbls: DiamondColors, diamonds, sqlite_stat1
```

With versions of **dplyr** beyond 0.6.0 this can also be performed using **DBI** directly.

```
> diaDBSource2 <- DBI::dbConnect(RSQLite::SQLite(), "data/diamonds.db")
> diaDBSource2

<SQLiteConnection>
  Path: C:\Users\jared\Documents\Consulting\book\book\
             FasterGroupManipulation\data\diamonds.db
  Extensions: TRUE
```

Now that we have a connection to the database we need to point to a specific table. In this example, the database has two data tables called `diamonds` and `DiamondColors` and a metadata table called `sqlite_stat1`. Each table in the database needs to be pointed to individually. For our purposes we are only concerned with the `diamonds` table.

```
> diaTab <- tbl(diaDBSource, "diamonds")
> diaTab
```

```
Source:   query [?? x 10]
Database: sqlite 3.11.1 [data/diamonds.db]

   carat       cut color clarity depth table price     x     y     z
   <dbl>     <chr> <chr>   <chr> <dbl> <dbl> <int> <dbl> <dbl> <dbl>
1   0.23     Ideal     E     SI2  61.5    55   326  3.95  3.98  2.43
2   0.21   Premium     E     SI1  59.8    61   326  3.89  3.84  2.31
3   0.23      Good     E     VS1  56.9    65   327  4.05  4.07  2.31
4   0.29   Premium     I     VS2  62.4    58   334  4.20  4.23  2.63
5   0.31      Good     J     SI2  63.3    58   335  4.34  4.35  2.75
6   0.24 Very Good     J    VVS2  62.8    57   336  3.94  3.96  2.48
7   0.24 Very Good     I    VVS1  62.3    57   336  3.95  3.98  2.47
8   0.26 Very Good     H     SI1  61.9    55   337  4.07  4.11  2.53
9   0.22      Fair     E     VS2  65.1    61   337  3.87  3.78  2.49
10  0.23 Very Good     H     VS1  59.4    61   338  4.00  4.05  2.39
# ... with more rows
```

This looks like a regular `data.frame` but is actually a table in the database, and just the first few rows are queried and displayed. Most calculations on this `tbl` are actually performed by the database itself.

```
> diaTab %>% group_by(cut) %>% dplyr::summarize(Price=mean(price))

Source:   query [?? x 2]
Database: sqlite 3.11.1 [data/diamonds.db]

        cut    Price
      <chr>    <dbl>
1      Fair 4358.758
2      Good 3928.864
3     Ideal 3457.542
4   Premium 4584.258
5 Very Good 3981.760

> diaTab %>% group_by(cut) %>%
+     dplyr::summarize(Price=mean(price), Carat=mean(Carat))

Source:   query [?? x 3]
Database: sqlite 3.11.1 [data/diamonds.db]

        cut    Price     Carat
      <chr>    <dbl>     <dbl>
1      Fair 4358.758 1.0461366
2      Good 3928.864 0.8491847
3     Ideal 3457.542 0.7028370
4   Premium 4584.258 0.8919549
5 Very Good 3981.760 0.8063814
```

12.12 Conclusion

This next generation package from Hadley Wickham makes data manipulation both easier to code and faster to execute. Its syntax is built around the verbs of data manipulation such as **select**, **filter**, **arrange** and **group_by** and is designed for writing for highly readable and fast code.

13

Iterating with purrr

R has numerous ways to iterate over the elements of a `list` (or `vector`), and Hadley Wickham aimed to improve on and standardize that experience with the **purrr** package. R has its roots in functional programming, and **purrr** is designed to utilize that computing paradigm. For our purposes, functional programming is when a function depends only on its input arguments. This enables us to iterate over a `list` and apply a function to each element independently. This is mainly aimed at operating on `list`s, though it can also be used to apply non-vectorized functions to `vector`s.

The name **purrr** has many meanings. It is primarily meant to convey that this package enforces pure programming. It is also a pun on a cat purring and has five letters, so it is the same length as many of Hadley Wickham's other packages such as **dplyr**, **readr** and **tidyr**.

13.1 map

The foundation of **purrr** is the **map** function, which applies a function to each element of a `list`, independently, and returns the results in a `list` of the same length. This works the same as **lapply**, detailed in Section 11.1.2, though it is designed with pipes in mind.

Returning to the example in Section 11.1.2, we build a `list` with four elements and then apply the **sum** function to each element with **lapply**.

```
> theList <- list(A=matrix(1:9, 3), B=1:5, C=matrix(1:4, 2), D=2)
> lapply(theList, sum)

$A
[1] 45

$B
[1] 15

$C
[1] 10

$D
[1] 2
```

Identical results are achieved using **map**.

```
> library(purrr)
> theList %>% map(sum)

$A
[1] 45

$B
[1] 15

$C
[1] 10

$D
[1] 2

> identical(lapply(theList, sum), theList %>% map(sum))

[1] TRUE
```

If elements of `theList` had missing values (`NA`), the **sum** function would need to have the `na.rm` argument set to `TRUE`. This can be accomplished by wrapping **sum** in an anonymous function defined directly inside the **map** call or by passing `na.rm=TRUE` as an additional argument to **sum** through **map**.

To illustrate we first set elements of `theList` to `NA`.

```
> theList2 <- theList
> theList2[[1]][2, 1] <- NA
> theList2[[2]][4] <- NA
```

Applying **sum** through **map** now results in two elements with `NA` as the sum.

```
> theList2 %>% map(sum)

$A
[1] NA

$B
[1] NA

$C
[1] 10

$D
[1] 2
```

We first solve this by using an anonymous function, which is simply a wrapper around **sum**.

```
> theList2 %>% map(function(x) sum(x, na.rm=TRUE))

$A
[1] 43

$B
[1] 11

$C
[1] 10

$D
[1] 2
```

We then solve it by passing `na.rm=TRUE` as an additional argument to **sum** through the `dot-dot-dot` argument (...) of **map**.

```
> theList2 %>% map(sum, na.rm=TRUE)

$A
[1] 43

$B
[1] 11

$C
[1] 10

$D
[1] 2
```

Writing the anonymous function may seem like a lot of work for such a simple operation, though it is very common in functional programming and is helpful when function arguments are not in an order conducive to being used with **map**.

13.2 map with Specified Types

Using **map** always results in a `list` as the returned result, which is highly generalized but not always desired. In base R **sapply** attempts to simplify the results to a `vector` when possible and reverts to a `list` if necessary. It is nice to have the simplification, but this removes certainty about the type of results. To eliminate this uncertainty while returning simplified results in a `vector`, **purrr** offers a number of `map` functions that specify the type of expected result. If the specified type cannot be returned, then the functions return an error. An error may seem undesired, but an early error is preferred over an unexpected result that causes an error further down in the code. All of these functions take the form of **map_***, where * specifies the type of return value. The possible return values, and their corresponding functions, are laid out in Table 13.1.

Table 13.1 **purrr functions and their corresponding outputs**

Function	Output Type
map	list
map_int	integer
map_dbl	numeric
map_chr	character
map_lgl	logical
map_df	data.frame

It is important to note that each of these **map_*** functions expects a vector of length one for each element. A result greater than length one for even a single element will result in an error.

13.2.1 map_int

map_int can be used when the result will be an integer. To illustrate this we apply the **NROW** function to each element of theList, which will return the length if the element is one-dimensional or the number of rows if the element is two-dimensional.

```
> theList %>% map_int(NROW)

A B C D
3 5 2 1
```

Applying a function that returns a numeric, such as **mean**, will result in an error.

```
> theList %>% map_int(mean)

Error: Can't coerce element 1 from a double to a integer
```

13.2.2 map_dbl

In order to apply a function that returns a numeric we use **map_dbl**.[1]

```
> theList %>% map_dbl(mean)

  A   B   C   D
5.0 3.0 2.5 2.0
```

1. The function refers to a double, which is the underlying type of a numeric.

13.2.3 map_chr

Applying a function that returns character data necessitates **map_chr**.

```
> theList %>% map_chr(class)

        A         B         C         D
 "matrix" "integer"  "matrix" "numeric"
```

If one of the elements of theList had multiple classes, **map_chr** would have returned an error. This is because the result of the function must be a vector of length one for each element of the input list. This can be seen if we add another element that is an ordered factor.

```
> theList3 <- theList
> theList3[['E']] <- factor(c('A', 'B', 'C'), ordered=TRUE)
>
> class(theList3$E)

[1] "ordered" "factor"
```

The class for this new element has a length of two, which will cause an error with **map_chr**.

```
> theList3 %>% map_chr(class)

Error: Result 5 is not a length 1 atomic vector
```

The simplest solution is to return a list using **map**. It will no longer be a simple vector, but there will be no error. Any operation that works with a **map_*** works with **map**.

```
> theList3 %>% map(class)

$A
[1] "matrix"

$B
[1] "integer"

$C
[1] "matrix"

$D
[1] "numeric"

$E
[1] "ordered" "factor"
```

13.2.4 map_lgl

The results of logical operations can be stored in a logical vector using **map_lgl**.

```
> theList %>% map_lgl(function(x) NROW(x) < 3)

    A     B     C     D
FALSE FALSE  TRUE  TRUE
```

13.2.5 map_df

A popular function in **plyr** is **ldply**, which iterates over a list, applies a function and combines the results into a data.frame. The equivalent in **purrr** is **map_df**.

As a contrived example we create a function that builds a data.frame with two columns, the length of which is decided by an argument to the function. We also create a list of numbers to provide those lengths.

```
> buildDF <- function(x)
+ {
+     data.frame(A=1:x, B=x:1)
+ }
>
> listOfLengths <- list(3, 4, 1, 5)
```

We iterate over that list, building a data.frame for each element. Using **map** would result in a list of length four, with each element being a data.frame.

```
> listOfLengths %>% map(buildDF)

[[1]]
  A B
1 1 3
2 2 2
3 3 1

[[2]]
  A B
1 1 4
2 2 3
3 3 2
4 4 1

[[3]]
  A B
1 1 1

[[4]]
  A B
1 1 5
```

```
2 2 4
3 3 3
4 4 2
5 5 1
```

This result would be more convenient as a `data.frame`, which is accomplished with **map_df**.

```
> listOfLengths %>% map_df(buildDF)

   A B
1  1 3
2  2 2
3  3 1
4  1 4
5  2 3
6  3 2
7  4 1
8  1 1
9  1 5
10 2 4
11 3 3
12 4 2
13 5 1
```

13.2.6 map_if

There are times when elements of a list should only be modified if a logical condition holds true. This is accomplished with **map_if**, where only the elements meeting the criteria are modified, and the rest are returned unmodified. We illustrate by multiplying the `matrix` elements of `theList` by two.

```
> theList %>% map_if(is.matrix, function(x) x*2)

$A
     [,1] [,2] [,3]
[1,]    2    8   14
[2,]    4   10   16
[3,]    6   12   18

$B
[1] 1 2 3 4 5

$C
     [,1] [,2]
[1,]    2    6
[2,]    4    8
```

```
$D
[1] 2
```

This was easily accomplished using an anonymous function, though **purrr** provides yet another way to specify a function inline. We could have supplied a `formula` rather than a function, and **map_if** (or any of the **map** functions) would create an anonymous function for us. Up to two arguments can be supplied and must be of the form `.x` and `.y`.

```
> theList %>% map_if(is.matrix, ~ .x*2)

$A
     [,1] [,2] [,3]
[1,]    2    8   14
[2,]    4   10   16
[3,]    6   12   18

$B
[1] 1 2 3 4 5

$C
     [,1] [,2]
[1,]    2    6
[2,]    4    8

$D
[1] 2
```

13.3 Iterating over a `data.frame`

Iterating over a `data.frame` is just as simple, because `data.frames` are technically `lists`. To see this we calculate the means of the `numeric` columns in the `diamonds` data.

```
> data(diamonds, package='ggplot2')
> diamonds %>% map_dbl(mean)

    carat         cut       color     clarity       depth
0.7979397          NA          NA          NA  61.7494049
    table       price           x           y           z
57.4571839 3932.7997219   5.7311572   5.7345260   3.5387338
```

This returns the means of the `numeric` columns and `NA`s for the non-`numeric` columns. It also displays a warning telling us that it cannot calculate the means of the non-`numeric` columns.

This operation can be similarly calculated using **summarize_each** in **dplyr**. Numerically, they are the same, but **map_dbl** returns a `numeric vector` and **mutate_each** returns a single-row `data.frame`.

```
> library(dplyr)
> diamonds %>% summarize_each(funs(mean))

Warning in mean.default(structure(c(5L, 4L, 2L, 4L, 2L, 3L, 3L, 3L, 1L, :
argument is not numeric or logical: returning NA
Warning in mean.default(structure(c(2L, 2L, 2L, 6L, 7L, 7L, 6L, 5L, 2L, :
argument is not numeric or logical: returning NA
Warning in mean.default(structure(c(2L, 3L, 5L, 4L, 2L, 6L, 7L, 3L, 4L, :
argument is not numeric or logical: returning NA

# A tibble: 1 × 10
      carat   cut color clarity   depth    table  price        x
      <dbl> <dbl> <dbl>   <dbl>   <dbl>    <dbl>  <dbl>    <dbl>
1 0.7979397    NA    NA      NA 61.7494 57.45718 3932.8 5.731157
# ... with 2 more variables: y <dbl>, z <dbl>
```

A `warning` was generated for each non-`numeric` column informing that **mean** cannot be used on non-`numeric` data. Even with the `warning` the function still completes, returning `NA` for each non-`numeric` column.

13.4 map with Multiple Inputs

In Section 11.1.3 we learned how to use **mapply** to apply a function that took two arguments to corresponding elements of two `lists`. The **purrr** analog is **pmap**, with **map2** as a special case when the function takes exactly two arguments.

```
> ## build two lists
> firstList <- list(A=matrix(1:16, 4), B=matrix(1:16, 2), C=1:5)
> secondList <- list(A=matrix(1:16, 4), B=matrix(1:16, 8), C=15:1)
>
> ## adds the number of rows (or length) of corresponding elements
> simpleFunc <- function(x, y)
+ {
+     NROW(x) + NROW(y)
+ }

>
> # apply the function to the two lists
> map2(firstList, secondList, simpleFunc)

$A
[1] 8

$B
[1] 10
```

```
$C
[1] 20

> # apply the function to the two lists and return an integer
> map2_int(firstList, secondList, simpleFunc)

 A  B  C
 8 10 20
```

The more general **pmap** requires that the `lists` being iterated over are stored in a `list`.

```
> # use the more general pmap
> pmap(list(firstList, secondList), simpleFunc)

$A
[1] 8

$B
[1] 10

$C
[1] 20

> pmap_int(list(firstList, secondList), simpleFunc)

 A  B  C
 8 10 20
```

13.5 Conclusion

Iterating over `lists` is easier than ever with **purrr**. Most of what can be done in **purrr** can already be accomplished using base R functions such as **lapply**, but it is quicker with **purrr**, both in terms of computation and programming time. In addition to the speed improvements, **purrr** ensures the results returned are as the programmer expected and was designed to work with pipes, which further enhances the user experience.

14

Data Reshaping

As noted in Chapter 11, manipulating the data takes a great deal of effort before serious analysis can begin. In this chapter we will consider when the data need to be rearranged from column-oriented to row-oriented (or the opposite) and when the data are in multiple, separate sets and need to be combined into one.

There are base functions to accomplish these tasks, but we will focus on those in **plyr**, **reshape2** and **data.table**.

While the tools covered in this chapter still form the backbone of data reshaping, newer packages like **tidyr** and **dplyr** are starting to supercede them. Chapter 15 is an analog to this chapter using these new packages.

14.1 `cbind` and `rbind`

The simplest case is when we have a two datasets with either identical columns (both the number of and names) or the same number of rows. In this case, either **rbind** or **cbind** work greatly.

As a first trivial example, we create two simple `data.frames` by combining a few `vectors` with **cbind**, and then stack them using **rbind**.

```
> # make three vectors and combine them as columns in a data.frame
> sport <- c("Hockey", "Baseball", "Football")
> league <- c("NHL", "MLB", "NFL")
> trophy <- c("Stanley Cup", "Commissioner's Trophy",
+             "Vince Lombardi Trophy")
> trophies1 <- cbind(sport, league, trophy)
>
> # make another data.frame using data.frame()
> trophies2 <- data.frame(sport=c("Basketball", "Golf"),
+                         league=c("NBA", "PGA"),
+                         trophy=c("Larry O'Brien Championship Trophy",
+                                  "Wanamaker Trophy"),
+                         stringsAsFactors=FALSE)
>
> # combine them into one data.frame with rbind
> trophies <- rbind(trophies1, trophies2)
```

Both **cbind** and **rbind** can take multiple arguments to combine an arbitrary number of objects. Note that it is possible to assign new column names to vectors in **cbind**.

```
> cbind(Sport=sport, Association=league, Prize=trophy)

     Sport      Association Prize
[1,] "Hockey"   "NHL"       "Stanley Cup"
[2,] "Baseball" "MLB"       "Commissioner's Trophy"
[3,] "Football" "NFL"       "Vince Lombardi Trophy"
```

14.2 Joins

Data do not always come so nicely aligned for combing using **cbind** and need to be joined together using a common key. This concept should be familiar to SQL users. Joins in R are not as flexible as SQL joins, but are still an essential operation in the data analysis process.

The three most commonly used functions for joins are **merge** in base R, **join** in **plyr** and the merging functionality in **data.table**. Each has pros and cons with some pros outweighing their respective cons.

To illustrate these functions we have prepared data originally made available as part of the USAID Open Government initiative.[1] The data have been chopped into eight separate files so that they can be joined together. They are all available in a zip file at `http://jaredlander.com/data/US_Foreign_Aid.zip`. These should be downloaded and unzipped to a folder on our computer. This can be done a number of ways (including using a mouse!) but we show how to download and unzip using R.

```
> download.file(url="http://jaredlander.com/data/US_Foreign_Aid.zip",
+               destfile="data/ForeignAid.zip")
> unzip("data/ForeignAid.zip", exdir="data")
```

To load all of these files programatically, we utilize a **for** loop as seen in Section 10.1. We get a list of the files using **dir**, and then loop through that list, assigning each dataset to a name specified using **assign**. The function **str_sub** extracts individual `characters` from a `character vector` and is explained in Section 16.3.

```
> library(stringr)
> # first get a list of the files
> theFiles <- dir("data/", pattern="\\.csv")
> ## loop through those files
> for(a in theFiles)
+ {
+     # build a good name to assign to the data
+     nameToUse <- str_sub(string=a, start=12, end=18)
+     # read in the csv using read.table
+     # file.path is a convenient way to specify a folder and file name
+     temp <- read.table(file=file.path("data", a),
+                        header=TRUE, sep=",", stringsAsFactors=FALSE)
```

1. More information about the data is available at `http://gbk.eads.usaidallnet.gov/`.

```
+      # assign them into the workspace
+      assign(x=nameToUse, value=temp)
+ }
```

14.2.1 merge

R comes with a built-in function, called **merge**, to merge two `data.frames`.

```
> Aid90s00s <- merge(x=Aid_90s, y=Aid_00s,
+                    by.x=c("Country.Name", "Program.Name"),
+                    by.y=c("Country.Name", "Program.Name"))
> head(Aid90s00s)

  Country.Name                                       Program.Name
1  Afghanistan                         Child Survival and Health
2  Afghanistan          Department of Defense Security Assistance
3  Afghanistan                             Development Assistance
4  Afghanistan Economic Support Fund/Security Support Assistance
5  Afghanistan                                 Food For Education
6  Afghanistan                   Global Health and Child Survival
  FY1990 FY1991 FY1992   FY1993  FY1994 FY1995 FY1996 FY1997 FY1998
1     NA     NA     NA       NA      NA     NA     NA     NA     NA
2     NA     NA     NA       NA      NA     NA     NA     NA     NA
3     NA     NA     NA       NA      NA     NA     NA     NA     NA
4     NA     NA     NA 14178135 2769948     NA     NA     NA     NA
5     NA     NA     NA       NA      NA     NA     NA     NA     NA
6     NA     NA     NA       NA      NA     NA     NA     NA     NA
  FY1999 FY2000  FY2001   FY2002    FY2003     FY2004     FY2005
1     NA     NA      NA  2586555  56501189   40215304   39817970
2     NA     NA      NA  2964313        NA   45635526  151334908
3     NA     NA 4110478  8762080  54538965  180539337  193598227
4     NA     NA   61144 31827014 341306822 1025522037 1157530168
5     NA     NA      NA       NA   3957312    2610006    3254408
6     NA     NA      NA       NA        NA         NA         NA
      FY2006     FY2007     FY2008     FY2009
1   40856382   72527069   28397435         NA
2  230501318  214505892  495539084  552524990
3  212648440  173134034  150529862    3675202
4 1357750249 1266653993 1400237791 1418688520
5     386891         NA         NA         NA
6         NA         NA   63064912    1764252
```

The `by.x` specifies the key column(s) in the left `data.frame` and `by.y` does the same for the right `data.frame`. The ability to specify different column names for each `data.frame` is the most useful feature of **merge**. The biggest drawback, however, is that **merge** can be much slower than the alternatives.

14.2.2 plyr join

Returning to Hadley Wickham's **plyr** package, we see it includes a **join** function, which works similarly to **merge** but is much faster. The biggest drawback, though, is that the key column(s) in each table must have the same name. We use the same data used previously to illustrate.

```
> library(plyr)
> Aid90s00sJoin <- join(x=Aid_90s, y=Aid_00s,
+                       by=c("Country.Name", "Program.Name"))
> head(Aid90s00sJoin)

  Country.Name                                      Program.Name
1  Afghanistan                         Child Survival and Health
2  Afghanistan         Department of Defense Security Assistance
3  Afghanistan                            Development Assistance
4  Afghanistan Economic Support Fund/Security Support Assistance
5  Afghanistan                                Food For Education
6  Afghanistan                  Global Health and Child Survival
  FY1990 FY1991 FY1992   FY1993  FY1994 FY1995 FY1996 FY1997 FY1998
1     NA     NA     NA       NA      NA     NA     NA     NA     NA
2     NA     NA     NA       NA      NA     NA     NA     NA     NA
3     NA     NA     NA       NA      NA     NA     NA     NA     NA
4     NA     NA     NA 14178135 2769948     NA     NA     NA     NA
5     NA     NA     NA       NA      NA     NA     NA     NA     NA
6     NA     NA     NA       NA      NA     NA     NA     NA     NA
  FY1999 FY2000  FY2001   FY2002    FY2003     FY2004     FY2005
1     NA     NA      NA  2586555  56501189   40215304   39817970
2     NA     NA      NA  2964313        NA   45635526  151334908
3     NA     NA 4110478  8762080  54538965  180539337  193598227
4     NA     NA   61144 31827014 341306822 1025522037 1157530168
5     NA     NA      NA       NA   3957312    2610006    3254408
6     NA     NA      NA       NA        NA         NA         NA
      FY2006     FY2007     FY2008     FY2009
1   40856382   72527069   28397435         NA
2  230501318  214505892  495539084  552524990
3  212648440  173134034  150529862    3675202
4 1357750249 1266653993 1400237791 1418688520
5     386891         NA         NA         NA
6         NA         NA   63064912    1764252
```

join has an argument for specifying a left, right, inner or full (outer) join.

We have eight `data.frames` containing foreign assistance data that we would like to combine into one `data.frame` without hand coding each join. The best way to do this is put all the `data.frames` into a `list`, and then successively join them together using **Reduce**.

```
> # first figure out the names of the data.frames
> frameNames <- str_sub(string=theFiles, start=12, end=18)
```

```
> # build an empty list
> frameList <- vector("list", length(frameNames))
> names(frameList) <- frameNames
> # add each data.frame into the list
> for(a in frameNames)
+ {
+     frameList[[a]] <- eval(parse(text=a))
+ }
```

A lot happened in that section of code, so let's go over it carefully. First we reconstructed the names of the `data.frame`s using **str_sub** from Hadley Wickham's **stringr** package, which is shown in more detail in Chapter 16. Then we built an empty `list` with as many elements as there are `data.frame`s, in this case eight, using **vector** and assigning its mode to "list". We then set appropriate names to the `list`.

Now that the `list` is built and named, we looped through it, assigning to each element the appropriate `data.frame`. The problem is that we have the names of the `data.frame`s as characters but the <- operator requires a variable, not a character. So we parse and evaluate the character, which realizes the actual variable. Inspecting, we see that the list does indeed contain the appropriate `data.frame`s.

```
> head(frameList[[1]])

  Country.Name                                        Program.Name
1  Afghanistan                          Child Survival and Health
2  Afghanistan          Department of Defense Security Assistance
3  Afghanistan                            Development Assistance
4  Afghanistan Economic Support Fund/Security Support Assistance
5  Afghanistan                                Food For Education
6  Afghanistan                  Global Health and Child Survival
  FY2000  FY2001   FY2002    FY2003     FY2004     FY2005     FY2006
1     NA      NA  2586555  56501189   40215304   39817970   40856382
2     NA      NA  2964313        NA   45635526  151334908  230501318
3     NA 4110478  8762080  54538965  180539337  193598227  212648440
4     NA   61144 31827014 341306822 1025522037 1157530168 1357750249
5     NA      NA       NA   3957312    2610006    3254408     386891
6     NA      NA       NA        NA         NA         NA         NA
      FY2007     FY2008     FY2009
1   72527069   28397435         NA
2  214505892  495539084  552524990
3  173134034  150529862    3675202
4 1266653993 1400237791 1418688520
5         NA         NA         NA
6         NA   63064912    1764252

> head(frameList[["Aid_00s"]])

  Country.Name                                        Program.Name
1  Afghanistan                          Child Survival and Health
2  Afghanistan          Department of Defense Security Assistance
```

```
3  Afghanistan                            Development Assistance
4  Afghanistan Economic Support Fund/Security Support Assistance
5  Afghanistan                                Food For Education
6  Afghanistan                  Global Health and Child Survival
  FY2000  FY2001   FY2002    FY2003     FY2004     FY2005     FY2006
1     NA      NA  2586555  56501189   40215304   39817970   40856382
2     NA      NA  2964313        NA   45635526  151334908  230501318
3     NA 4110478  8762080  54538965  180539337  193598227  212648440
4     NA   61144 31827014 341306822 1025522037 1157530168 1357750249
5     NA      NA       NA   3957312    2610006    3254408     386891
6     NA      NA       NA        NA         NA         NA         NA
     FY2007     FY2008     FY2009
1  72527069   28397435         NA
2 214505892  495539084  552524990
3 173134034  150529862    3675202
4 1266653993 1400237791 1418688520
5        NA         NA         NA
6        NA   63064912    1764252
> head(frameList[[5]])
  Country.Name                                      Program.Name
1  Afghanistan                         Child Survival and Health
2  Afghanistan         Department of Defense Security Assistance
3  Afghanistan                            Development Assistance
4  Afghanistan Economic Support Fund/Security Support Assistance
5  Afghanistan                                Food For Education
6  Afghanistan                  Global Health and Child Survival
  FY1960 FY1961    FY1962 FY1963 FY1964 FY1965 FY1966 FY1967 FY1968
1     NA     NA        NA     NA     NA     NA     NA     NA     NA
2     NA     NA        NA     NA     NA     NA     NA     NA     NA
3     NA     NA        NA     NA     NA     NA     NA     NA     NA
4     NA     NA 181177853     NA     NA     NA     NA     NA     NA
5     NA     NA        NA     NA     NA     NA     NA     NA     NA
6     NA     NA        NA     NA     NA     NA     NA     NA     NA
  FY1969
1     NA
2     NA
3     NA
4     NA
5     NA
6     NA
> head(frameList[["Aid_60s"]])
  Country.Name                                      Program.Name
1  Afghanistan                         Child Survival and Health
2  Afghanistan         Department of Defense Security Assistance
3  Afghanistan                            Development Assistance
4  Afghanistan Economic Support Fund/Security Support Assistance
5  Afghanistan                                Food For Education
```

```
6  Afghanistan                    Global Health and Child Survival
  FY1960 FY1961    FY1962 FY1963 FY1964 FY1965 FY1966 FY1967 FY1968
1     NA     NA        NA     NA     NA     NA     NA     NA     NA
2     NA     NA        NA     NA     NA     NA     NA     NA     NA
3     NA     NA        NA     NA     NA     NA     NA     NA     NA
4     NA     NA 181177853     NA     NA     NA     NA     NA     NA
5     NA     NA        NA     NA     NA     NA     NA     NA     NA
6     NA     NA        NA     NA     NA     NA     NA     NA     NA
  FY1969
1     NA
2     NA
3     NA
4     NA
5     NA
6     NA
```

Having all the `data.frames` in a `list` allows us to iterate through the `list`, joining all the elements together (or applying any function to the elements iteratively). Rather than using a loop, we use the **Reduce** function to speed up the operation.

```
> allAid <- Reduce(function(...){
+     join(..., by=c("Country.Name", "Program.Name"))},
+     frameList)
> dim(allAid)

[1] 2453   67

> library(useful)
> corner(allAid, c=15)

  Country.Name                                            Program.Name
1  Afghanistan                            Child Survival and Health
2  Afghanistan         Department of Defense Security Assistance
3  Afghanistan                                Development Assistance
4  Afghanistan Economic Support Fund/Security Support Assistance
5  Afghanistan                                    Food For Education
  FY2000  FY2001   FY2002    FY2003     FY2004     FY2005     FY2006
1     NA      NA  2586555  56501189   40215304   39817970   40856382
2     NA      NA  2964313        NA   45635526  151334908  230501318
3     NA 4110478  8762080  54538965  180539337  193598227  212648440
4     NA   61144 31827014 341306822 1025522037 1157530168 1357750249
5     NA      NA       NA   3957312    2610006    3254408     386891
      FY2007     FY2008     FY2009     FY2010 FY1946 FY1947
1   72527069   28397435         NA         NA     NA     NA
2  214505892  495539084  552524990  316514796     NA     NA
3  173134034  150529862    3675202         NA     NA     NA
4 1266653993 1400237791 1418688520 2797488331     NA     NA
5         NA         NA         NA         NA     NA     NA

> bottomleft(allAid, c=15)
```

```
     Country.Name              Program.Name  FY2000  FY2001   FY2002
2449     Zimbabwe Other State Assistance 1341952  322842       NA
2450     Zimbabwe Other USAID Assistance 3033599 8464897  6624408
2451     Zimbabwe               Peace Corps 2140530 1150732   407834
2452     Zimbabwe                   Title I      NA      NA       NA
2453     Zimbabwe                  Title II      NA      NA 31019776
       FY2003   FY2004   FY2005  FY2006    FY2007    FY2008    FY2009
2449       NA   318655    44553  883546   1164632   2455592   2193057
2450 11580999 12805688 10091759 4567577  10627613  11466426  41940500
2451       NA       NA       NA      NA        NA        NA        NA
2452       NA       NA       NA      NA        NA        NA        NA
2453       NA       NA       NA  277468 100053600 180000717 174572685
       FY2010 FY1946 FY1947
2449  1605765     NA     NA
2450 30011970     NA     NA
2451       NA     NA     NA
2452       NA     NA     NA
2453 79545100     NA     NA
```

Reduce can be a difficult function to grasp, so we illustrate it with a simple example. Let's say we have a `vector` of the first ten integers, `1:10`, and want to sum them (forget for a moment that `sum(1:10)` will work perfectly). We can call `Reduce(sum, 1:10)`, which will first add 1 and 2. It will then add 3 to that result, then 4 to that result and so on, resulting in 55.

Likewise, we passed a `list` to a function that joins its inputs, which in this case was simply ..., meaning that anything could be passed. Using ... is an advanced trick of R programming that can be difficult to get right. **Reduce** passed the first two `data.frames` in the `list`, which were then joined. That result was then joined to the next `data.frame` and so on until they were all joined together.

14.2.3 `data.table` merge

Like many other operations in **data.table**, joining data requires a different syntax, and possibly a different way of thinking. To start, we convert two of our foreign aid datasets' `data.frames` into `data.tables`.

```
> library(data.table)
> dt90 <- data.table(Aid_90s, key=c("Country.Name", "Program.Name"))
> dt00 <- data.table(Aid_00s, key=c("Country.Name", "Program.Name"))
```

Then, doing the join is a simple operation. Note that the join requires specifying the keys for the `data.tables`, which we did during their creation.

```
> dt0090 <- dt90[dt00]
```

In this case `dt90` is the left side, `dt00` is the right side and a left join was performed.

14.3 reshape2

The next most common munging need is either melting data (going from column orientation to row orientation) or casting data (going from row orientation to column orientation). As with most other procedures in R, there are multiple functions available to accomplish these tasks, but we will focus on Hadley Wickham's **reshape2** package. We talk about Wickham a lot, but that is because his products have become so fundamental to the R developer's toolbox.

14.3.1 melt

Looking at the `Aid_00s` `data.frame`, we see that each year is stored in its own column. That is, the dollar amount for a given country and program is found in a different column for each year. This is called a cross table, which while nice for human consumption, is not ideal for graphing with **ggplot2** or for some analysis algorithms.

```
> head(Aid_00s)

  Country.Name                                             Program.Name
1  Afghanistan                          Child Survival and Health
2  Afghanistan          Department of Defense Security Assistance
3  Afghanistan                             Development Assistance
4  Afghanistan Economic Support Fund/Security Support Assistance
5  Afghanistan                                 Food For Education
6  Afghanistan                    Global Health and Child Survival
  FY2000  FY2001   FY2002    FY2003     FY2004     FY2005     FY2006
1     NA      NA  2586555  56501189   40215304   39817970   40856382
2     NA      NA  2964313        NA   45635526  151334908  230501318
3     NA 4110478  8762080  54538965  180539337  193598227  212648440
4     NA   61144 31827014 341306822 1025522037 1157530168 1357750249
5     NA      NA       NA   3957312    2610006    3254408     386891
6     NA      NA       NA        NA         NA         NA         NA
      FY2007     FY2008     FY2009
1   72527069   28397435         NA
2  214505892  495539084  552524990
3  173134034  150529862    3675202
4 1266653993 1400237791 1418688520
5         NA         NA         NA
6         NA   63064912    1764252
```

We want it set up so that each row represents a single country-program-year entry with the dollar amount stored in one column. To achieve this we melt the data using **melt** from **reshape2**.

```
> library(reshape2)
> melt00 <- melt(Aid_00s, id.vars=c("Country.Name", "Program.Name"),
+                variable.name="Year", value.name="Dollars")
> tail(melt00, 10)
```

```
      Country.Name
24521     Zimbabwe
24522     Zimbabwe
24523     Zimbabwe
24524     Zimbabwe
24525     Zimbabwe
24526     Zimbabwe
24527     Zimbabwe
24528     Zimbabwe
24529     Zimbabwe
24530     Zimbabwe
                                                    Program.Name   Year
24521                       Migration and Refugee Assistance FY2009
24522                                      Narcotics Control FY2009
24523 Nonproliferation, Anti-Terrorism, Demining and Related FY2009
24524                            Other Active Grant Programs FY2009
24525                                Other Food Aid Programs FY2009
24526                                 Other State Assistance FY2009
24527                                 Other USAID Assistance FY2009
24528                                            Peace Corps FY2009
24529                                                Title I FY2009
24530                                               Title II FY2009
        Dollars
24521   3627384
24522        NA
24523        NA
24524   7951032
24525        NA
24526   2193057
24527  41940500
24528        NA
24529        NA
24530 174572685
```

The `id.vars` argument specifies which columns uniquely identify a row.

After some manipulation of the `Year` column and aggregating, this is now prime for plotting, as shown in Figure 14.1. The plot uses faceting, enabling us to quickly see and understand the funding for each program over time.

```
> library(scales)
> # strip the "FY" out of the year column and convert it to numeric
> melt00$Year <- as.numeric(str_sub(melt00$Year, start=3, 6))
> # aggregate the data so we have yearly numbers by program
> meltAgg <- aggregate(Dollars ~ Program.Name + Year, data=melt00,
+                      sum, na.rm=TRUE)
> # just keep the first 10 characters of program name
> # then it will fit in the plot
> meltAgg$Program.Name <- str_sub(meltAgg$Program.Name, start=1,
+                                 end=10)
>
```

```
> ggplot(meltAgg, aes(x=Year, y=Dollars)) +
+     geom_line(aes(group=Program.Name)) +
+     facet_wrap(~ Program.Name) +
+     scale_x_continuous(breaks=seq(from=2000, to=2009, by=2)) +
+     theme(axis.text.x=element_text(angle=90, vjust=1, hjust=0)) +
+     scale_y_continuous(labels=multiple_format(extra=dollar,
+                                                multiple="B"))
```

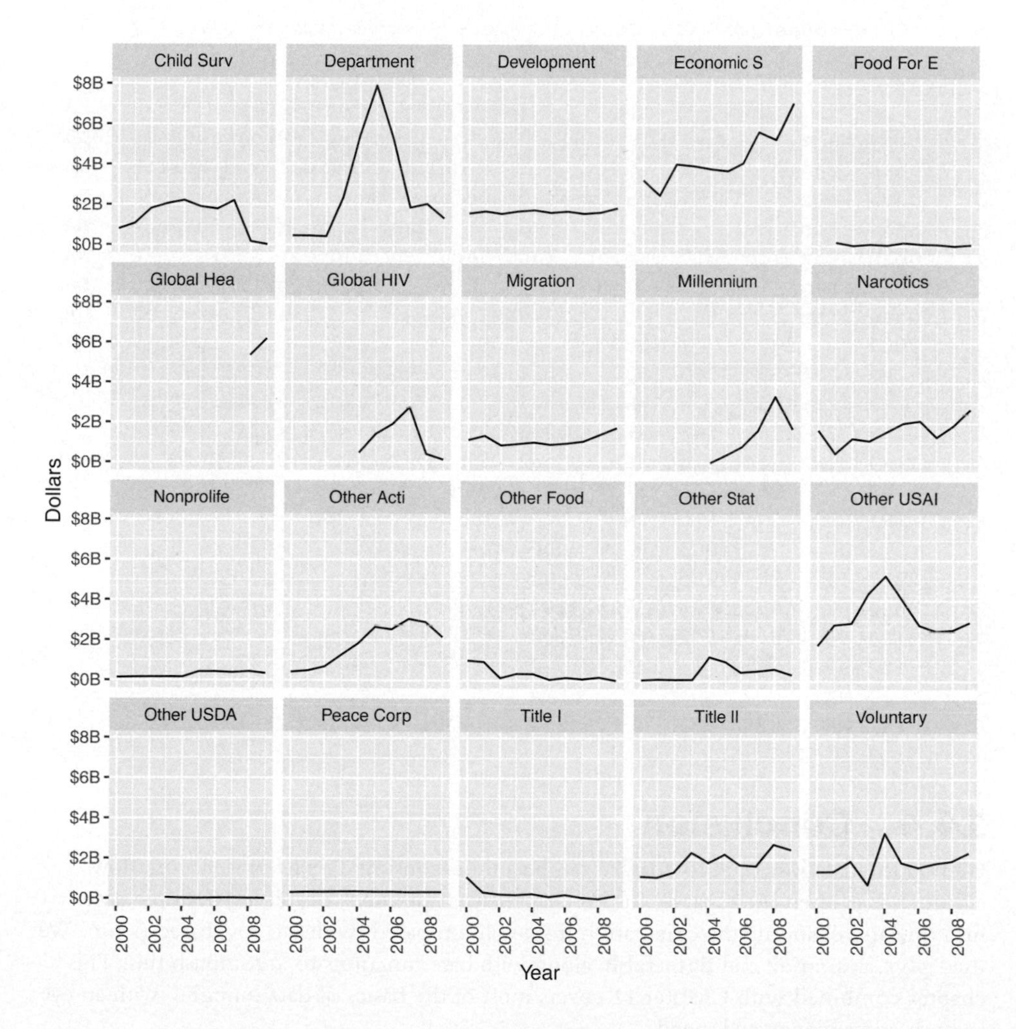

Figure 14.1 Plot of foreign assistance by year for each of the programs.

14.3.2 dcast

Now that we have the foreign aid data melted, we cast it back into the wide format for illustration purposes. The function for this is **dcast**, and it has trickier arguments than **melt**. The first is the data to be used, in our case `melt00`. The second argument is a `formula` where the left side specifies the columns that should remain columns and the right side specifies the columns that should become column names. The third argument is the column (as a character) that holds the values to be populated into the new columns, representing the unique values of the right side of the `formula` argument.

```
> cast00 <- dcast(melt00, Country.Name + Program.Name ~ Year,
+                 value.var="Dollars")
> head(cast00)

  Country.Name                                  Program.Name 2000
1  Afghanistan                    Child Survival and Health   NA
2  Afghanistan         Department of Defense Security Assistance   NA
3  Afghanistan                       Development Assistance   NA
4  Afghanistan Economic Support Fund/Security Support Assistance   NA
5  Afghanistan                           Food For Education   NA
6  Afghanistan                Global Health and Child Survival   NA
     2001     2002      2003       2004       2005       2006
1      NA  2586555  56501189   40215304   39817970   40856382
2      NA  2964313        NA   45635526  151334908  230501318
3 4110478  8762080  54538965  180539337  193598227  212648440
4   61144 31827014 341306822 1025522037 1157530168 1357750249
5      NA       NA   3957312    2610006    3254408     386891
6      NA       NA        NA         NA         NA         NA
        2007       2008       2009
1   72527069   28397435         NA
2  214505892  495539084  552524990
3  173134034  150529862    3675202
4 1266653993 1400237791 1418688520
5         NA         NA         NA
6         NA   63064912    1764252
```

14.4 Conclusion

Getting the data just right to analyze can be a time-consuming part of our work flow, although it is often inescapable. In this chapter we examined combining multiple datasets into one and changing the orientation from column based (wide) to row based (long). We used **plyr**, **reshape2** and **data.table** along with base functions to accomplish this. This chapter combined with Chapter 11 covers most of the basics of data munging with an eye to both convenience and speed.

15

Reshaping Data in the Tidyverse

Chapter 14 covered numerous ways to reshape data such as binding rows and columns, joins, and converting between wide and long data formats. All the described functions from base R and packages such as **plyr**, **data.table** and **reshape2** are still great options, but the newer packages **dplyr** and **tidyr** are designed to work with pipes and may be easier to use for some people. In some cases, these packages will see speed improvements but not always. These, and other packages written mainly by Hadley Wickham, make up the Tidyverse.

15.1 Binding Rows and Columns

The **dplyr** analogs to **rbind** and **cbind** are **bind_rows** and **bind_cols**, respectively. These functions do not behave exactly like their base R counterparts; they only work with `data.frames` (and `tibbles`), whereas **cbind** and **rbind** work with `data.frames` and `matrices` and can also bind `vectors` into `matrices` and `data.frames`. So **bind_rows** and **bind_cols** may be more limited, but they work well when given `data.frames`.

We return to the example in Section 14.1 and adapt it for use with **tibble** and **dplyr**.

```
> # load dplyr
> library(dplyr)
> library(tibble)
>
> # create a two-column tibble
> sportLeague <- tibble(sport=c("Hockey", "Baseball", "Football"),
+                       league=c("NHL", "MLB", "NFL"))
> # create a one-column tibble
> trophy <- tibble(trophy=c("Stanley Cup", "Commissioner's Trophy",
+                           "Vince Lombardi Trophy"))
>
> # combine them into one tibble
> trophies1 <- bind_cols(sportLeague, trophy)
>
> # make another tibble using tribble as a shortcut to build it row-wise
> trophies2 <- tribble(
```

```
+     ~sport, ~league, ~trophy,
+     "Basketball", "NBA", "Larry O'Brien Championship Trophy",
+     "Golf", "PGA", "Wanamaker Trophy"
+ )
>
> # combine them into one tibble
> trophies <- bind_rows(trophies1, trophies2)
>
> trophies

# A tibble: 5 × 3
       sport league                           trophy
       <chr>  <chr>                            <chr>
1     Hockey    NHL                      Stanley Cup
2   Baseball    MLB           Commissioner's Trophy
3   Football    NFL           Vince Lombardi Trophy
4 Basketball    NBA Larry O'Brien Championship Trophy
5       Golf    PGA                 Wanamaker Trophy
```

Both **bind_cols** and **bind_rows** can bind multiple `tibbles` (or `data.frames`) together.

15.2 Joins with `dplyr`

Joins are an important part of the data manipulation process. In Section 14.2 we used functions from base R, **plyr** and **data.table** to perform joins. Another option is to use the join functions in **dplyr**: **left_join**, **right_join**, **inner_join**, **full_join**, **semi_join** and **anti_join**. As our motivating example we look at the `diamonds` data, which has a column about the color of the diamonds. This column indicates the color as a single letter, one of D, E, F, G, H, I and J, that depends on prior knowledge of these codes. Fortunately, we have more data that details the color specification that we can join to the `diamonds` data.

First, we read the data using **read_csv** from the **readr** package. This package was introduced in Section 6.1.1 and enables fast reading of data stored in plain text files. The result is a `tibble` object.

```
> library(readr)
> colorsURL <- 'http://www.jaredlander.com/data/DiamondColors.csv'
> diamondColors <- read_csv(colorsURL)
> diamondColors

# A tibble: 10 × 3
   Color          Description                     Details
   <chr>                <chr>                       <chr>
1      D Absolutely Colorless                    No color
2      E            Colorless       Minute traces of color
3      F            Colorless       Minute traces of color
4      G       Near Colorless Color is dificult to detect
5      H       Near Colorless Color is dificult to detect
6      I       Near Colorless   Slightly detectable color
7      J       Near Colorless   Slightly detectable color
8      K          Faint Color            Noticeable color
```

```
9       L           Faint Color                   Noticeable color
10      M           Faint Color                   Noticeable color
```

Looking at the diamonds data, we see the following values for color.

```
> # load the diamonds data without loading the ggplot2 package
> data(diamonds, package='ggplot2')
> unique(diamonds$color)

[1] E I J H F G D
Levels: D < E < F < G < H < I < J
```

We perform a left join that will result in a combined tibble (which we refer to interchangeably with data.frame) that has all the information from each of the datasets. We use the diamonds data as the left-hand table and the Diamond Colors data as the right-hand table. Since the key column is not the same in both data.frames (color with a lower-case 'c' in diamonds and Color with an upper-case 'C' in diamondColors), we specify it using the by argument. The input to this argument should be a named vector, where the names of the vector are the keys in the left-hand table and the values of the vector are the keys in the right-hand table. In this case there is only one key per table, but there can be multiple keys, and each would be specified in the vector of keys.

```
> library(dplyr)
> left_join(diamonds, diamondColors, by=c('color'='Color'))

Warning in left_join_impl(x, y, by$x, by$y, suffix$x, suffix$y): joining character
vector and factor, coercing into character vector

# A tibble: 53,940 × 12
   carat       cut color clarity depth table price     x     y     z
   <dbl>     <ord> <chr>   <ord> <dbl> <dbl> <int> <dbl> <dbl> <dbl>
1   0.23     Ideal     E     SI2  61.5    55   326  3.95  3.98  2.43
2   0.21   Premium     E     SI1  59.8    61   326  3.89  3.84  2.31
3   0.23      Good     E     VS1  56.9    65   327  4.05  4.07  2.31
4   0.29   Premium     I     VS2  62.4    58   334  4.20  4.23  2.63
5   0.31      Good     J     SI2  63.3    58   335  4.34  4.35  2.75
6   0.24 Very Good     J    VVS2  62.8    57   336  3.94  3.96  2.48
7   0.24 Very Good     I    VVS1  62.3    57   336  3.95  3.98  2.47
8   0.26 Very Good     H     SI1  61.9    55   337  4.07  4.11  2.53
9   0.22      Fair     E     VS2  65.1    61   337  3.87  3.78  2.49
10  0.23 Very Good     H     VS1  59.4    61   338  4.00  4.05  2.39
# ... with 53,930 more rows, and 2 more variables: Description <chr>,
#   Details <chr>
```

There is a warning message saying the the key columns in the two data.frames are of different types (factor and character). Given the underlying structure of factors, this would be akin to joining a character column on an integer column, so **left_join** automatically coerced the factor into a character.

Due to space limitations in the layout of this book, several columns were not displayed. In interactive usage the number of columns displayed is determined by the width of the console, so this can vary. To fully see the results of the join, we execute it again and then use **select** to choose just some columns to be displayed.

```
> left_join(diamonds, diamondColors, by=c('color'='Color')) %>%
+     select(carat, color, price, Description, Details)

Warning in left_join_impl(x, y, by$x, by$y, suffix$x, suffix$y): joining
character vector and factor, coercing into character vector

# A tibble: 53,940 × 5
   carat color price     Description                     Details
   <dbl> <chr> <int>           <chr>                       <chr>
1   0.23     E   326       Colorless       Minute traces of color
2   0.21     E   326       Colorless       Minute traces of color
3   0.23     E   327       Colorless       Minute traces of color
4   0.29     I   334  Near Colorless   Slightly detectable color
5   0.31     J   335  Near Colorless   Slightly detectable color
6   0.24     J   336  Near Colorless   Slightly detectable color
7   0.24     I   336  Near Colorless   Slightly detectable color
8   0.26     H   337  Near Colorless Color is dificult to detect
9   0.22     E   337       Colorless       Minute traces of color
10  0.23     H   338  Near Colorless Color is dificult to detect
# ... with 53,930 more rows
```

This was a left join, so all of the rows of the left-hand table (diamonds) are kept and only the rows of the right-hand table (diamondColors) that have matches are kept. We can see that diamondColors has more unique values for Color and Description than the joined results.

```
> left_join(diamonds, diamondColors, by=c('color'='Color')) %>%
+     distinct(color, Description)

Warning in left_join_impl(x, y, by$x, by$y, suffix$x, suffix$y): joining
character vector and factor, coercing into character vector

# A tibble: 7 × 2
  color          Description
  <chr>                <chr>
1     E            Colorless
2     I       Near Colorless
3     J       Near Colorless
4     H       Near Colorless
5     F            Colorless
6     G       Near Colorless
7     D Absolutely Colorless

> diamondColors %>% distinct(Color, Description)

# A tibble: 10 × 2
   Color          Description
   <chr>                <chr>
1      D Absolutely Colorless
2      E            Colorless
3      F            Colorless
4      G       Near Colorless
5      H       Near Colorless
6      I       Near Colorless
7      J       Near Colorless
```

```
8       K           Faint Color
9       L           Faint Color
10      M           Faint Color
```

A right join keeps all the rows of the right-hand table and the matching rows of the left-hand table. Since diamondColors has more unique values of Color than diamonds, the resulting join will also have more rows than diamonds by itself.

```
> right_join(diamonds, diamondColors, by=c('color'='Color')) %>% nrow

Warning in right_join_impl(x, y, by$x, by$y, suffix$x, suffix$y):
joining factor and character vector, coercing into character vector

[1] 53943

> diamonds %>% nrow

[1] 53940
```

An inner join returns rows from both tables that have matching keys. If a row in either table is not matched by a row in the other table, then that row is not returned. For the data we are using, an inner join will be equivalent to a left join.

```
> all.equal(
+     left_join(diamonds, diamondColors, by=c('color'='Color')),
+     inner_join(diamonds, diamondColors, by=c('color'='Color'))
+ )

Warning in left_join_impl(x, y, by$x, by$y, suffix$x, suffix$y): joining
character vector and factor, coercing into character vector
Warning in inner_join_impl(x, y, by$x, by$y, suffix$x, suffix$y): joining
factor and character vector, coercing into character vector

[1] TRUE
```

A full join (also known as an outer join) returns all rows of both tables, even those that do not have a match with rows in the other table. For the data we are using, a full join will be equivalent to a right join.

```
> all.equal(
+     right_join(diamonds, diamondColors, by=c('color'='Color')),
+     full_join(diamonds, diamondColors, by=c('color'='Color'))
+ )

Warning in right_join_impl(x, y, by$x, by$y, suffix$x, suffix$y): joining
factor and character vector, coercing into character vector
Warning in full_join_impl(x, y, by$x, by$y, suffix$x, suffix$y): joining
character vector and factor, coercing into character vector

[1] TRUE
```

A semi-join does not join two tables together but rather returns the first rows in the left-hand table that have matches in the right-hand table. This is a form of row filtering. If a row in the left-hand table is matched by multiple rows in the right-hand table, only the first matched rows are returned. Using diamondColors as the left-hand table, only the colors E, I, J, H, F, G, D are found in diamonds, so those are returned, along with the rest of the data.frame.

```
> semi_join(diamondColors, diamonds, by=c('Color'='color'))

# A tibble: 7 × 3
  Color          Description                       Details
  <chr>                <chr>                         <chr>
1     E            Colorless        Minute traces of color
2     I       Near Colorless     Slightly detectable color
3     J       Near Colorless     Slightly detectable color
4     H       Near Colorless Color is dificult to detect
5     F            Colorless        Minute traces of color
6     G       Near Colorless Color is dificult to detect
7     D Absolutely Colorless                      No color
```

An anti-join is the opposite of a semi-join and returns the rows of the left-hand table that are not matched to any rows in the right-hand table. Within diamondColors, the colors K, L, M are not matched in diamonds, so those rows are returned.

```
> anti_join(diamondColors, diamonds, by=c('Color'='color'))

# A tibble: 3 × 3
  Color Description           Details
  <chr>       <chr>             <chr>
1     K Faint Color Noticeable color
2     L Faint Color Noticeable color
3     M Faint Color Noticeable color
```

The same results for both **semi_join** and **anti_join** could have been achieved using **filter** and **unique**. This alternative coding is easy when the data are in data.frames, but **semi_join** and **anti_join** are better options when using **dplyr** to operate on databases.

```
> diamondColors %>% filter(Color %in% unique(diamonds$color))

# A tibble: 7 × 3
  Color          Description                       Details
  <chr>                <chr>                         <chr>
1     D Absolutely Colorless                      No color
2     E            Colorless        Minute traces of color
3     F            Colorless        Minute traces of color
4     G       Near Colorless Color is dificult to detect
5     H       Near Colorless Color is dificult to detect
6     I       Near Colorless     Slightly detectable color
7     J       Near Colorless     Slightly detectable color

> diamondColors %>% filter(!Color %in% unique(diamonds$color))
```

```
# A tibble: 3 × 3
  Color Description            Details
  <chr>       <chr>              <chr>
1     K Faint Color Noticeable color
2     L Faint Color Noticeable color
3     M Faint Color Noticeable color
```

15.3 Converting Data Formats

Converting data between long and wide formats is handled well using base functions and **melt** and **dcast** in Hadley Wickham's **reshape2** package, as seen in Section 14.3. Much as **dplyr** is the next generation of **plyr**, **tidyr** is the next generation of **reshape2**. Hadley Wickham wrote **tidyr** to be easier to use than **reshape2** (and to work with pipes) rather than to be computationally faster, so speed gains should be minimal.

For an example we look at data from an experiment conducted at Columbia University about emotion reactivity and regulation. The dataset has been annonymized and random noise has been added so that no personal information is shared. The file is tab separated and saved as text, so we use **read_tsv** from the **readr** package to read the data into memory as a `tibble`. By default, data reading functions in **readr** display a message that indicates the types of data stored in each column.

```
> library(readr)
> emotion <- read_tsv('http://www.jaredlander.com/data/reaction.txt')

Parsed with column specification:
  cols(
    ID = col_integer(),
    Test = col_integer(),
    Age = col_double(),
    Gender = col_character(),
    BMI = col_double(),
    React = col_double(),
    Regulate = col_double()
  )

> emotion

# A tibble: 99 × 7
     ID  Test   Age Gender   BMI React Regulate
  <int> <int> <dbl>  <chr> <dbl> <dbl>    <dbl>
1     1     1  9.69      F 14.71  4.17     3.15
2     1     2 12.28      F 14.55  3.89     2.55
3     2     1 15.72      F 19.48  4.39     4.41
4     2     2 17.62      F 19.97  2.03     2.20
5     3     1  9.52      F 20.94  3.38     2.65
6     3     2 11.84      F 23.97  4.00     3.63
7     4     1 16.29      M 25.13  3.15     3.59
8     4     2 18.85      M 27.96  3.02     3.54
```

```
9      5     1 15.78     M 28.35  3.08      2.64
10     5     2 18.25     M 19.57  3.17      2.29
# ... with 89 more rows
```

We see that the `tibble` is in wide format, so we use **gather**[1] (the corollary to **melt** from **reshape2**) to make it long. We put the `Age`, `React` and `Regulate` columns into one column that we call `Measurement` and the column names. Another column, `Type`, holds the original column names. The first argument to **gather** is the `tibble` (or `data.frame`) of interest. The `key` is the name of the newly created column that holds the original column names or keys. The `Value` argument is the name of the newly created column that holds the actual data from the columns being gathered. Both of these arguments are specified without quotes. Following arguments provide the names, unquoted, of columns that are to be gathered and pivoted into the value column.

```
> library(tidyr)
> emotion %>%
+     gather(key=Type, value=Measurement, Age, BMI, React, Regulate)

# A tibble: 396 × 5
      ID  Test Gender  Type Measurement
   <int> <int>  <chr> <chr>       <dbl>
1      1     1      F   Age        9.69
2      1     2      F   Age       12.28
3      2     1      F   Age       15.72
4      2     2      F   Age       17.62
5      3     1      F   Age        9.52
6      3     2      F   Age       11.84
7      4     1      M   Age       16.29
8      4     2      M   Age       18.85
9      5     1      M   Age       15.78
10     5     2      M   Age       18.25
# ... with 386 more rows
```

The data are sorted by the newly created `Type` column, so it is hard to see what happened to the data. To make it easier we sort the data by `ID`.

```
> library(tidyr)
> emotionLong <- emotion %>%
+     gather(key=Type, value=Measurement, Age, BMI, React, Regulate) %>%
+     arrange(ID)
>
> head(emotionLong, 20)

# A tibble: 20 × 5
     ID  Test Gender    Type Measurement
  <int> <int>  <chr>   <chr>       <dbl>
1     1     1      F     Age        9.69
2     1     2      F     Age       12.28
3     1     1      F     BMI       14.71
```

1. It gathers data from multiple columns and stores them in a single column.

```
4     1     2     F      BMI       14.55
5     1     1     F    React        4.17
6     1     2     F    React        3.89
7     1     1     F Regulate        3.15
8     1     2     F Regulate        2.55
9     2     1     F      Age       15.72
10    2     2     F      Age       17.62
11    2     1     F      BMI       19.48
12    2     2     F      BMI       19.97
13    2     1     F    React        4.39
14    2     2     F    React        2.03
15    2     1     F Regulate        4.41
16    2     2     F Regulate        2.20
17    3     1     F      Age        9.52
18    3     2     F      Age       11.84
19    3     1     F      BMI       20.94
20    3     2     F      BMI       23.97
```

In the original data each `ID` was contained in two rows, each of which had `Age`, `BMI`, `React` and `Regulate columns`. The transformed data has four rows for each original row with a column (`Type`) specifying `Age`, `BMI`, `React` and `Regulate columns` and another (`Measurement`) holding the actual values. The remaining columns—`ID`, `Test` and `Gender`—were not pivoted.

While we specified the columns we wanted to gather, we could also specify the columns we do not want to pivot by preceding them with a minus sign (-).

```
> emotion %>%
+     gather(key=Type, value=Measurement, -ID, -Test, -Gender)

# A tibble: 396 × 5
      ID  Test Gender  Type Measurement
   <int> <int>  <chr> <chr>       <dbl>
1      1     1      F   Age        9.69
2      1     2      F   Age       12.28
3      2     1      F   Age       15.72
4      2     2      F   Age       17.62
5      3     1      F   Age        9.52
6      3     2      F   Age       11.84
7      4     1      M   Age       16.29
8      4     2      M   Age       18.85
9      5     1      M   Age       15.78
10     5     2      M   Age       18.25
# ... with 386 more rows

> identical(
+     emotion %>%
+         gather(key=Type, value=Measurement, -ID, -Test, -Gender),
+     emotion %>%
+         gather(key=Type, value=Measurement, Age, BMI, React, Regulate)
+ )

[1] TRUE
```

The opposite of **gather** is **spread**[2] (the corollary to **dcast** from **reshape2**), which makes the data wide. The `key` argument specifies the column that will become the new column names and `value` specifies the column that holds the values that will populate the new columns.

```
> emotionLong %>%
+     spread(key=Type, value=Measurement)

# A tibble: 99 × 7
      ID  Test Gender   Age   BMI React Regulate
*  <int> <int>  <chr> <dbl> <dbl> <dbl>    <dbl>
1      1     1      F  9.69 14.71  4.17     3.15
2      1     2      F 12.28 14.55  3.89     2.55
3      2     1      F 15.72 19.48  4.39     4.41
4      2     2      F 17.62 19.97  2.03     2.20
5      3     1      F  9.52 20.94  3.38     2.65
6      3     2      F 11.84 23.97  4.00     3.63
7      4     1      M 16.29 25.13  3.15     3.59
8      4     2      M 18.85 27.96  3.02     3.54
9      5     1      M 15.78 28.35  3.08     2.64
10     5     2      M 18.25 19.57  3.17     2.29
# ... with 89 more rows
```

15.4 Conclusion

Over the past few years many advancements in data reshaping have made munging easier. In particular, **bind_rows**, **bind_cols**, **left_join** and **inner_join** from **dplyr** and **gather** and **spread** from **tidyr** have improved the state of data manipulation. While their functionality already exists in base R and other packages, these are meant to be easier to use, and **dplyr** is meant to be faster too.

2. It spreads data from one column into multiple columns.

16 Manipulating Strings

Strings (character data) often need to be constructed or deconstructed to identify observations, preprocess text, combine information or satisfy any number of other needs. R offers functions for building strings, like **paste** and **sprintf**. It also provides a number of functions for using regular expressions and examining text data, although for those purposes it is better to use Hadley Wickham's **stringr** package.

16.1 paste

The first function new R users reach for when putting together strings is **paste**. This function takes a series of strings, or expressions that evaluate to strings, and puts them together into one string. We start by putting together three simple strings.

```
> paste("Hello", "Jared", "and others")

[1] "Hello Jared and others"
```

Notice that spaces were put between the two words. This is because **paste** has a third argument, sep, that determines what to put in between entries. This can be any valid text, including empty text ("").

```
> paste("Hello", "Jared", "and others", sep="/")

[1] "Hello/Jared/and others"
```

Like many functions in R, **paste** is vectorized. This means each element can be a vector of data to be put together.

```
> paste(c("Hello", "Hey", "Howdy"), c("Jared", "Bob", "David"))

[1] "Hello Jared" "Hey Bob"     "Howdy David"
```

In this case each vector had the same number of entries so they paired one-to-one. When the vectors do not have the same length, they are recycled.

```
> paste("Hello", c("Jared", "Bob", "David"))

[1] "Hello Jared" "Hello Bob"   "Hello David"
```

```
> paste("Hello", c("Jared", "Bob", "David"), c("Goodbye", "Seeya"))

[1] "Hello Jared Goodbye" "Hello Bob Seeya"     "Hello David Goodbye"
```

paste also has the ability to collapse a `vector` of text into one `vector` containing all the elements with any arbitrary separator, using the `collapse` argument.

```
> vectorOfText <- c("Hello", "Everyone", "out there", ".")
> paste(vectorOfText, collapse=" ")

[1] "Hello Everyone out there ."

> paste(vectorOfText, collapse="*")

[1] "Hello*Everyone*out there*."
```

16.2 sprintf

While **paste** is convenient for putting together short bits of text, it can become unwieldy when piecing together long pieces of text, such as when inserting a number of variables into a long piece of text. For instance, we might have a lengthy sentence that has a few spots that require the insertion of special variables. An example is "Hello Jared, your party of eight will be seated in 25 minutes" where "Jared," "eight" and "25" could be replaced with other information.

Reforming this with **paste** can make reading the line in code difficult.

To start, we make some variables to hold the information.

```
> person <- "Jared"
> partySize <- "eight"
> waitTime <- 25
```

Now we build the **paste** expression.

```
> paste("Hello ", person, ", your party of ", partySize,
+       " will be seated in ", waitTime, " minutes.", sep="")

[1] "Hello Jared, your party of eight will be seated in 25 minutes."
```

Making even a small change to this sentence would require putting the commas in just the right places.

A good alternative is the **sprintf** function. With this function we build one long string with special markers indicating where to insert values.

```
> sprintf("Hello %s, your party of %s will be seated in %s minutes",
+         person, partySize, waitTime)

[1] "Hello Jared, your party of eight will be seated in 25 minutes"
```

Here, each `%s` was replaced with its corresponding variable. While the long sentence is easier to read in code, we must maintain the order of `%s`'s and variables.

sprintf is also vectorized. Note that the `vector` lengths must be multiples of each other.

```
> sprintf("Hello %s, your party of %s will be seated in %s minutes",
+         c("Jared", "Bob"), c("eight", 16, "four", 10), waitTime)

[1] "Hello Jared, your party of eight will be seated in 25 minutes"
[2] "Hello Bob, your party of 16 will be seated in 25 minutes"
[3] "Hello Jared, your party of four will be seated in 25 minutes"
[4] "Hello Bob, your party of 10 will be seated in 25 minutes"
```

16.3 Extracting Text

Often text needs to be ripped apart to be made useful, and while R has a number of functions for doing so, the **stringr** package is much easier to use.

First we need some data, so we use the **XML** package to download a table of United States Presidents from Wikipedia.

```
> library(XML)
```

Then we use **readHTMLTable** to parse the table.

```
> load("data/presidents.rdata")

> theURL <- "http://www.loc.gov/rr/print/list/057_chron.html"
> presidents <- readHTMLTable(theURL, which=3, as.data.frame=TRUE,
+                             skip.rows=1, header=TRUE,
+                             stringsAsFactors=FALSE)
```

Now we take a look at the data.

```
> head(presidents)

       YEAR         PRESIDENT
1 1789-1797 George Washington
2 1797-1801        John Adams
3 1801-1805  Thomas Jefferson
4 1805-1809  Thomas Jefferson
5 1809-1812     James Madison
6 1812-1813     James Madison
                                         FIRST LADY    VICE PRESIDENT
1                                Martha Washington        John Adams
2                                    Abigail Adams Thomas Jefferson
3 Martha Wayles Skelton Jefferson\n    (no image)        Aaron Burr
4 Martha Wayles Skelton Jefferson\n    (no image)    George Clinton
5                                   Dolley Madison   George Clinton
6                                   Dolley Madison    office vacant
```

Examining it more closely, we see that the last few rows contain information we do not want, so we keep only the first 64 rows.

```
> tail(presidents$YEAR)

[1] "2001-2009"
[2] "2009-"
[3] "Presidents: Introduction (Rights/Ordering\n        Info.) |
        Adams\n       - Cleveland |
        Clinton - Harding Harrison\n       - Jefferson |
        Johnson - McKinley |
        Monroe\n                         - Roosevelt |
        Taft - Truman |
        Tyler\n                          - WilsonList of names, Alphabetically"
[4] "First Ladies: Introduction\n                  (Rights/Ordering Info.) |
        Adams\n                  - Coolidge |
        Eisenhower - HooverJackson\n                 - Pierce   |
        \n                                  Polk - Wilson |
        List\n               of names, Alphabetically"
[5] "Vice Presidents: Introduction (Rights/Ordering Info.) |
        Adams - Coolidge | Curtis - Hobart Humphrey - Rockefeller |
        Roosevelt - WilsonList of names, Alphabetically"
[6] "Top\n              of Page"

> presidents <- presidents[1:64, ]
```

To start, we create two new columns, one for the beginning of the term and one for the end of the term. To do this we need to split the `Year` column on the hyphen ("-"). The **stringr** package has the **str_split** function that splits a string based on some value. It returns a `list` with an element for each element of the input `vector`. Each of these elements has as many elements as necessary for the split, in this case either two (a start and stop year) or one (when the president served less than one year).

```
> library(stringr)
> # split the string
> yearList <- str_split(string=presidents$YEAR, pattern="-")
> head(yearList)

[[1]]
[1] "1789" "1797"

[[2]]
[1] "1797" "1801"

[[3]]
[1] "1801" "1805"

[[4]]
[1] "1805" "1809"

[[5]]
[1] "1809" "1812"

[[6]]
[1] "1812" "1813"
```

```
> # combine them into one matrix
> yearMatrix <- data.frame(Reduce(rbind, yearList))
> head(yearMatrix)

    X1   X2
1 1789 1797
2 1797 1801
3 1801 1805
4 1805 1809
5 1809 1812
6 1812 1813

> # give the columns good names
> names(yearMatrix) <- c("Start", "Stop")
> # bind the new columns onto the data.frame
> presidents <- cbind(presidents, yearMatrix)
> # convert the start and stop columns into numeric
> presidents$Start <- as.numeric(as.character(presidents$Start))
> presidents$Stop <- as.numeric(as.character(presidents$Stop))
> # view the changes
> head(presidents)

       YEAR         PRESIDENT
1 1789-1797 George Washington
2 1797-1801        John Adams
3 1801-1805  Thomas Jefferson
4 1805-1809  Thomas Jefferson
5 1809-1812     James Madison
6 1812-1813     James Madison
                                          FIRST LADY   VICE PRESIDENT
1                                  Martha Washington       John Adams
2                                      Abigail Adams Thomas Jefferson
3 Martha Wayles Skelton Jefferson\n    (no image)       Aaron Burr
4 Martha Wayles Skelton Jefferson\n    (no image)   George Clinton
5                                     Dolley Madison   George Clinton
6                                     Dolley Madison    office vacant
  Start Stop
1  1789 1797
2  1797 1801
3  1801 1805
4  1805 1809
5  1809 1812
6  1812 1813

> tail(presidents)

        YEAR      PRESIDENT                FIRST LADY    VICE PRESIDENT
59 1977-1981   Jimmy Carter            Rosalynn Carter Walter F. Mondale
60 1981-1989 Ronald Reagan               Nancy Reagan       George Bush
61 1989-1993    George Bush              Barbara Bush        Dan Quayle
62 1993-2001   Bill Clinton Hillary Rodham Clinton         Albert Gore
```

```
63 2001-2009 George W. Bush              Laura Bush    Richard Cheney
64      2009-   Barack Obama         Michelle Obama   Joseph R. Biden
   Start Stop
59  1977 1981
60  1981 1989
61  1989 1993
62  1993 2001
63  2001 2009
64  2009   NA
```

In the preceding example there was a quirk of R that can be frustrating at first pass. In order to convert the factor `presidents$Start` into a `numeric`, we first had to convert it into a `character`. That is because `factors` are simply labels on top of integers, as seen in Section 4.4.2. So when applying **as.numeric** to a `factor`, it is converted to the underlying integers.

Just like in Excel, it is possible to select specified characters from text using **str_sub**.

```
> # get the first 3 characters
> str_sub(string=presidents$PRESIDENT, start=1, end=3)

 [1] "Geo" "Joh" "Tho" "Tho" "Jam" "Jam" "Jam" "Jam" "Jam" "Joh" "And"
[12] "And" "Mar" "Wil" "Joh" "Jam" "Zac" "Mil" "Fra" "Fra" "Jam" "Abr"
[23] "Abr" "And" "Uly" "Uly" "Uly" "Rut" "Jam" "Che" "Gro" "Gro" "Ben"
[34] "Gro" "Wil" "Wil" "Wil" "The" "The" "Wil" "Wil" "Woo" "War" "Cal"
[45] "Cal" "Her" "Fra" "Fra" "Fra" "Har" "Har" "Dwi" "Joh" "Lyn" "Lyn"
[56] "Ric" "Ric" "Ger" "Jim" "Ron" "Geo" "Bil" "Geo" "Bar"

> # get the 4rd through 8th characters
> str_sub(string=presidents$PRESIDENT, start=4, end=8)

 [1] "rge W" "n Ada" "mas J" "mas J" "es Ma" "es Ma" "es Ma" "es Ma"
 [9] "es Mo" "n Qui" "rew J" "rew J" "tin V" "liam " "n Tyl" "es K."
[17] "hary " "lard " "nklin" "nklin" "es Bu" "aham " "aham " "rew J"
[25] "sses " "sses " "sses " "herfo" "es A." "ster " "ver C" "ver C"
[33] "jamin" "ver C" "liam " "liam " "liam " "odore" "odore" "liam "
[41] "liam " "drow " "ren G" "vin C" "vin C" "bert " "nklin" "nklin"
[49] "nklin" "ry S." "ry S." "ght D" "n F. " "don B" "don B" "hard "
[57] "hard " "ald R" "my Ca" "ald R" "rge B" "l Cli" "rge W" "ack O"
```

This is good for finding a president whose term started in a year ending in 1, which means he got elected in a year ending in 0, a preponderance of which ones died in office.

```
> presidents[str_sub(string=presidents$Start, start=4, end=4) == 1,
+            c("YEAR", "PRESIDENT", "Start", "Stop")]

        YEAR                PRESIDENT Start Stop
3  1801-1805         Thomas Jefferson  1801 1805
14      1841 William Henry Harrison  1841 1841
15 1841-1845               John Tyler  1841 1845
22 1861-1865          Abraham Lincoln  1861 1865
29      1881        James A. Garfield  1881 1881
```

```
30 1881-1885        Chester A. Arthur  1881 1885
37        1901      William McKinley   1901 1901
38 1901-1905     Theodore Roosevelt   1901 1905
43 1921-1923        Warren G. Harding  1921 1923
48 1941-1945 Franklin D. Roosevelt   1941 1945
53 1961-1963          John F. Kennedy  1961 1963
60 1981-1989           Ronald Reagan  1981 1989
63 2001-2009          George W. Bush  2001 2009
```

16.4 Regular Expressions

Sifting through text often requires searching for patterns, and usually these patterns have to be general and flexible. This is where regular expressions are very useful. We will not make an exhaustive lesson of regular expressions but will illustrate how to use them within R.

Let's say we want to find any president with "John" in his name, either first or last. Since we do not know where in the name "John" would occur, we cannot simply use **str_sub**. Instead we use **str_detect**.

```
> # returns TRUE/FALSE if John was found in the name
> johnPos <- str_detect(string=presidents$PRESIDENT, pattern="John")
> presidents[johnPos, c("YEAR", "PRESIDENT", "Start", "Stop")]

        YEAR           PRESIDENT Start Stop
2  1797-1801          John Adams  1797 1801
10 1825-1829 John Quincy Adams  1825 1829
15 1841-1845          John Tyler  1841 1845
24 1865-1869      Andrew Johnson  1865 1869
53 1961-1963     John F. Kennedy  1961 1963
54 1963-1965 Lyndon B. Johnson  1963 1965
55 1963-1969 Lyndon B. Johnson  1963 1969
```

This found John Adams, John Quincy Adams, John Tyler, Andrew Johnson, John F. Kennedy and Lyndon B. Johnson. Note that regular expressions are case sensitive, so to ignore case we have to put the pattern in **ignore.case**.

```
> badSearch <- str_detect(presidents$PRESIDENT, "john")
> goodSearch <- str_detect(presidents$PRESIDENT, ignore.case("John"))
> sum(badSearch)

[1] 0

> sum(goodSearch)

[1] 7
```

To show off some more interesting regular expressions we will make use of yet another table from Wikipedia, the list of United States Wars. Because we only care about one column, which has some encoding issues, we put an `RData` file of just that one column at `http://www.jaredlander.com/data/warTimes.rdata`. We load that file using **load**, and we then see a new object in our session named `warTimes`.

For some odd reason, loading `RData` files from a URL is not as straightforward as reading in a CSV file from a URL. A connection must first be made using **url**; then that connection is loaded with **load**, and then the connection must be closed with **close**.

```
> con <- url("http://www.jaredlander.com/data/warTimes.rdata")
> load(con)
> close(con)
```

This `vector` holds the starting and stopping dates of the wars. Sometimes it has just years; sometimes it also includes months and possibly days. There are instances where it has only one year. Because of this, it is a good dataset to comb through with various text functions. The first few entries follow.

```
> head(warTimes, 10)

 [1] "September 1, 1774 ACAEA September 3, 1783"
 [2] "September 1, 1774 ACAEA March 17, 1776"
 [3] "1775ACAEA1783"
 [4] "June 1775 ACAEA October 1776"
 [5] "July 1776 ACAEA March 1777"
 [6] "June 14, 1777 ACAEA October 17, 1777"
 [7] "1777ACAEA1778"
 [8] "1775ACAEA1782"
 [9] "1776ACAEA1794"
[10] "1778ACAEA1782"
```

We want to create a new column that contains information for the start of the war. To get at this information we need to split the `Time` column. Thanks to Wikipedia's encoding, the separator is generally "ACAEA", which was originally "Ã¢Â€Â ' '" and converted to these characters to make life easier. There are two instances where the "-" appears, once as a separator and once to make a hyphenated word. This is seen in the following code.

```
> warTimes[str_detect(string=warTimes, pattern="-")]

[1] "6 June 1944 ACAEA mid-July 1944"
[2] "25 August-17 December 1944"
```

So when we are splitting our string, we need to search for either "ACAEA" or "-". In **str_split** the `pattern` argument can take a regular expression. In this case it will be "(ACAEA)|-", which tells the engine to search for either "(ACAEA)" or (denoted by the vertical pipe) "-" in the string. To avoid the instance seen before, where the hyphen is used in "mid-July," we set the argument `n` to 2 so it returns at most only two pieces for each element of the input `vector`. The parentheses are not matched but rather act to group the characters "ACAEA" in the search.[1] This grouping capability will prove important for advanced replacement of text, which will be demonstrated later in this section.

1. To match parentheses, they should be prefixed with a backslash (\).

```
> theTimes <- str_split(string=warTimes, pattern="(ACAEA)|-", n=2)
> head(theTimes)

[[1]]
[1] "September 1, 1774 " " September 3, 1783"

[[2]]
[1] "September 1, 1774 " " March 17, 1776"

[[3]]
[1] "1775" "1783"

[[4]]
[1] "June 1775 "      " October 1776"

[[5]]
[1] "July 1776 "   " March 1777"

[[6]]
[1] "June 14, 1777 "     " October 17, 1777"
```

Seeing that this worked for the first few entries, we also check on the two instances where a hyphen was the separator.

```
> which(str_detect(string=warTimes, pattern="-"))

[1] 147 150

> theTimes[[147]]

[1] "6 June 1944 "     " mid-July 1944"

> theTimes[[150]]

[1] "25 August"           "17 December 1944"
```

This looks correct, as the first entry shows "mid-July" still intact while the second entry shows the two dates split apart.

For our purposes we only care about the start date of the wars, so we need to build a function that extracts the first (in some cases only) element of each `vector` in the `list`.

```
> theStart <- sapply(theTimes, FUN=function(x) x[1])
> head(theStart)

[1] "September 1, 1774 " "September 1, 1774 " "1775"
[4] "June 1775 "          "July 1776 "          "June 14, 1777 "
```

The original text sometimes had spaces around the separators and sometimes did not, meaning that some of our text has trailing white spaces. The easiest way to get rid of them is with the **str_trim** function.

```
> theStart <- str_trim(theStart)
> head(theStart)

[1] "September 1, 1774" "September 1, 1774" "1775"
[4] "June 1775"         "July 1776"         "June 14, 1777"
```

To extract the word "January" wherever it might occur, use **str_extract**. In places where it is not found will be `NA`.

```
> # pull out "January" anywhere it's found, otherwise return NA
> str_extract(string=theStart, pattern="January")

  [1] NA        NA        NA        NA        NA        NA
  [7] NA        NA        NA        NA        NA        NA
 [13] "January" NA        NA        NA        NA        NA
 [19] NA        NA        NA        NA        NA        NA
 [25] NA        NA        NA        NA        NA        NA
 [31] NA        NA        NA        NA        NA        NA
 [37] NA        NA        NA        NA        NA        NA
 [43] NA        NA        NA        NA        NA        NA
 [49] NA        NA        NA        NA        NA        NA
 [55] NA        NA        NA        NA        NA        NA
 [61] NA        NA        NA        NA        NA        NA
 [67] NA        NA        NA        NA        NA        NA
 [73] NA        NA        NA        NA        NA        NA
 [79] NA        NA        NA        NA        NA        NA
 [85] NA        NA        NA        NA        NA        NA
 [91] NA        NA        NA        NA        NA        NA
 [97] NA        NA        "January" NA        NA        NA
[103] NA        NA        NA        NA        NA        NA
[109] NA        NA        NA        NA        NA        NA
[115] NA        NA        NA        NA        NA        NA
[121] NA        NA        NA        NA        NA        NA
[127] NA        NA        NA        NA        "January" NA
[133] NA        NA        "January" NA        NA        NA
[139] NA        NA        NA        NA        NA        NA
[145] "January" "January" NA        NA        NA        NA
[151] NA        NA        NA        NA        NA        NA
[157] NA        NA        NA        NA        NA        NA
[163] NA        NA        NA        NA        NA        NA
[169] "January" NA        NA        NA        NA        NA
[175] NA        NA        NA        NA        NA        NA
[181] "January" NA        NA        NA        NA        "January"
[187] NA        NA
```

Contrarily, to find elements that contain "January" and return the entire entry—not just "January"—use **str_detect** and subset `theStart` with the results.

```
> # just return elements where "January" was detected
> theStart[str_detect(string=theStart, pattern="January")]
```

```
[1] "January"          "January 21"         "January 1942"
[4] "January"          "January 22, 1944" "22 January 1944"
[7] "January 4, 1989"  "15 January 2002"   "January 14, 2010"
```

To extract the year, we search for an occurrence of four numbers together. Because we do not know specific numbers, we have to use a pattern. In a regular expression search, "[0-9]" searches for any number. We use "[0-9][0-9][0-9][0-9]" to search for four consecutive numbers.

```
> # get incidents of 4 numeric digits in a row
> head(str_extract(string=theStart, "[0-9][0-9][0-9][0-9]"), 20)
```

```
 [1] "1774" "1774" "1775" "1775" "1776" "1777" "1777" "1775" "1776"
[10] "1778" "1775" "1779" NA     "1785" "1798" "1801" NA     "1812"
[19] "1812" "1813"
```

Writing "[0-9]" repeatedly is inefficient, especially when searching for many occurences of a number. Putting "4" after "[0-9]" causes the engine to search for any set of four numbers.

```
> # a smarter way to search for four numbers
> head(str_extract(string=theStart, "[0-9]{4}"), 20)
```

```
 [1] "1774" "1774" "1775" "1775" "1776" "1777" "1777" "1775" "1776"
[10] "1778" "1775" "1779" NA     "1785" "1798" "1801" NA     "1812"
[19] "1812" "1813"
```

Even writing "[0-9]" can be inefficient, so there is a shortcut to denote any integer. In most other languages the shortcut is "\d" but in R there needs to be two backslashes ("\\d").

```
> # "\\d" is a shortcut for "[0-9]"
> head(str_extract(string=theStart, "\\d{4}"), 20)
```

```
 [1] "1774" "1774" "1775" "1775" "1776" "1777" "1777" "1775" "1776"
[10] "1778" "1775" "1779" NA     "1785" "1798" "1801" NA     "1812"
[19] "1812" "1813"
```

The curly braces offer even more functionality, for instance, searching for a number one to three times.

```
> # this looks for any digit that occurs either once, twice or thrice
> str_extract(string=theStart, "\\d{1,3}")
```

```
 [1] "1"   "1"   "177" "177" "177" "14"  "177" "177" "177" "177"
[11] "177" "177" NA    "178" "179" "180" NA    "18"  "181" "181"
[21] "181" "181" "181" "181" "181" "181" "181" "181" "181" "181"
[31] "22"  "181" "181" "5"   "182" "182" "182" NA    "6"   "183"
[41] "23"  "183" "19"  "11"  "25"  "184" "184" "184" "184" "184"
[51] "185" "184" "28"  "185" "13"  "4"   "185" "185" "185" "185"
[61] "185" "185" "6"   "185" "6"   "186" "12"  "186" "186" "186"
```

```
 [71] "186" "186" "17"  "31"  "186" "20"  "186" "186" "186" "186"
 [81] "186" "17"  "1"   "6"   "12"  "27"  "187" "187" "187" "187"
 [91] "187" "187" NA    "30"  "188" "189" "22"  "189" "21"  "189"
[101] "25"  "189" "189" "189" "189" "189" "189" "2"   "189" "28"
[111] "191" "21"  "28"  "191" "191" "191" "191" "191" "191" "191"
[121] "191" "191" "191" "7"   "194" "194" NA    NA    "3"   "7"
[131] "194" "194" NA    "20"  NA    "1"   "16"  "194" "8"   "194"
[141] "17"  "9"   "194" "3"   "22"  "22"  "6"   "6"   "15"  "25"
[151] "25"  "16"  "8"   "6"   "194" "195" "195" "195" "195" "197"
[161] "28"  "25"  "15"  "24"  "19"  "198" "15"  "198" "4"   "20"
[171] "2"   "199" "199" "199" "19"  "20"  "24"  "7"   "7"   "7"
[181] "15"  "7"   "6"   "20"  "16"  "14"  "200" "19"
```

Regular expressions can search for text with anchors indicating the beginning of a line ("^") and the end of a line ("$").

```
> # extract 4 digits at the beginning of the text
> head(str_extract(string=theStart, pattern="^\\d{4}"), 30)

 [1] NA     NA     "1775" NA     NA     NA     "1777" "1775" "1776"
[10] "1778" "1775" "1779" NA     "1785" "1798" "1801" NA     NA
[19] "1812" "1813" "1812" "1812" "1813" "1813" "1813" "1814" "1813"
[28] "1814" "1813" "1815"

> # extract 4 digits at the end of the text
> head(str_extract(string=theStart, pattern="\\d{4}$"), 30)

 [1] "1774" "1774" "1775" "1775" "1776" "1777" "1777" "1775" "1776"
[10] "1778" "1775" "1779" NA     "1785" "1798" "1801" NA     "1812"
[19] "1812" "1813" "1812" "1812" "1813" "1813" "1813" "1814" "1813"
[28] "1814" "1813" "1815"

> # extract 4 digits at the beginning AND the end of the text
> head(str_extract(string=theStart, pattern="^\\d{4}$"), 30)

 [1] NA     NA     "1775" NA     NA     NA     "1777" "1775" "1776"
[10] "1778" "1775" "1779" NA     "1785" "1798" "1801" NA     NA
[19] "1812" "1813" "1812" "1812" "1813" "1813" "1813" "1814" "1813"
[28] "1814" "1813" "1815"
```

Replacing text selectively is another powerful feature of regular expressions. We start by simply replacing numbers with a fixed value.

```
> # replace the first digit seen with "x"
> head(str_replace(string=theStart, pattern="\\d", replacement="x"), 30)

 [1] "September x, 1774" "September x, 1774" "x775"
 [4] "June x775"         "July x776"         "June x4, 1777"
 [7] "x777"              "x775"              "x776"
[10] "x778"              "x775"              "x779"
[13] "January"           "x785"              "x798"
```

```
[16] "x801"              "August"            "June x8, 1812"
[19] "x812"              "x813"              "x812"
[22] "x812"              "x813"              "x813"
[25] "x813"              "x814"              "x813"
[28] "x814"              "x813"              "x815"

> # replace all digits seen with "x"
> # this means "7" -> "x" and "382" -> "xxx"
> head(str_replace_all(string=theStart, pattern="\\d", replacement="x"),
+      30)

 [1] "September x, xxxx" "September x, xxxx" "xxxx"
 [4] "June xxxx"         "July xxxx"         "June xx, xxxx"
 [7] "xxxx"              "xxxx"              "xxxx"
[10] "xxxx"              "xxxx"              "xxxx"
[13] "January"           "xxxx"              "xxxx"
[16] "xxxx"              "August"            "June xx, xxxx"
[19] "xxxx"              "xxxx"              "xxxx"
[22] "xxxx"              "xxxx"              "xxxx"
[25] "xxxx"              "xxxx"              "xxxx"
[28] "xxxx"              "xxxx"              "xxxx"

> # replace any strings of digits from 1 to 4 in length with "x"
> # this means "7" -> "x" and "382" -> "x"
> head(str_replace_all(string=theStart, pattern="\\d{1,4}",
+                      replacement="x"), 30)

 [1] "September x, x" "September x, x" "x"
 [4] "June x"         "July x"         "June x, x"
 [7] "x"              "x"              "x"
[10] "x"              "x"              "x"
[13] "January"        "x"              "x"
[16] "x"              "August"         "June x, x"
[19] "x"              "x"              "x"
[22] "x"              "x"              "x"
[25] "x"              "x"              "x"
[28] "x"              "x"              "x"
```

Not only can regular expressions substitute fixed values into a string, they can also substitute part of the search pattern. To see this, we create a vector of some HTML commands.

```
> # create a vector of HTML commands
> commands <- c("<a href=index.html>The Link is here</a>",
+               "<b>This is bold text</b>")
```

Now we would like to extract the text between the HTML tags. The pattern is a set of opening and closing angle brackets with something in between ("<.+?>"), some text (".+?") and another set of opening and closing brackets ("<.+?>"). The "." indicates a search for anything, while the "+" means to search for it one or more times with the "?", meaning it is not a greedy search. Because we do not know what the text between the tags will be, and that is what we want to substitute back into the text, we group it inside parentheses and use a back reference to reinsert it using "\\1", which indicates use of the first grouping. Subsequent groupings are referenced using subsequent numerals, up to nine. In other languages a "$" is used instead of "\\."

```
> # get the text between the HTML tags
> # the content in (.+?) is substituted using \\1
> str_replace(string=commands, pattern="<.+?>(.+?)<.+>",
+             replacement="\\1")

[1] "The Link is here"  "This is bold text"
```

Since R has its own regular expression peculiarities, there is a handy help file that can be accessed with `?regex`.

16.5 Conclusion

R has many facilities for dealing with text, whether creating, extracting or manipulating it. For creating text, it is best to use **sprintf** and if necessary **paste**. For all other text needs, it is best to use Hadley Wickham's **stringr** package. This includes pulling out text specified by character position (**str_sub**), regular expressions (**str_detect**, **str_extract** and **str_replace**) and splitting strings (**str_split**).

17

Probability Distributions

Being a statistical programming language, R easily handles all the basic necessities of statistics, including drawing random numbers and calculating distribution values (the focus of this chapter), means, variances, maxmima and minima, correlation and t-tests (the focus of Chapter 18).

Probability distributions lie at the heart of statistics, so naturally R provides numerous functions for making use of them. These include functions for generating random numbers and calculating the distribution and quantile.

17.1 Normal Distribution

Perhaps the most famous, and most used, statistical distribution is the normal distribution, sometimes referred to as the Gaussian distribution, which is defined as

$$f(x; \mu, \sigma) = \frac{1}{\sqrt{2\pi}\sigma} e^{(-(x-\mu)^2)/(2\sigma^2)} \tag{17.1}$$

where μ is the mean and σ the standard deviation. This is the famous bell curve that describes so many phenomena in life. To draw random numbers from the normal distribution use the **rnorm** function, which optionally allows the specification of the mean and standard deviation.

```
> # 10 draws from the standard 0-1 normal distribution
> rnorm(n=10)

 [1]  0.4385627  1.1969098  1.0130680  0.0053413 -0.6086422 -1.5829601
 [7]  0.9106169 -1.9663997  1.0108341  0.1931879

> # 10 draws from the 100-20 distribution
> rnorm(n=10, mean=100, sd=20)

 [1] 114.99418 121.15465  95.35524  95.73121  86.45346 106.73548
 [7] 104.05061 113.61679 101.40346  61.48190
```

The density (the probability of a particular value) for the normal distribution is calculated using **dnorm**.

```
> randNorm10 <- rnorm(10)
> randNorm10

 [1]  1.9125749 -0.5822831  0.5553026 -2.3583206  0.7638454  1.1312883
 [7] -0.1721544  1.8832073  0.5361347 -1.2932703

> dnorm(randNorm10)

 [1] 0.06406161 0.33673288 0.34194033 0.02472905 0.29799802 0.21037889
 [7] 0.39307411 0.06773357 0.34553589 0.17287050

> dnorm(c(-1, 0, 1))

[1] 0.2419707 0.3989423 0.2419707
```

dnorm returns the probability of a specific number occuring. While it is technically mathematically impossible to find the exact probability of a number from a continuous distribution, this is an estimate of the probability. Like with **rnorm**, a mean and standard deviation can be specified for **dnorm**.

To see this visually we generate a number of normal random variables, calculate their distributions and then plot them. This should result in a nicely shaped bell curve, as seen in Figure 17.1.

```
> # generate the normal variables
> randNorm <- rnorm(30000)
> # calcualte their distributions
> randDensity <- dnorm(randNorm)
> # load ggplot2
> library(ggplot2)
> # plot them
> ggplot(data.frame(x=randNorm, y=randDensity)) + aes(x=x, y=y) +
+     geom_point() + labs(x="Random Normal Variables", y="Density")
```

Similarly, **pnorm** calculates the distribution of the normal distribution, that is, the cumulative probability that a given number, or smaller number, occurs. This is defined as

$$\Phi(a) = P\{X <= a\} = \int_{-\infty}^{a} \frac{1}{\sqrt{2\pi}\sigma} e^{\frac{-(x-\mu)^2}{2\sigma^2}} \mathrm{d}x \tag{17.2}$$

```
> pnorm(randNorm10)

 [1] 0.972098753 0.280188016 0.710656152 0.009178915 0.777520317
 [6] 0.871033114 0.431658071 0.970163858 0.704067283 0.097958799

> pnorm(c(-3, 0, 3))

[1] 0.001349898 0.500000000 0.998650102

> pnorm(-1)

[1] 0.1586553
```

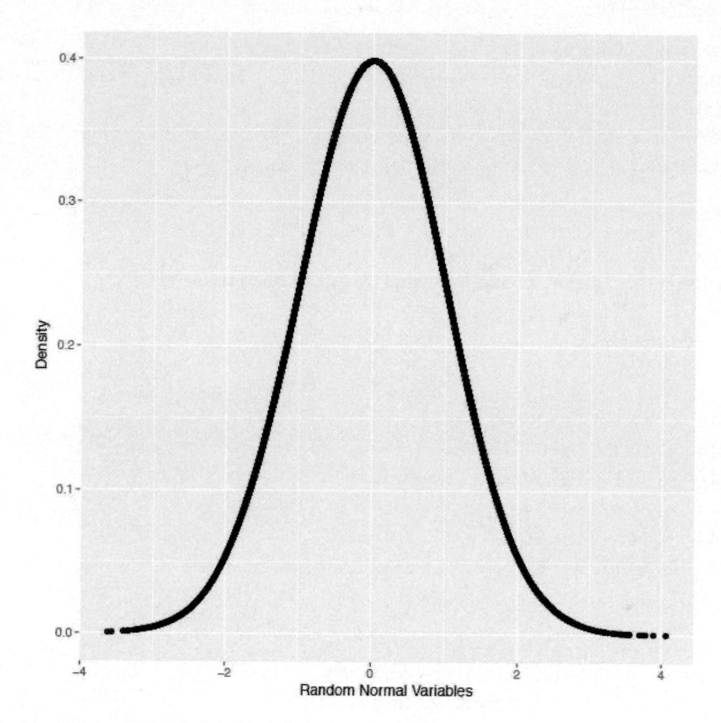

Figure 17.1 Plot of random normal variables and their densities, which results in a bell curve.

By default this is left-tailed. To find the probability that the variable falls between two points, we must calculate the two probabilities and subtract them from each other.

```
> pnorm(1) - pnorm(0)

[1] 0.3413447

> pnorm(1) - pnorm(-1)

[1] 0.6826895
```

This probability is represented by the area under the curve and illustrated in Figure 17.2, which is drawn by the following code.

```
> # a few things happen with this first line of code
> # the idea is to build a ggplot2 object that we can build upon later
> # that is why it is saved to p
> # we take randNorm and randDensity and put them into a data.frame
> # we declare the x and y aes outside of any other function
> # this just gives more flexibility
> # we add lines with geom_line()
> # x- and y-axis labels with labs(x="x", y="Density")
> p <- ggplot(data.frame(x=randNorm, y=randDensity)) + aes(x=x, y=y) +
+     geom_line() + labs(x="x", y="Density")
>
```

```
> # plotting p will print a nice distribution
> # to create a shaded area under the curve we first calcu-
late that area
> # generate a sequence of numbers going from the far left to -1
> neg1Seq <- seq(from=min(randNorm), to=-1, by=.1)
>
> # build a data.frame of that sequence as x
> # the distribution values for that sequence as y
> lessThanNeg1 <- data.frame(x=neg1Seq, y=dnorm(neg1Seq))
>
> head(lessThanNeg1)

          x            y
1 -4.164144 6.847894e-05
2 -4.064144 1.033313e-04
3 -3.964144 1.543704e-04
4 -3.864144 2.283248e-04
5 -3.764144 3.343484e-04
6 -3.664144 4.847329e-04

> # combine this with endpoints at the far left and far right
> # the height is 0
> lessThanNeg1 <- rbind(c(min(randNorm), 0),
+                       lessThanNeg1,
+                       c(max(lessThanNeg1$x), 0))
>
> # use that shaded region as a polygon
> p + geom_polygon(data=lessThanNeg1, aes(x=x, y=y))
>
> # create a similar sequence going from -1 to 1
> neg1Pos1Seq <- seq(from=-1, to=1, by=.1)
>
> # build a data.frame of that sequence as x
> # the distribution values for that sequence as y
> neg1To1 <- data.frame(x=neg1Pos1Seq, y=dnorm(neg1Pos1Seq))
>
> head(neg1To1)

     x         y
1 -1.0 0.2419707
2 -0.9 0.2660852
3 -0.8 0.2896916
4 -0.7 0.3122539
5 -0.6 0.3332246
6 -0.5 0.3520653

> # combine this with endpoints at the far left and far right
> # the height is 0
> neg1To1 <- rbind(c(min(neg1To1$x), 0),
+                  neg1To1,
+                  c(max(neg1To1$x), 0))
```

```
>
> # use that shaded region as a polygon
> p + geom_polygon(data=neg1To1, aes(x=x, y=y))
```

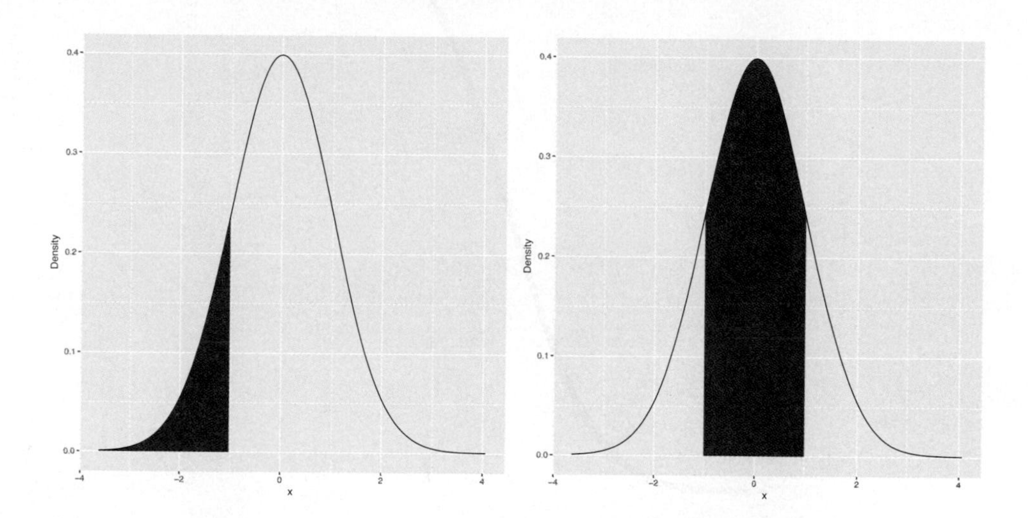

Figure 17.2 Area under a normal curve. The plot on the left shows the area to the left of -1, while the plot on the right shows the area between -1 and 1.

The distribution has a non-decreasing shape, as shown in Figure 17.3. The information displayed here is the same as in Figure 17.2 but it is shown differently. Instead of the cumulative probability being shown as a shaded region, it is displayed as a single point along the y-axis.

```
> randProb <- pnorm(randNorm)
> ggplot(data.frame(x=randNorm, y=randProb)) + aes(x=x, y=y) +
+     geom_point() + labs(x="Random Normal Variables", y="Probability")
```

The opposite of **pnorm** is **qnorm**. Given a cumulative probability it returns the quantile.

```
> randNorm10

 [1]  1.9125749 -0.5822831  0.5553026 -2.3583206  0.7638454  1.1312883
 [7] -0.1721544  1.8832073  0.5361347 -1.2932703

> qnorm(pnorm(randNorm10))

 [1]  1.9125749 -0.5822831  0.5553026 -2.3583206  0.7638454  1.1312883
 [7] -0.1721544  1.8832073  0.5361347 -1.2932703

> all.equal(randNorm10, qnorm(pnorm(randNorm10)))

[1] TRUE
```

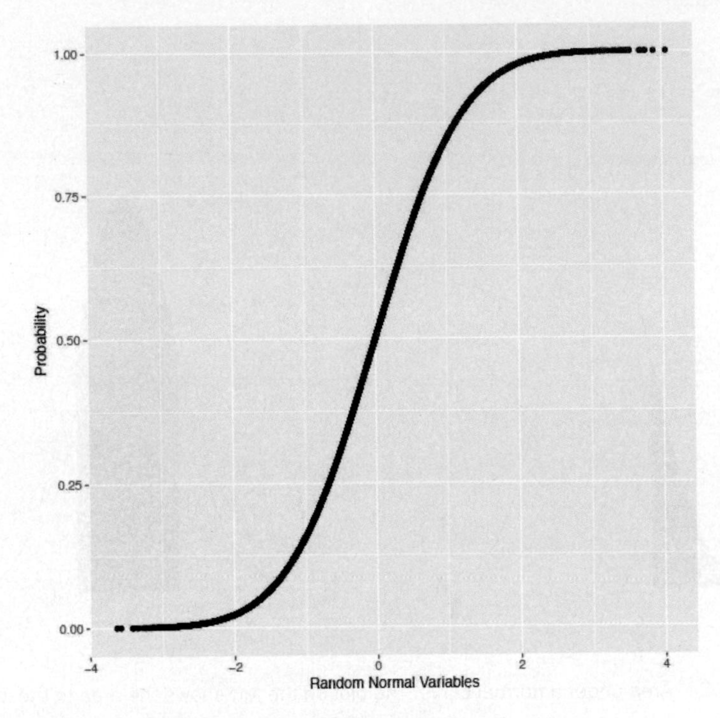

Figure 17.3 Normal distribution function.

17.2 Binomial Distribution

Like the normal distribution, the binomial distribution is well represented in R. Its probability mass function is

$$p(x; n, p) = \binom{n}{x} p^x (1-p)^{n-x} \tag{17.3}$$

where

$$\binom{n}{x} = \frac{n!}{x!(n-x)!} \tag{17.4}$$

and n is the number of trials and p is the probability of success of a trial. The mean is np and the variance is $np(1-p)$. When $n = 1$ this reduces to the Bernoulli distribution.

Generating random numbers from the binomial distribution is not simply generating random numbers but rather generating the number of successes of independent trials. To simulate the number of successes out of ten trials with probability 0.4 of success, we run **rbinom** with `n=1` (only one run of the trials), `size=10` (trial size of 10) and `prob=0.4` (probability of success is 0.4).

```
> rbinom(n=1, size=10, prob=.4)

[1] 1
```

That is to say that ten trials were conducted, each with 0.4 probability of success, and the number generated is the number that succeeded. As this is random, different numbers will be generated each time.

By setting `n` to anything greater than 1, R will generate the number of successes for each of the `n` sets of `size` trials.

```
> rbinom(n=1, size=10, prob=.4)

[1] 3

> rbinom(n=5, size=10, prob=.4)

[1] 4 3 5 2 5

> rbinom(n=10, size=10, prob=.4)

 [1] 5 2 7 4 7 3 2 3 3 3
```

Setting `size` to 1 turns the numbers into a Bernoulli random variable, which can take on only the value 1 (success) or 0 (failure).

```
> rbinom(n=1, size=1, prob=.4)

[1] 0

> rbinom(n=5, size=1, prob=.4)

[1] 1 1 0 0 0

> rbinom(n=10, size=1, prob=.4)

 [1] 0 0 0 0 0 0 0 1 0 0
```

To visualize the binomial distribution, we randomly generate 10,000 experiments, each with 10 trials and 0.3 probability of success. This is seen in Figure 17.4, which shows that the most common number of successes is 3, as expected.

```
> binomData <- data.frame(Successes=rbinom(n=10000, size=10, prob=.3))
> ggplot(binomData, aes(x=Successes)) + geom_histogram(binwidth=1)
```

To see how the binomial distribution is well approximated by the normal distribution as the number of trials grows large, we run similar experiments with differing numbers of trials and graph the results, as shown in Figure 17.5.

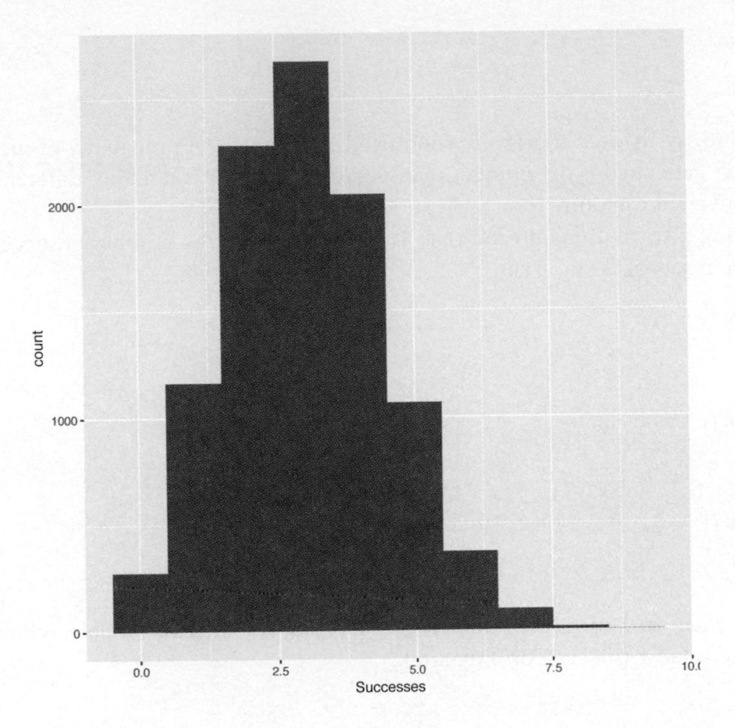

Figure 17.4 Ten thousand runs of binomial experiments with ten trials each and probability of success of 0.3.

```
> # create a data.frame with Successes being the 10,000 random draws
> # Size equals 5 for all 10,000 rows

> binom5 <- data.frame(Successes=rbinom(n=10000, size=5, prob=.3), Size=5)
> dim(binom5)

[1] 10000     2

> head(binom5)

  Successes Size
1         2    5
2         1    5
3         2    5
4         1    5
5         2    5
6         2    5

> # similar as before, still 10,000 rows
> # numbers are drawn from a distribution with a different size
> # Size now equals 10 for all 10,000 rows
> binom10 <- data.frame(Successes=rbinom(n=10000, size=10, prob=.3), Size=10)
> dim(binom10)
```

```
[1] 10000     2

> head(binom10)

  Successes Size
1         2   10
2         2   10
3         1   10
4         2   10
5         4   10
6         1   10

> binom100 <- data.frame(Successes=rbinom(n=10000, size=100, prob=.3), Size=100)
>
> binom1000 <- data.frame(Successes=rbinom(n=10000, size=1000, prob=.3), Size=1000)
>
> # combine them all into one data.frame
> binomAll <- rbind(binom5, binom10, binom100, binom1000)
> dim(binomAll)

[1] 40000     2

> head(binomAll, 10)

   Successes Size
1          2    5
2          1    5
3          2    5
4          1    5
5          2    5
6          2    5
7          1    5
8          1    5
9          2    5
10         1    5

> tail(binomAll, 10)

      Successes Size
39991       288 1000
39992       289 1000
39993       297 1000
39994       327 1000
39995       336 1000
39996       290 1000
39997       310 1000
39998       328 1000
39999       281 1000
40000       307 1000
```

```
> # build the plot
> # histograms only need an x aesthetic
> # it is faceted (broken up) based on the values of Size
> # these are 5, 10, 100, 1000
> ggplot(binomAll, aes(x=Successes)) + geom_histogram() +
+     facet_wrap(~ Size, scales="free")
```

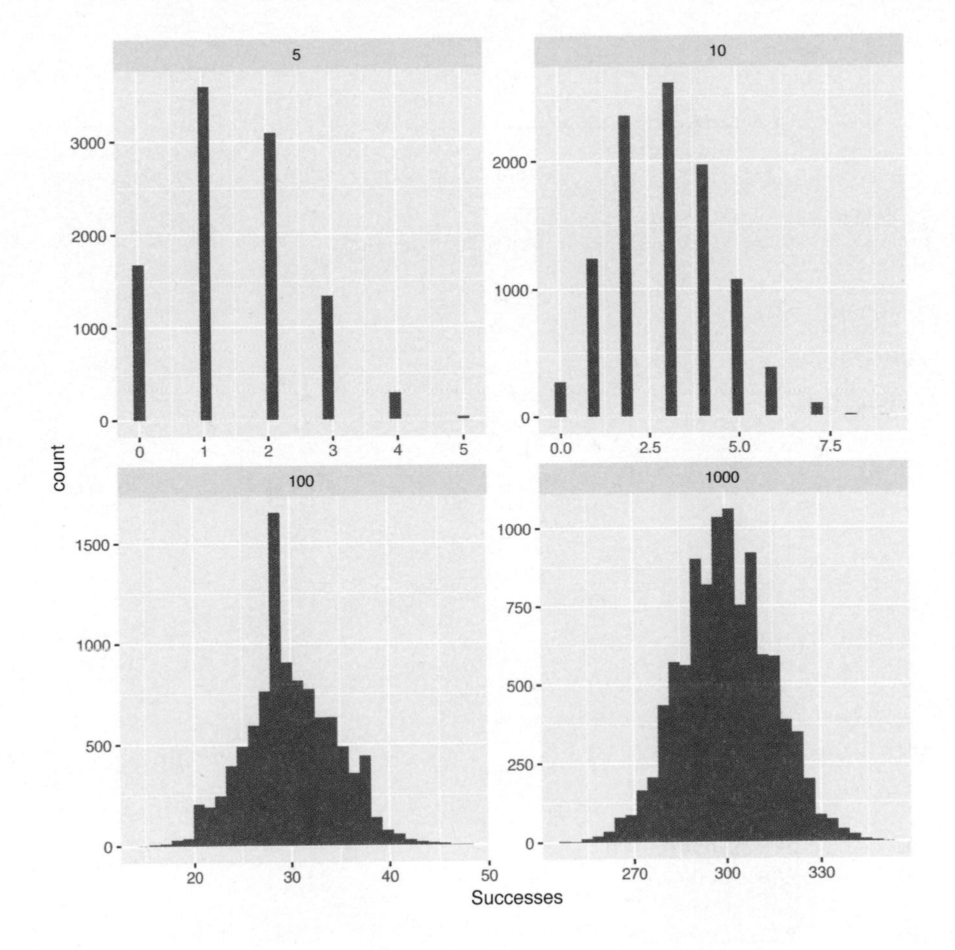

Figure 17.5 Random binomial histograms faceted by trial size. Notice that while not perfect, as the number of trials increases the distribution appears more normal. Also note the differing scales in each pane.

The cumulative distribution function is

$$F(a; n, p) = P\{X <= a\} = \sum_{i=0}^{a} \binom{n}{i} p^i (1-p)^{n-i} \tag{17.5}$$

where n and p are the number of trials and the probability of success, respectively, as before.

Similar to the normal distribution functions, **dbinom** and **pbinom** provide the density (probability of an exact value) and distribution (cumulative probability), respectively, for the binomial distribution.

```
> # probability of 3 successes out of 10
> dbinom(x=3, size=10, prob=.3)

[1] 0.2668279

> # probability of 3 or fewer successes out of 10
> pbinom(q=3, size=10, prob=.3)

[1] 0.6496107

> # both functions can be vectorized
> dbinom(x=1:10, size=10, prob=.3)

 [1] 0.1210608210 0.2334744405 0.2668279320 0.2001209490 0.1029193452
 [6] 0.0367569090 0.0090016920 0.0014467005 0.0001377810 0.0000059049

> pbinom(q=1:10, size=10, prob=.3)

 [1] 0.1493083 0.3827828 0.6496107 0.8497317 0.9526510 0.9894079
 [7] 0.9984096 0.9998563 0.9999941 1.0000000
```

Given a certain probability, **qbinom** returns the quantile, which for this distribution is the number of successes.

```
> qbinom(p=.3, size=10, prob=.3)

[1] 2

> qbinom(p=c(.3, .35, .4, .5, .6), size=10, prob=.3)

[1] 2 2 3 3 3
```

17.3 Poisson Distribution

Another popular distribution is the Poisson distribution, which is for count data. Its probability mass function is

$$p(x; \lambda) = \frac{\lambda^x e^{-\lambda}}{x!} \tag{17.6}$$

and the cumulative distribution is

$$F(a; \lambda) = P\{X <= a\} = \sum_{i=0}^{a} \frac{\lambda^i e^{-\lambda}}{i!} \tag{17.7}$$

where λ is both the mean and variance.

To generate random counts, the density, the distribution and quantiles use **rpois**, **dpois**, **ppois** and **qpois**, respectively.

As λ grows large the Poisson distribution begins to resemble the normal distribution. To see this we will simulate 10,000 draws from the Poisson distribution and plot their histograms to see the shape.

```
> # generate 10,000 random counts from 5 different Poisson distributions
> pois1 <- rpois(n=10000, lambda=1)
> pois2 <- rpois(n=10000, lambda=2)
> pois5 <- rpois(n=10000, lambda=5)
> pois10 <- rpois(n=10000, lambda=10)
> pois20 <- rpois(n=10000, lambda=20)
> pois <- data.frame(Lambda.1=pois1, Lambda.2=pois2,
+                    Lambda.5=pois5, Lambda.10=pois10, Lambda.20=pois20)
> # load reshape2 package to melt the data to make it easier to plot
> library(reshape2)
> # melt the data into a long format
> pois <- melt(data=pois, variable.name="Lambda", value.name="x")
> # load the stringr package to help clean up the new column name
> library(stringr)
> # clean up the Lambda to just show the value for that lambda
> pois$Lambda <- as.factor(as.numeric(str_extract(string=pois$Lambda,
+                                                  pattern="\\d+")))
> head(pois)

  Lambda x
1      1 1
2      1 1
3      1 1
4      1 2
5      1 2
6      1 0

> tail(pois)

      Lambda  x
49995     20 22
49996     20 15
49997     20 24
49998     20 23
49999     20 20
50000     20 23
```

Now we will plot a separate histogram for each value of λ, as shown in Figure 17.6.

```
> library(ggplot2)
> ggplot(pois, aes(x=x)) + geom_histogram(binwidth=1) +
+     facet_wrap(~ Lambda) + ggtitle("Probability Mass Function")
```

Another, perhaps more compelling, way to visualize this convergence to normality is within overlaid density plots, as seen in Figure 17.7.

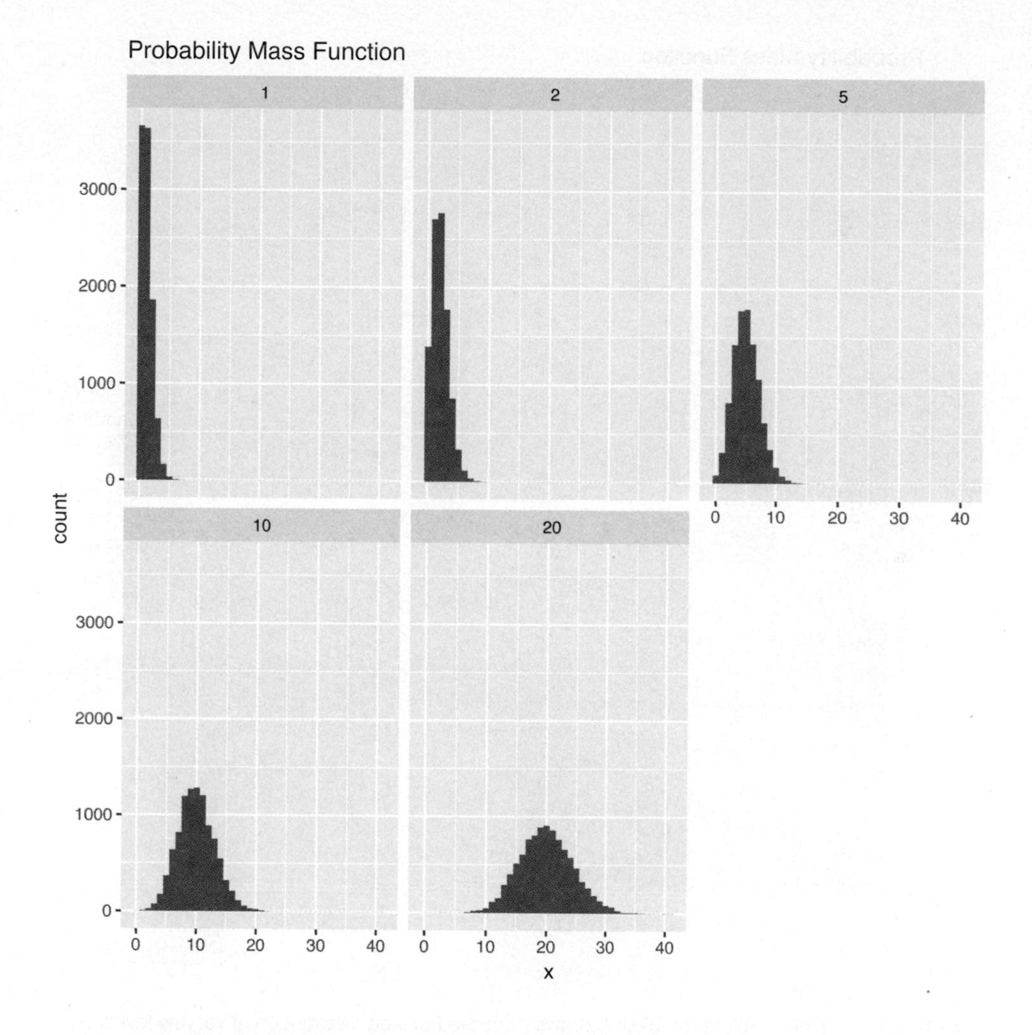

Figure 17.6 Histograms for 10,000 draws from the Poisson distribution at varying levels of λ. Notice how the histograms become more like the normal distribution.

```
> ggplot(pois, aes(x=x)) +
+     geom_density(aes(group=Lambda, color=Lambda, fill=Lambda),
+                  adjust=4, alpha=1/2) +
+     scale_color_discrete() + scale_fill_discrete() +
+     ggtitle("Probability Mass Function")
```

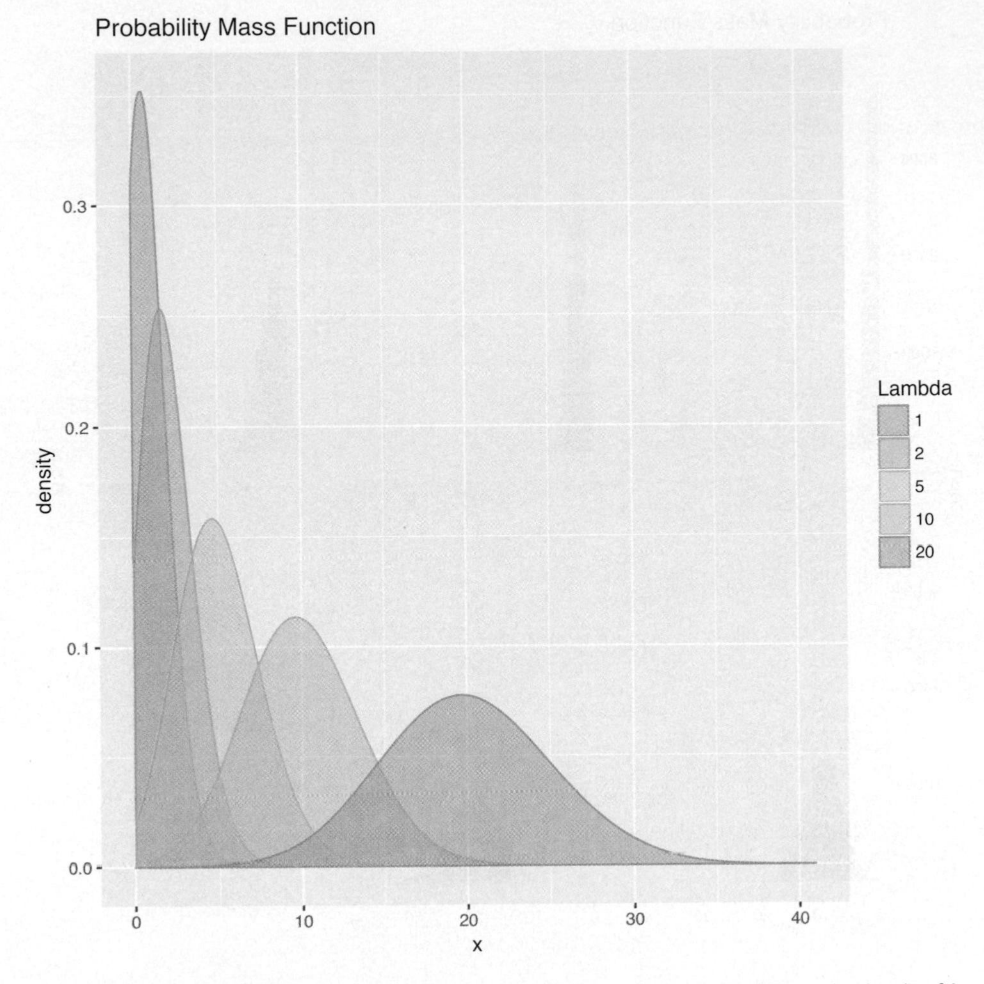

Figure 17.7 Density plots for 10,000 draws from the Poisson distribution at varying levels of λ. Notice how the density plots become more like the normal distribution.

17.4 Other Distributions

R supports many distributions, some of which are very common, while others are quite obscure. They are listed in Table 17.1; the mathematical formulas, means and variances are in Table 17.2.

Table 17.1 **Statistical distributions and their functions**

Distribution	Random Number	Density	Distribution	Quantile
Normal	rnorm	dnorm	pnorm	qnorm
Binomial	rbinom	dbinom	pbinom	qbinom
Poisson	rpois	dpois	ppois	qpois
t	rt	dt	pt	qt
F	rf	df	pf	qf
Chi-Squared	rchisq	dchisq	pchisq	qchisq
Gamma	rgamma	dgamma	pgamma	qgamma
Geometric	rgeom	dgeom	pgeom	qgeom
Negative Binomial	rnbinom	dnbinom	pnbinom	qnbinom
Exponential	rexp	dexp	pexp	qexp
Weibull	rweibull	dweibull	pweibull	qweibull
Uniform (Continuous)	runif	dunif	punif	qunif
Beta	rbeta	dbeta	pbeta	qbeta
Cauchy	rcauchy	dcauchy	pcauchy	qcauchy
Multinomial	rmultinom	dmultinom	pmultinom	qmultinom
Hypergeometric	rhyper	dhyper	phyper	qhyper
Log normal	rlnorm	dlnorm	plnorm	qlnorm
Logistic	rlogis	dlogis	plogis	qlogis

Table 17.2 **Formulas, means and variances for various statistical distributions. The B in the F distribution is the Beta function, $B(x,y)=\int_0^1 t^{x-1}(1-t)^{y-1}\mathrm{d}t$.**

Distribution	Formula	Mean	Variance
Normal	$f(x;\mu,\sigma)=\frac{1}{\sqrt{2\pi}\sigma}e^{\frac{-(x-\mu)^2}{2\sigma^2}}$	μ	σ^2
Binomial	$p(x;n,p)=\binom{n}{x}p^x(1-p)^{n-x}$	np	$np(1-p)$
Poisson	$p(i)=\binom{n}{i}p^i(1-p)^{n-i}$	λ	λ
t	$f(x;n)=\frac{\Gamma(\frac{n+1}{2})}{\sqrt{n\pi}\Gamma(\frac{n}{2})}\left(1+\frac{x^2}{n}\right)^{-\frac{n+1}{2}}$	0	$\frac{n}{n-2}$
F	$f(x;\lambda,s)=\frac{\sqrt{\frac{(n_1x)^{n_1}n_2^{n_2}}{(n_1x+n_2)^{n_1+n_2}}}}{xB(\frac{n_1}{2},\frac{n_2}{2})}$	$\frac{n_2}{n_2-2}$	$\frac{2n_2^2(n_1+n_2-2)}{n_1(n_2-2)^2(n_2-4)}$
Chi-Squared	$f(x;n)=\frac{e^{-\frac{y}{2}}y^{(\frac{n}{2})-1}}{2^{\frac{n}{2}}\Gamma(\frac{n}{2})}$	n	$2n$

(Continued)

Distribution	Formula	Mean	Variance
Gamma	$f(x;\lambda,s)=\frac{\lambda e^{-\lambda x}(\lambda x)^{s-1}}{\Gamma(s)}$	$\frac{s}{\lambda}$	$\frac{s}{\lambda^2}$
Geometric	$p(x;p)=p(1-p)^{x-1}$	$\frac{1}{\lambda}$	$\frac{1}{\lambda^2}$
Negative Binomial	$p(x;r,p)=\binom{x-1}{r-1}p^r(1-p)^{x-r}$	$\frac{r}{p}$	$\frac{r(1-p)}{p^2}$
Exponential	$f(x;\lambda)=\lambda e^{-\lambda x}$	$\frac{1}{\lambda}$	$\frac{1}{\lambda^2}$
Weibull	$f(x;\lambda,k)=\frac{k}{\lambda}(\frac{x}{\lambda})^{k-1}e^{-(x/\lambda)^k}$	$\lambda\Gamma(1+\frac{1}{k})$	$\lambda^2\Gamma(1+\frac{2}{k})-\mu^2$
Uniform	$f(x;a,b)=\frac{1}{b-a}$	$\frac{a+b}{2}$	$\frac{(b-a)^2}{12}$
Beta	$f(x;\alpha,\beta)=\frac{1}{B(\alpha,\beta)}x^{\alpha-1}(1-x)^{\beta-1}$	$\frac{\alpha}{\alpha+\beta}$	$\frac{\alpha\beta}{(\alpha+\beta)^2(\alpha+\beta+1)}$
Cauchy	$f(x;s,t)=\frac{s}{\pi\left(s^2+(x-t)^2\right)}$	Undefined	Undefined
Multinomial	$p(x_1,\ldots,x_k;n,p_1,\ldots,p_k)=\frac{n!}{x_1!\cdots x_k!}p_1^{x_1}\cdots p_k^{x_k}$	np_i	$np_i(1-p_i)$
Hypergeometric	$p(x;N,n,m)=\frac{\binom{m}{x}\binom{N-m}{n-x}}{\binom{N}{n}}$	$\frac{nm}{N}$	$\frac{nm}{N}\left[\frac{(n-1)(m-1)}{N-1}+1-\frac{nm}{N}\right]$
Log-normal	$f(x;\mu,\sigma)=\frac{1}{x\sigma\sqrt{2\pi}}e^{-\frac{(\ln x-\mu)^2}{2\sigma^2}}$	$e^{\mu+\frac{\sigma^2}{2}}$	$\left(e^{\sigma^2}-1\right)e^{2\mu+\sigma^2}$
Logistic	$f(x;\mu,s)=\frac{e^{-\frac{x-\mu}{s}}}{s\left(1+e^{-\frac{x-\mu}{s}}\right)^2}$	μ	$\frac{1}{3}s^2\pi^2$

17.5 Conclusion

R facilitates the use of many different probability distributions through the various random number, density, distribution and quantile functions outlined in Table 17.1. We focused on three distributions—normal, Bernoulli and Poisson—in detail as they are the most commonly used. The formulas for every distribution available in the base packages of R, along with their means and variances, are listed in Table 17.2.

18
Basic Statistics

Some of the most common tools used in statistics are means, variances, correlations and t-tests. These are all well represented in R with easy-to-use functions such as **mean**, **var**, **cor** and **t.test**.

18.1 Summary Statistics

The first thing many people think of in relation to statistics is the average, or mean, as it is properly called. We start by looking at some simple numbers and later in the chapter play with bigger datasets. First we generate a random sampling of 100 numbers between 1 and 100.

```
> x <- sample(x=1:100, size=100, replace=TRUE)
> x

  [1] 53 89 28 97 35 51 21 55 47  3 46 35 86 66 51 20 41 15 10 22 31
 [22] 86 19 13 10 59 60 58 90 11 54 79 45 49 23 91 80 30 83 69 20 76
 [43]  2 42 35 51 76 77 90 84 12 36 79 38 68 87 72 17 20 57 61 83 23
 [64] 61 64 41 31 74 35 20 85 89 64 73 11 36 12 81 10 64 39  4 69 42
 [85] 41 85 84 66 76 23 47 56 50 82 21 67 89 57  6 13
```

sample uniformly draws `size` entries from `x`. Setting `replace=TRUE` means that the same number can be drawn multiple times.

Now that we have a `vector` of data we can calculate the mean.

```
> mean(x)

[1] 49.85
```

This is the simple arithmetic mean.

$$E[X] = \frac{\sum_{i=1}^{N} x_i}{N} \tag{18.1}$$

Simple enough. Because this is statistics, we need to consider cases where some data is missing. To create this we take `x` and randomly set 20 percent of the elements to `NA`.

```
> # copy x
> y <- x
```

```
> # choose a random 20 elements, using sample, to set to NA
> y[sample(x=1:100, size=20, replace=FALSE)] <- NA
> y

  [1] 53 89 28 97 35 51 21 55 47 NA 46 35 86 NA NA NA 41 15 10 22 31
 [22] NA 19 13 NA 59 60 NA 90 11 NA 79 45 NA 23 91 80 30 83 69 20 76
 [43]  2 42 35 51 76 77 NA 84 NA 36 79 38 NA 87 72 17 20 57 61 83 NA
 [64] 61 64 41 31 74 NA 20 NA 89 64 73 NA 36 12 NA 10 64 39  4 NA 42
 [85] 41 85 84 66 76 23 47 56 50 82 21 67 NA NA  6 13
```

Using **mean** on `y` will return `NA`. This is because, by default, if **mean** encounters even one element that is `NA` it will return `NA`. This is to avoid providing misleading information.

```
> mean(y)

[1] NA
```

To have the `NA`s removed before calculating the mean, set `na.rm` to `TRUE`.

```
> mean(y, na.rm=TRUE)

[1] 49.6
```

To calculate the weighted mean of a set of numbers, the function **weighted.mean** takes a `vector` of numbers and a `vector` of weights. It also has an optional argument, `na.rm`, to remove `NA`s before calculating; otherwise, a `vector` with `NA` values will return `NA`.

```
> grades <- c(95, 72, 87, 66)
> weights <- c(1/2, 1/4, 1/8, 1/8)
> mean(grades)

[1] 80

> weighted.mean(x=grades, w=weights)

[1] 84.625
```

The formula for **weighted.mean** is in Equation 18.2, which is the same as the expected value of a random variable.

$$E[X] = \frac{\sum_{i=1}^{N} w_i x_i}{\sum_{i=1}^{N} w_i} = \sum_{i=1}^{N} p_i x_i \tag{18.2}$$

Another vitally important metric is the variance, which is calculated with **var**.

```
> var(x)

[1] 724.5328
```

This calculates variance as

$$Var(x) = \frac{\sum_{i=1}^{N} (x_i - \bar{x})^2}{N-1} \tag{18.3}$$

which can be verified in R.

```
> var(x)

[1] 724.5328

> sum((x - mean(x))^2) / (length(x) - 1)

[1] 724.5328
```

Standard deviation is the square root of variance and is calculated with **sd**. Like **mean** and **var**, **sd** has the `na.rm` argument to remove `NA`s before computation; otherwise, any `NA`s will cause the answer to be `NA`.

```
> sqrt(var(x))

[1] 26.91715

> sd(x)

[1] 26.91715

> sd(y)

[1] NA

> sd(y, na.rm-TRUE)

[1] 26.48506
```

Other commonly used functions for summary statistics are **min**, **max** and **median**. Of course, all of these also have `na.rm` arguments.

```
> min(x)

[1] 2

> max(x)

[1] 97

> median(x)

[1] 51

> min(y)

[1] NA

> min(y, na.rm=TRUE)

[1] 2
```

The median, as calculated before, is the middle of an ordered set of numbers. For instance, the median of 5, 2, 1, 8 and 6 is 5. In the case when there are an even amount of numbers, the median is the mean of the middle two numbers. For 5, 1, 7, 4, 3, 8, 6 and 2, the median is 4.5.

A helpful function that computes the mean, minimum, maximum and median is **summary**. There is no need to specify `na.rm` because if there are `NA`s, they are automatically removed and their count is included in the results.

```
> summary(x)

   Min. 1st Qu.  Median    Mean 3rd Qu.    Max.
   2.00   23.00   51.00   49.85   74.50   97.00

> summary(y)

   Min. 1st Qu.  Median    Mean 3rd Qu.    Max.    NA's
   2.00   26.75   48.50   49.60   74.50   97.00      20
```

This summary also displayed the first and third quantiles. These can be computed using **quantile**.

```
> # calculate the 25th and 75th quantile
> quantile(x, probs=c(.25, .75))

 25%  75%
23.0 74.5

> # try the same on y
> quantile(y, probs=c(.25, .75))

Error in quantile.default(y, probs = c(0.25, 0.75)): missing
values and NaN's not allowed if 'na.rm' is FALSE

> # this time use na.rm=TRUE
> quantile(y, probs=c(.25, .75), na.rm=TRUE)

  25%   75%
26.75 74.50

> # compute other quantiles
> quantile(x, probs=c(.1, .25, .5, .75, .99))

  10%   25%   50%   75%   99%
12.00 23.00 51.00 74.50 91.06
```

Quantiles are numbers in a set where a certain percentage of the numbers are smaller than that quantile. For instance, of the numbers one through 200, the 75th quantile—the number that is larger than 75 percent of the numbers—is 150.25.

18.2 Correlation and Covariance

When dealing with more than one variable, we need to test their relationship with each other. Two simple, straightforward methods are correlation and covariance. To examine these concepts we look at the `economics` data from **ggplot2**.

```
> library(ggplot2)
> head(economics)

# A tibble: 6 × 8
        date   pce    pop psavert uempmed unemploy  year month
      <date> <dbl>  <int>   <dbl>   <dbl>    <int> <dbl> <ord>
1 1967-07-01 507.4 198712    12.5     4.5     2944  1967   Jul
2 1967-08-01 510.5 198911    12.5     4.7     2945  1967   Aug
3 1967-09-01 516.3 199113    11.7     4.6     2958  1967   Sep
4 1967-10-01 512.9 199311    12.5     4.9     3143  1967   Oct
5 1967-11-01 518.1 199498    12.5     4.7     3066  1967   Nov
6 1967-12-01 525.8 199657    12.1     4.8     3018  1967   Dec
```

In the `economics` dataset, `pce` is personal consumption expenditures and `psavert` is the personal savings rate. We calculate their correlation using **cor**.

```
> cor(economics$pce, economics$psavert)

[1] -0.837069
```

This very low correlation makes sense because spending and saving are opposites of each other. Correlation is defined as

$$r_{xy} = \frac{\sum_{i=1}^{n}(x_i - \bar{x})(y_i - \bar{y})}{(n-1)s_x s_y} \tag{18.4}$$

where $\bar{x}$ and $\bar{y}$ are the means of x and y, and s_x and s_y are the standard deviations of x and y. It can range between -1 and 1, with higher positive numbers meaning a closer relationship between the two variables, lower negative numbers meaning an inverse relationship and numbers near zero meaning no relationship. This can be easily checked by computing Equation 18.4.

```
> # use cor to calculate correlation
> cor(economics$pce, economics$psavert)

[1] -0.837069

> ## calculate each part of correlation
> xPart <- economics$pce - mean(economics$pce)
> yPart <- economics$psavert - mean(economics$psavert)
> nMinusOne <- (nrow(economics) - 1)
> xSD <- sd(economics$pce)
> ySD <- sd(economics$psavert)
> # use correlation formula
> sum(xPart * yPart) / (nMinusOne * xSD * ySD)

[1] -0.837069
```

To compare multiple variables at once, use **cor** on a `matrix` (only for `numeric` variables).

```
> cor(economics[, c(2, 4:6)])
                pce    psavert    uempmed   unemploy
pce       1.0000000 -0.8370690  0.7273492  0.6139997
psavert  -0.8370690  1.0000000 -0.3874159 -0.3540073
uempmed   0.7273492 -0.3874159  1.0000000  0.8694063
unemploy  0.6139997 -0.3540073  0.8694063  1.0000000
```

Because this is just a table of numbers, it would be helpful to also visualize the information using a plot. For this we use the **ggpairs** function from the **GGally** package (a collection of helpful plots built on **ggplot2**) shown in Figure 18.1. This shows a scatterplot of every variable in the data against every other variable. Loading **GGally** also loads the **reshape** package, which causes `namespace` issues with the newer **reshape2** package. So rather than load **GGally**, we call its function using the `::` operator, which allows access to functions within a package without loading it.

```
> GGally::ggpairs(economics[, c(2, 4:6)])
```

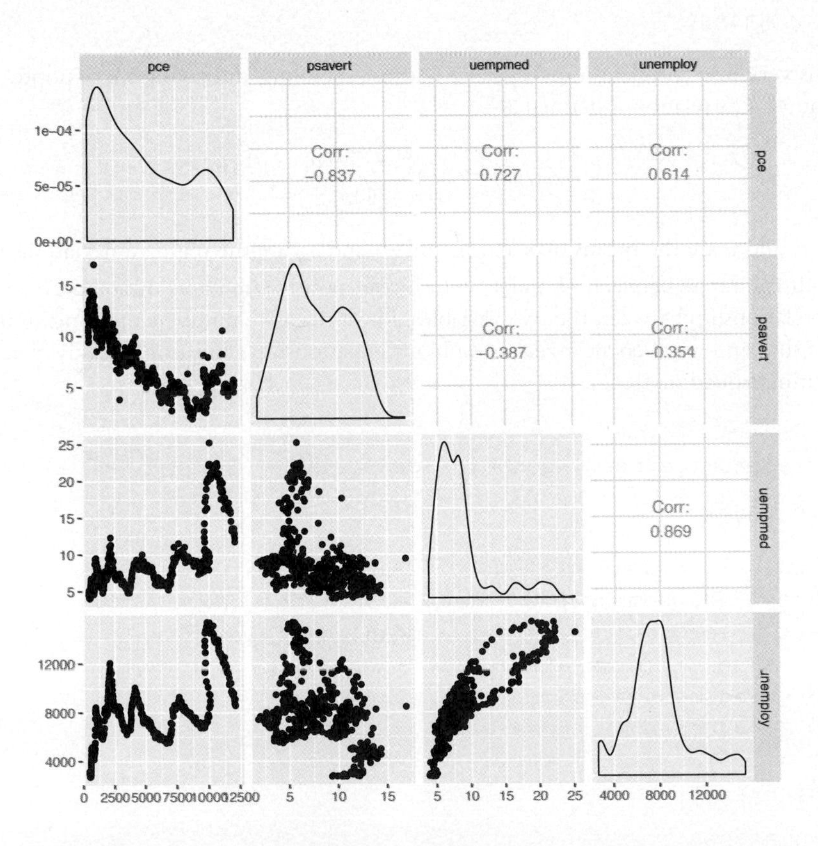

Figure 18.1 Pairs plot of economics data showing the relationship between each pair of variables as a scatterplot with the correlations printed as numbers.

This is similar to a small multiples plot except that each pane has different x- and y-axes. While this shows the original data, it does not actually show the correlation. To show that we build a heatmap of the correlation numbers, as shown in Figure 18.2. High positive correlation indicates a positive relationship between the variables, high negative correlation indicates a negative relationship between the variables and near zero correlation indicates no strong relationship.

```
> # load the reshape package for melting the data
> library(reshape2)
> # load the scales package for some extra plotting features
> library(scales)
> # build the correlation matrix
> econCor <- cor(economics[, c(2, 4:6)])
> # melt it into the long format
> econMelt <- melt(econCor, varnames=c("x", "y"), value.name="Correlation")
> # order it according to the correlation
> econMelt <- econMelt[order(econMelt$Correlation), ]
> # display the melted data
> econMelt

          x        y Correlation
2   psavert      pce  -0.8370690
5       pce  psavert  -0.8370690
7   uempmed  psavert  -0.3874159
10  psavert  uempmed  -0.3874159
8  unemploy  psavert  -0.3540073
14  psavert unemploy  -0.3540073
4  unemploy      pce   0.6139997
13      pce unemploy   0.6139997
3   uempmed      pce   0.7273492
9       pce  uempmed   0.7273492
12 unemploy  uempmed   0.8694063
15  uempmed unemploy   0.8694063
1       pce      pce   1.0000000
6   psavert  psavert   1.0000000
11  uempmed  uempmed   1.0000000
16 unemploy unemploy   1.0000000

> ## plot it with ggplot
> # initialize the plot with x and y on the x and y axes
> ggplot(econMelt, aes(x=x, y=y)) +
+     # draw tiles filling the color based on Correlation
+     geom_tile(aes(fill=Correlation)) +
+     # make the fill (color) scale a three color gradient with muted
+     # red for the low point, white for the middle and steel blue
+     # for the high point
+     # the guide should be a colorbar with no ticks, whose height is
+     # 10 lines
+     # limits indicates the scale should be filled from -1 to 1
```

```
+       scale_fill_gradient2(low=muted("red"), mid="white",
+           high="steelblue",
+           guide=guide_colorbar(ticks=FALSE, barheight=10),
+           limits=c(-1, 1)) +
+       # use the minimal theme so there are no extras in the plot
+       theme_minimal() +
+       # make the x and y labels blank
+       labs(x=NULL, y=NULL)
```

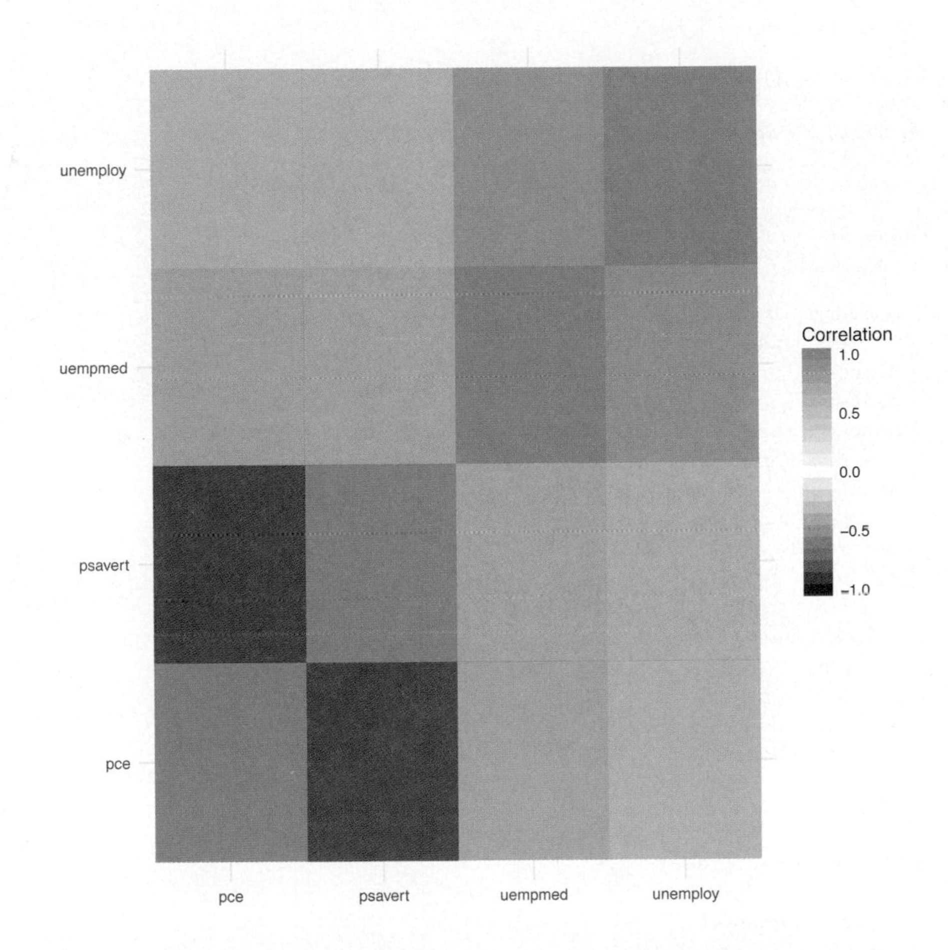

Figure 18.2 Heatmap of the correlation of the economics data. The diagonal has elements with correlation 1 because every element is perfectly correlated with itself. Red indicates highly negative correlation, blue indicates highly positive correlation and white is no correlation.

Missing data is just as much a problem with **cor** as it is with **mean** and **var**, but is dealt with differently because multiple columns are being considered simultaneously. Instead of specifying `na.rm=TRUE` to remove `NA` entries, one of `"all.obs"`, `"complete.obs"`, "pairwise.complete.obs", "everything" or "na.or.complete" is used. To illustrate this we first make a five-column matrix where only the fourth and fifth columns have no NA values; the other columns have one or two NAs.

```
> m <- c(9, 9, NA, 3, NA, 5, 8, 1, 10, 4)
> n <- c(2, NA, 1, 6, 6, 4, 1, 1, 6, 7)
> p <- c(8, 4, 3, 9, 10, NA, 3, NA, 9, 9)
> q <- c(10, 10, 7, 8, 4, 2, 8, 5, 5, 2)
> r <- c(1, 9, 7, 6, 5, 6, 2, 7, 9, 10)
> # combine them together
> theMat <- cbind(m, n, p, q, r)
```

The first option for use is "everything", which means that the entirety of all columns must be free of NAs; otherwise the result is NA. Running this should generate a matrix of all NAs except ones on the diagonal—because a vector is always perfectly correlated with itself—and between q and r. With the second option—"all.obs"—even a single NA in any column will cause an error.

```
> cor(theMat, use="everything")

   m  n  p          q          r
m  1 NA NA         NA         NA
n NA  1 NA         NA         NA
p NA NA  1         NA         NA
q NA NA NA  1.0000000 -0.4242958
r NA NA NA -0.4242958  1.0000000

> cor(theMat, use="all.obs")

Error in cor(theMat, use = "all.obs"): missing observations in cov/cor
```

The third and fourth options—"complete.obs" and "na.or.complete"—work similarly to each other in that they keep only rows where every entry is not NA. That means our matrix will be reduced to rows 1, 4, 7, 9 and 10, and then have its correlation computed. The difference is that "complete.obs" will return an error if not a single complete row can be found, while "na.or.complete" will return NA in that case.

```
> cor(theMat, use="complete.obs")

           m          n          p          q          r
m  1.0000000 -0.5228840 -0.2893527  0.2974398 -0.3459470
n -0.5228840  1.0000000  0.8090195 -0.7448453  0.9350718
p -0.2893527  0.8090195  1.0000000 -0.3613720  0.6221470
q  0.2974398 -0.7448453 -0.3613720  1.0000000 -0.9059384
r -0.3459470  0.9350718  0.6221470 -0.9059384  1.0000000

> cor(theMat, use="na.or.complete")

           m          n          p          q          r
m  1.0000000 -0.5228840 -0.2893527  0.2974398 -0.3459470
```

```
n -0.5228840  1.0000000  0.8090195 -0.7448453  0.9350718
p -0.2893527  0.8090195  1.0000000 -0.3613720  0.6221470
q  0.2974398 -0.7448453 -0.3613720  1.0000000 -0.9059384
r -0.3459470  0.9350718  0.6221470 -0.9059384  1.0000000

> # calculate the correlation just on complete rows
> cor(theMat[c(1, 4, 7, 9, 10), ])

           m          n          p          q          r
m  1.0000000 -0.5228840 -0.2893527  0.2974398 -0.3459470
n -0.5228840  1.0000000  0.8090195 -0.7448453  0.9350718
p -0.2893527  0.8090195  1.0000000 -0.3613720  0.6221470
q  0.2974398 -0.7448453 -0.3613720  1.0000000 -0.9059384
r -0.3459470  0.9350718  0.6221470 -0.9059384  1.0000000

> # compare "complete.obs" and computing on select rows
> # should give the same result
> identical(cor(theMat, use="complete.obs"),
+           cor(theMat[c(1, 4, 7, 9, 10), ]))

[1] TRUE
```

The final option is "pairwise.complete", which is much more inclusive. It compares two columns at a time and keeps rows—for those two columns—where neither entry is NA. This is essentially the same as computing the correlation between every combination of two columns with use set to "complete.obs".

```
> # the entire correlation matrix
> cor(theMat, use="pairwise.complete.obs")

            m           n          p          q          r
m  1.00000000 -0.02511812 -0.3965859  0.4622943 -0.2001722
n -0.02511812  1.00000000  0.8717389 -0.5070416  0.5332259
p -0.39658588  0.87173889  1.0000000 -0.5197292  0.1312506
q  0.46229434 -0.50704163 -0.5197292  1.0000000 -0.4242958
r -0.20017222  0.53322585  0.1312506 -0.4242958  1.0000000

> # compare the entries for m vs n to this matrix
> cor(theMat[, c("m", "n")], use="complete.obs")

            m           n
m  1.00000000 -0.02511812
n -0.02511812  1.00000000

> # compare the entries for m vs p to this matrix
> cor(theMat[, c("m", "p")], use="complete.obs")

           m          p
m  1.0000000 -0.3965859
p -0.3965859  1.0000000
```

To see **ggpairs** in all its glory, look at tips data from the **reshape2** package in Figure 18.3. This shows every pair of variables in relation to each other building either histograms, boxplots or scatterplots depending on the combination of continuous and discrete variables. While a data dump like this looks really nice, it is not always the most informative form of exploratory data analysis.

```
> data(tips, package="reshape2")
> head(tips)

  total_bill  tip    sex smoker day   time size
1      16.99 1.01 Female     No Sun Dinner    2
2      10.34 1.66   Male     No Sun Dinner    3
3      21.01 3.50   Male     No Sun Dinner    3
4      23.68 3.31   Male     No Sun Dinner    2
5      24.59 3.61 Female     No Sun Dinner    4
6      25.29 4.71   Male     No Sun Dinner    4

> GGally::ggpairs(tips)
```

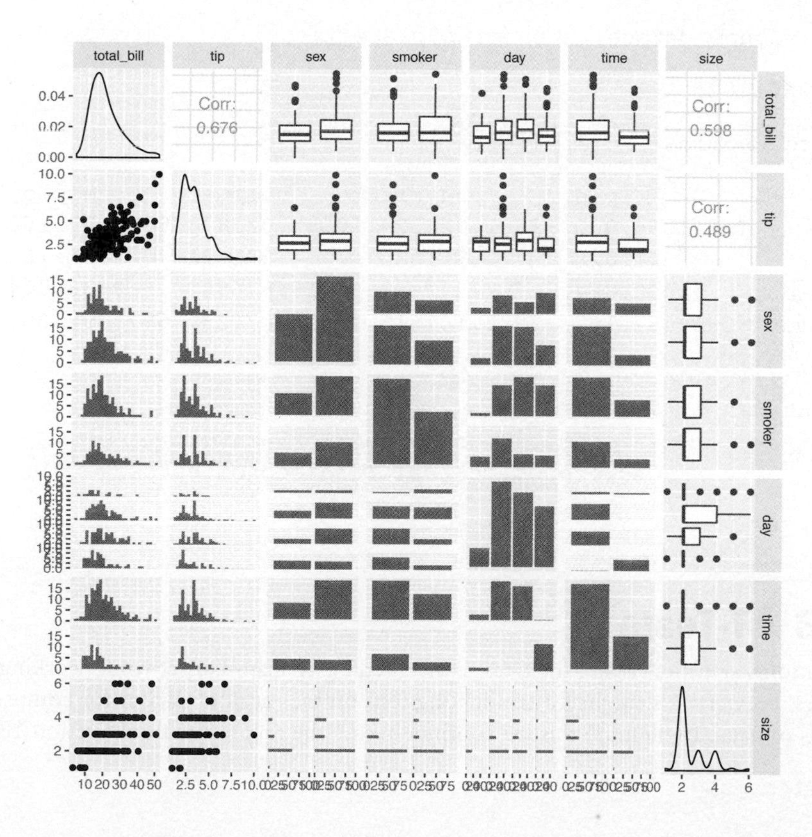

Figure 18.3 ggpairs plot of tips data using both continuous and categorial variables.

No discussion of correlation would be complete without the old refrain, "Correlation does not mean causation." In other words, just because two variables are correlated does not mean they have an effect on each other. This is exemplified in xkcd[1] comic number 552. There is even an R package, **RXKCD**, for downloading individual comics. Running the following code should generate a pleasant surprise.

```
> library(RXKCD)
> getXKCD(which="552")
```

Similar to correlation is covariance, which is like a variance between variables, its formula is in Equation 18.5. Notice the similarity to correlation in Equation 18.4 and variance in Equation 18.3.

$$cov(X,Y) = \frac{1}{N-1}\sum_{i=1}^{N}(x_i - \bar{x})(y_i - \bar{y}) \quad (18.5)$$

The **cov** function works similarly to the **cor** function, with the same arguments for dealing with missing data. In fact, `?cor` and `?cov` pull up the same help menu.

```
> cov(economics$pce, economics$psavert)

[1] -9361.028

> cov(economics[, c(2, 4:6)])

                 pce      psavert      uempmed     unemploy
pce      12811296.900 -9361.028324 10695.023873 5806187.162
psavert     -9361.028     9.761835    -4.972622   -2922.162
uempmed     10695.024    -4.972622    16.876582    9436.074
unemploy  5806187.162 -2922.161618  9436.074287 6979955.661

> # check that cov and cor*sd*sd are the same
> identical(cov(economics$pce, economics$psavert),
+           cor(economics$pce, economics$psavert) *
+               sd(economics$pce) * sd(economics$psavert))

[1] TRUE
```

18.3 T-Tests

In traditional statistics classes, the t-test—invented by William Gosset while working at the Guinness brewery—is taught for conducting tests on the mean of data or for comparing two sets of data. To illustrate this we continue to use the `tips` data from Section 18.2.

```
> head(tips)
```

1. xkcd is a Web comic by Randall Munroe, beloved by statisticians, physicists, mathematicians and the like. It can be found at `http://xkcd.com`.

```
  total_bill  tip    sex smoker day   time size
1      16.99 1.01 Female     No Sun Dinner    2
2      10.34 1.66   Male     No Sun Dinner    3
3      21.01 3.50   Male     No Sun Dinner    3
4      23.68 3.31   Male     No Sun Dinner    2
5      24.59 3.61 Female     No Sun Dinner    4
6      25.29 4.71   Male     No Sun Dinner    4

> # sex of the bill payer
> unique(tips$sex)

[1] Female Male
Levels: Female Male

> # day of the week
> unique(tips$day)

[1] Sun  Sat  Thur Fri
Levels: Fri Sat Sun Thur
```

18.3.1 One-Sample T-Test

First we conduct a one-sample t-test on whether the average tip is equal to $2.50. This test essentially calculates the mean of data and builds a confidence interval. If the value we are testing falls within that confidence interval, then we can conclude that it is the true value for the mean of the data; otherwise, we conclude that it is not the true mean.

```
> t.test(tips$tip, alternative="two.sided", mu=2.50)

One Sample t-test


data:  tips$tip
t = 5.6253, df = 243, p-value = 5.08e-08
alternative hypothesis: true mean is not equal to 2.5
95 percent confidence interval:
 2.823799 3.172758
sample estimates:
mean of x
 2.998279
```

The output very nicely displays the setup and results of the hypothesis test of whether the mean is equal to $2.50. It prints the t-statistic, the degrees of freedom and p-value. It also provides the 95 percent confidence interval and mean for the variable of interest. The p-value indicates that the null hypothesis[2] should be rejected, and we conclude that the mean is not equal to $2.50.

2. The null hypothesis is what is considered to be true; in this case that the mean is equal to $2.50.

We encountered a few new concepts here. The t-statistic is the ratio where the numerator is the difference between the estimated mean and the hypothesized mean and the denominator is the standard error of the estimated mean. It is defined in Equation 18.6.

$$\text{t-statistic} = \frac{(\bar{x} - \mu_0)}{s_{\bar{x}}/\sqrt{n}} \tag{18.6}$$

Here, $\bar{x}$ is the estimated mean, μ_0 is the hypothesized mean and $(s_{\bar{x}})/(\sqrt{n})$ is the standard error of $\bar{x}$.[3]

If the hypothesized mean is correct, then we expect the t-statistic to fall somewhere in the middle—about two standard deviations from the mean—of the t distribution. In Figure 18.4 we see that the thick black line, which represents the estimated mean, falls so far outside the distribution that we must conclude that the mean is not equal to $2.50.

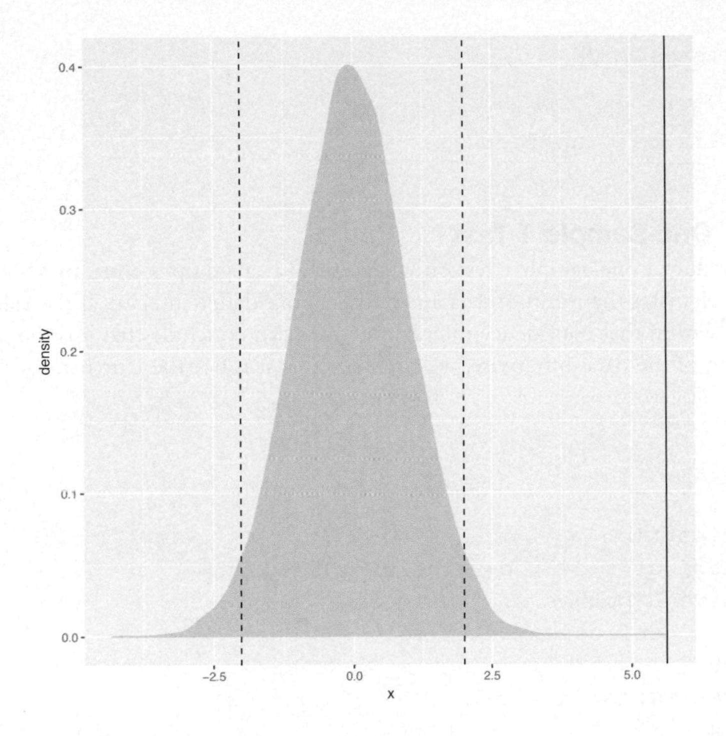

Figure 18.4 t-distribution and t-statistic for tip data. The dashed lines are two standard deviations from the mean in either direction. The thick black line, the t-statistic, is so far outside the distribution that we must reject the null hypothesis and conclude that the true mean is not $2.50.

```
> ## build a t distribution
> randT <- rt(30000, df=NROW(tips)-1)
>
```

3. $s_{\bar{x}}$ is the standard deviation of the data and n is the number of observations.

```
> # get t-statistic and other information
> tipTTest <- t.test(tips$tip, alternative="two.sided", mu=2.50)
>
> # plot it
> ggplot(data.frame(x=randT)) +
+     geom_density(aes(x=x), fill="grey", color="grey") +
+     geom_vline(xintercept=tipTTest$statistic) +
+     geom_vline(xintercept=mean(randT) + c(-2, 2)*sd(randT), linetype=2)
```

The p-value is an often misunderstood concept. Despite all the misinterpretations, a p-value is the probability, if the null hypothesis were correct, of getting as extreme, or more extreme, a result. It is a measure of how extreme the statistic—in this case, the estimated mean—is. If the statistic is too extreme, we conclude that the null hypothesis should be rejected. The main problem with p-values, however, is determining what should be considered too extreme. Ronald A. Fisher, the father of modern statistics, decided we should consider a p-value that is smaller than 0.10, 0.05 or 0.01 to be too extreme. While those p-values have been the standard for decades, they were arbitrarily chosen, leading some modern data scientists to question their usefulness. In this example, the p-value is 5.0799885×10^{-8}; this is smaller than 0.01, so we reject the null hypothesis.

Degrees of freedom is another difficult concept to grasp but is pervasive throughout statistics. It represents the effective number of observations. Generally, the degrees of freedom for some statistic or distribution is the number of observations minus the number of parameters being estimated. In the case of the t distribution, one parameter, the standard error, is being estimated. In this example, there are `nrow(tips)-1`=243 degrees of freedom.

Next we conduct a one-sided t-test to see if the mean is greater than $2.50.

```
> t.test(tips$tip, alternative="greater", mu=2.50)

One Sample t-test

data:  tips$tip
t = 5.6253, df = 243, p-value = 2.54e-08
alternative hypothesis: true mean is greater than 2.5
95 percent confidence interval:
 2.852023      Inf
sample estimates:
mean of x
 2.998279
```

Once again, the p-value indicates that we should reject the null hypothesis and conclude that the mean is greater than $2.50, which coincides nicely with the confidence interval.

18.3.2 Two-Sample T-Test

More often than not the t-test is used for comparing two samples. Continuing with the `tips` data, we compare how female and male diners tip. Before running the t-test, however, we first need to check the variance of each sample. A traditional t-test requires both groups to have the same variance, whereas the Welch two-sample t-test can handle groups with differing variances. We explore this both numerically and visually in Figure 18.5.

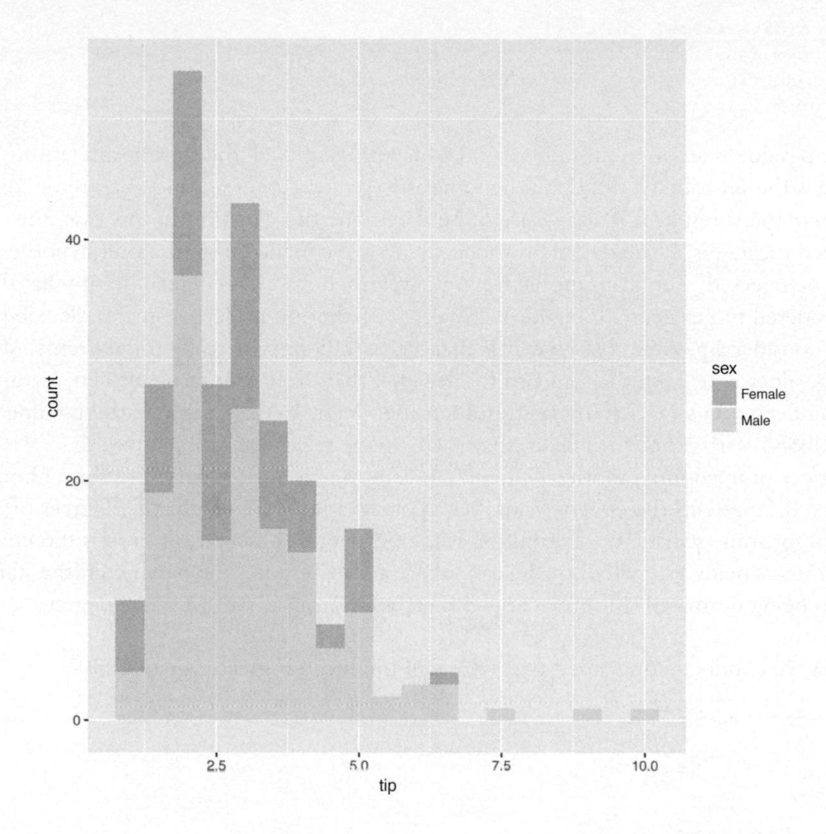

Figure 18.5 Histogram of tip amount by sex. Note that neither distribution appears to be normal.

```
> # first just compute the variance for each group
> # using the the formula interface
> # calculate the variance of tip for each level of sex
> aggregate(tip ~ sex, data=tips, var)

     sex      tip
1 Female 1.344428
2   Male 2.217424

> # now test for normality of tip distribution
> shapiro.test(tips$tip)
```

```
Shapiro-Wilk normality test

data:  tips$tip
W = 0.89781, p-value = 8.2e-12

> shapiro.test(tips$tip[tips$sex == "Female"])

Shapiro-Wilk normality test

data:  tips$tip[tips$sex == "Female"]
W = 0.95678, p-value = 0.005448

> shapiro.test(tips$tip[tips$sex == "Male"])

Shapiro-Wilk normality test

data:  tips$tip[tips$sex == "Male"]
W = 0.87587, p-value = 3.708e-10

> # all the tests fail so inspect visually
> ggplot(tips, aes(x=tip, fill=sex)) +
+     geom_histogram(binwidth=.5, alpha=1/2)
```

Since the data do not appear to be normally distributed, neither the standard F-test (via the **var.test** function) nor the Bartlett test (via the **bartlett.test** function) will suffice. So we use the nonparametric Ansari-Bradley test to examine the equality of variances.

```
> ansari.test(tip ~ sex, tips)

Ansari-Bradley test

data:  tip by sex
AB = 5582.5, p-value = 0.376
alternative hypothesis: true ratio of scales is not equal to 1
```

This test indicates that the variances are equal, meaning we can use the standard two-sample t-test.

```
> # setting var.equal=TRUE runs a standard two sample t-test
> # var.equal=FALSE (the default) would run the Welch test
> t.test(tip ~ sex, data=tips, var.equal=TRUE)

Two Sample t-test

data:  tip by sex
t = -1.3879, df = 242, p-value = 0.1665
alternative hypothesis: true difference in means is not equal to 0
95 percent confidence interval:
 -0.6197558  0.1074167
```

```
sample estimates:
mean in group Female   mean in group Male
            2.833448             3.089618
```

According to this test, the results were not significant, and we should conclude that female and male diners tip roughly equally. While all this statistical rigor is nice, a simple rule of thumb would be to see if the two means are within two standard deviations of each other.

```
> library(plyr)
> tipSummary <- ddply(tips, "sex", summarize,
+                      tip.mean=mean(tip), tip.sd=sd(tip),
+                      Lower=tip.mean - 2*tip.sd/sqrt(NROW(tip)),
+                      Upper=tip.mean + 2*tip.sd/sqrt(NROW(tip)))
> tipSummary

     sex tip.mean   tip.sd    Lower    Upper
1 Female 2.833448 1.159495 2.584827 3.082070
2   Male 3.089618 1.489102 2.851931 3.327304
```

A lot happened in that code. First, **ddply** was used to split the data according to the `levels` of `sex`. It then applied the **summarize** function to each subset of the data. This function applied the indicated functions to the data, creating a new `data.frame`.

As usual, we prefer visualizing the results rather than comparing numerical values. This requires reshaping the data a bit. The results, in Figure 18.6, clearly show the confidence intervals overlapping, suggesting that the means for the two sexes are roughly equivalent.

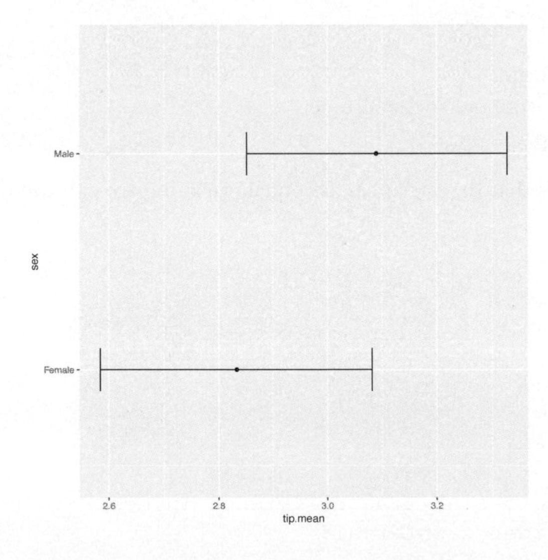

Figure 18.6 Plot showing the mean and two standard errors of tips broken down by the sex of the diner.

```
> ggplot(tipSummary, aes(x=tip.mean, y=sex)) + geom_point() +
+     geom_errorbarh(aes(xmin=Lower, xmax=Upper), height=.2)
```

18.3.3 Paired Two-Sample T-Test

For testing paired data (for example, measurements on twins, before and after treatment effects, father and son comparisons) a paired t-test should be used. This is simple enough to do by setting the `paired` argument in **t.test** to `TRUE`. To illustrate, we use data collected by Karl Pearson on the heights of fathers and sons that is located in the **UsingR** package. Heights are generally normally distributed, so we will forgo the tests of normality and equal variance.

```
> data(father.son, package='UsingR')
> head(father.son)

   fheight  sheight
1 65.04851 59.77827
2 63.25094 63.21404
3 64.95532 63.34242
4 65.75250 62.79238
5 61.13723 64.28113
6 63.02254 64.24221

> t.test(father.son$fheight, father.son$sheight, paired=TRUE)

Paired t-test

data:  father.son$fheight and father.son$sheight
t = -11.789, df = 1077, p-value < 2.2e-16
alternative hypothesis: true difference in means is not equal to 0
95 percent confidence interval:
 -1.1629160 -0.8310296
sample estimates:
mean of the differences
             -0.9969728
```

This test shows that we should reject the null hypothesis and conclude that fathers and sons (at least for this dataset) have different heights. We visualize this data using a density plot of the differences, as shown in Figure 18.7. In it we see a distribution with a mean not at zero and a confidence interval that barely excludes zero, which agrees with the test.

```
> heightDiff <- father.son$fheight - father.son$sheight
> ggplot(father.son, aes(x=fheight - sheight)) +
+     geom_density() +
+     geom_vline(xintercept=mean(heightDiff)) +
+     geom_vline(xintercept=mean(heightDiff) +
+                    2*c(-1, 1)*sd(heightDiff)/sqrt(nrow(father.son)),
+                linetype=2)
```

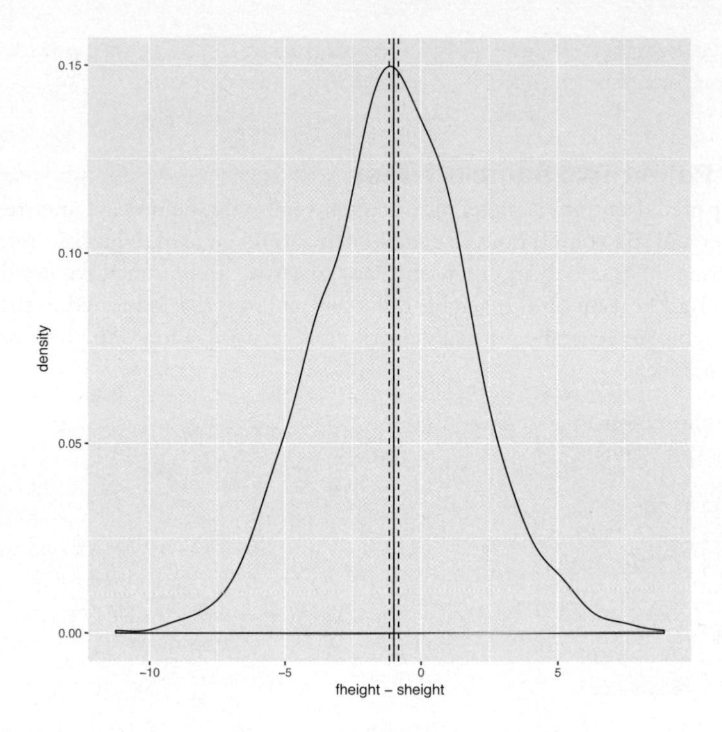

Figure 18.7 Density plot showing the difference of heights of fathers and sons.

18.4 ANOVA

After comparing two groups, the natural next step is comparing multiple groups. Every year, far too many students in introductory statistics classes are forced to learn the ANOVA (analysis of variance) test and memorize its formula, which is

$$F = \frac{\sum_i n_i(\overline{Y}_i - \overline{Y})^2/(K-1)}{\sum_{ij}(Y_{ij} - \overline{Y}_i)^2/(N-K)} \quad (18.7)$$

where n_i is the number of observations in group i, $\overline{Y}_i$ is the mean of group i, $\overline{Y}$ is the overall mean, Y_{ij} is observation j in group i, N is the total number of observations and K is the number of groups.

Not only is this a laborious formula that often turns off a lot of students from statistics; it is also a bit of an old-fashioned way of comparing groups. Even so, there is an R function—albeit rarely used—to conduct the ANOVA test. This also uses the `formula` interface where the left side is the variable of interest and the right side contains the variables that control grouping. To see this, we compare tips by day of the week, with `levels` Fri, Sat, Sun, Thur.

```
> tipAnova <- aov(tip ~ day - 1, tips)
```

In the `formula` the right side was `day - 1`. This might seem odd at first but will make more sense when comparing it to a call without `-1`.

```
> tipIntercept <- aov(tip ~ day, tips)
> tipAnova$coefficients

  dayFri   daySat   daySun  dayThur
2.734737 2.993103 3.255132 2.771452

> tipIntercept$coefficients

(Intercept)      daySat      daySun     dayThur
 2.73473684  0.25836661  0.52039474  0.03671477
```

Here we see that just using `tip ~ day` includes only Saturday, Sunday and Thursday, along with an intercept, while `tip ~ day - 1` compares Friday, Saturday, Sunday and Thursday with no intercept. The importance of the intercept is made clear in Chapter 19, but for now it suffices that having no intercept makes the analysis more straightforward.

The ANOVA tests whether any group is different from any other group but it does not specify which group is different. So printing a summary of the test just returns a single p-value.

```
> summary(tipAnova)

           Df Sum Sq Mean Sq F value Pr(>F)
day         4 2203.0   550.8   290.1 <2e-16 ***
Residuals 240  455.7     1.9
---
Signif. codes:  0 '***' 0.001 '**' 0.01 '*' 0.05 '.' 0.1 ' ' 1
```

Since the test had a significant p-value, we would like to see which group differed from the others. The simplest way is to make a plot of the group means and confidence intervals and see which overlap. Figure 18.8 shows that tips on Sunday differ (just barely, at the 90 percent confidence level) from both Thursday and Friday.

```
> tipsByDay <- ddply(tips, "day", plyr::summarize,
+                    tip.mean=mean(tip), tip.sd=sd(tip),
+                    Length=NROW(tip),
+                    tfrac=qt(p=.90, df=Length-1),
+                    Lower=tip.mean - tfrac*tip.sd/sqrt(Length),
+                    Upper=tip.mean + tfrac*tip.sd/sqrt(Length)
+                    )
>
> ggplot(tipsByDay, aes(x=tip.mean, y=day)) + geom_point() +
+     geom_errorbarh(aes(xmin=Lower, xmax=Upper), height=.3)
```

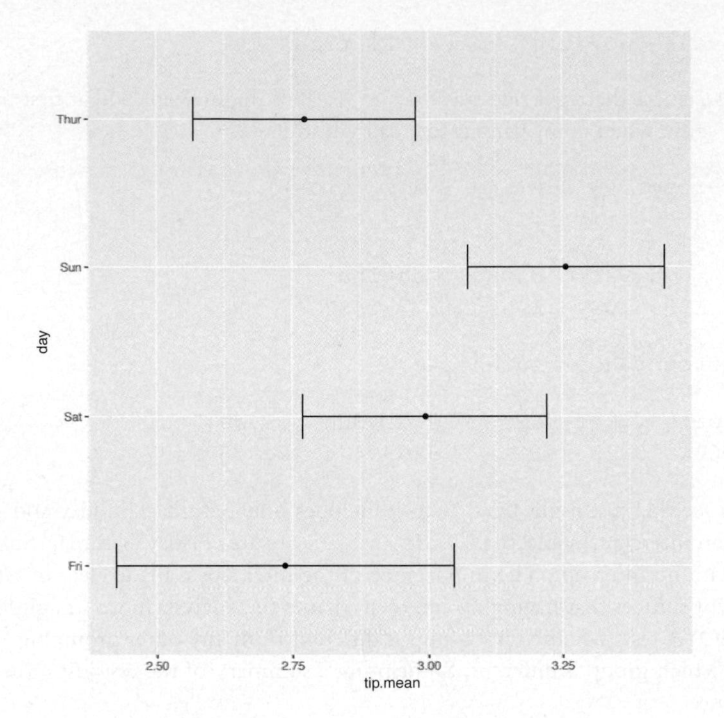

Figure 18.8 Means and confidence intervals of tips by day. This shows that Sunday tips differ from Thursday and Friday tips.

The use of **NROW** instead of **nrow** is to guarantee computation. Where **nrow** works only on `data.frames` and `matrices`, **NROW** returns the length of objects that have only one dimension.

```
> nrow(tips)

[1] 244

> NROW(tips)

[1] 244

> nrow(tips$tip)

NULL

> NROW(tips$tip)

[1] 244
```

To confirm the results from the ANOVA, individual t-tests could be run on each pair of groups, just like in Section 18.3.2. Traditional texts encourage adjusting the p-value to accommodate the multiple comparisons. However, some professors, including Andrew Gelman, suggest not worrying about adjustments for multiple comparisons.

An alternative to the ANOVA is to fit a linear regression with one categorical variable and no intercept. This is discussed in Section 19.1.1.

18.5 Conclusion

Whether computing simple numerical summaries or conducting hypothesis tests, R has functions for all of them. Means, variances and standard deviations are computed with **mean**, **var** and **sd**, respectively. Correlation and covariance are computed with **cor** and **cov**. For t-tests **t.test** is used, while **aov** is used for ANOVA.

19

Linear Models

The workhorse of statistical analysis is the linear model, particularly regression. Originally invented by Francis Galton to study the relationships between parents and children, which he described as regressing to the mean, it has become one of the most widely used modelling techniques and has spawned other models such as generalized linear models, regression trees, penalized regression and many others. In this chapter we focus on simple and multiple regression.

19.1 Simple Linear Regression

In its simplest form regression is used to determine the relationship between two variables. That is, given one variable, it tells us what we can expect from the other variable. This powerful tool, which is frequently taught and can accomplish a great deal of analysis with minimal effort, is called simple linear regression.

Before we go any further, we clarify some terminology. The outcome variable (what we are trying to predict) is called the response, and the input variable (what we are using to predict) is the predictor. Fields outside of statistics use other terms, such as measured variable, outcome variable and experimental variable for response, and covariate, feature and explanatory variable for predictor. Worst of all are the terms dependent (response) and independent (predictor) variables. These very names are misnomers. According to probability theory, if variable y is dependent on variable x, then variable x *cannot* be independent of variable y. So we stick with the terms response and predictor exclusively.

The general idea behind simple linear regression is using the predictor to come up with some average value of the response. The relationship is defined as

$$y = a + bx + \epsilon \tag{19.1}$$

where

$$b = \frac{\sum_{i=1}^{n}(x_i - \bar{x})(y_i - \bar{y})}{\sum_{i=1}^{n}(x_i - \bar{x})^2} \tag{19.2}$$

$$a = \bar{y} - b \tag{19.3}$$

and

$$\epsilon \sim \mathcal{N}(0, \sigma^2) \tag{19.4}$$

which is to say that there are normally distributed errors.

Equation 19.1 is essentially describing a straight line that goes through the data where a is the y-intercept and b is the slope. This is illustrated using fathers' and sons' height data, which are plotted in Figure 19.1. In this case we are using the fathers' heights as the predictor and the sons' heights as the response. The blue line running through the points is the regression line and the gray band around it represents the uncertainty in the fit.

```
> data(father.son, package='UsingR')
> library(ggplot2)
> head(father.son)

   fheight  sheight
1 65.04851 59.77827
2 63.25094 63.21404
3 64.95532 63.34242
4 65.75250 62.79238
5 61.13723 64.28113
6 63.02254 64.24221

> ggplot(father.son, aes(x=fheight, y=sheight)) + geom_point() +
+       geom_smooth(method="lm") + labs(x="Fathers", y="Sons")
```

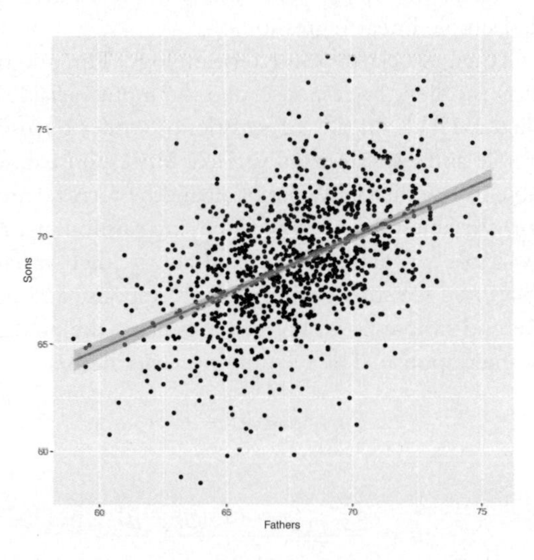

Figure 19.1 Using fathers' heights to predict sons' heights using simple linear regression. The fathers' heights are the predictors and the sons' heights are the responses. The blue line running through the points is the regression line and the gray band around it represents the uncertainty in the fit.

While that code generated a nice graph showing the results of the regression (generated with `geom_smooth(method="lm")`), it did not actually make those results available to us. To actually calculate a regression, use the **lm** function.

```
> heightsLM <- lm(sheight ~ fheight, data=father.son)
> heightsLM

Call:
lm(formula = sheight ~ fheight, data = father.son)

Coefficients:
(Intercept)      fheight
    33.8866       0.5141
```

Here we once again see the `formula` notation that specifies to regress `sheight` (the response) on `fheight` (the predictor), using the `father.son` data, and adds the intercept term automatically. The results show coefficients for `(Intercept)` and `fheight` which is the slope for the `fheight`, predictor. The interpretation of this is that, for every extra inch of height in a father, we expect an extra half inch in height for his son. The intercept in this case does not make much sense because it represents the height of a son whose father had zero height, which obviously cannot exist in reality.

While the point estimates for the coefficients are nice, they are not very helpful without the standard errors, which give the sense of uncertainty about the estimate and are similar to standard deviations. To quickly see a full report on the model, use **summary**.

```
> summary(heightsLM)

Call:
lm(formula = sheight ~ fheight, data = father.son)

Residuals:
    Min      1Q  Median      3Q     Max
-8.8772 -1.5144 -0.0079  1.6285  8.9685

Coefficients:
            Estimate Std. Error t value Pr(>|t|)
(Intercept) 33.88660    1.83235   18.49   <2e-16 ***
fheight      0.51409    0.02705   19.01   <2e-16 ***
---
Signif. codes:  0 '***' 0.001 '**' 0.01 '*' 0.05 '.' 0.1 ' ' 1

Residual standard error: 2.437 on 1076 degrees of freedom
Multiple R-squared:  0.2513,Adjusted R-squared:  0.2506
F-statistic: 361.2 on 1 and 1076 DF,  p-value: < 2.2e-16
```

This prints out a lot more information about the model, including the standard errors, t-test values and p-values for the coefficients, the degrees of freedom, residual summary statistics (seen in more detail in Section 21.1) and the results of an F-test. This is all diagnostic information to check the fit of the model, and is covered in more detail in Section 19.2 about multiple regression.

19.1.1 ANOVA Alternative

An alternative to running an ANOVA test (discussed in Section 18.4) is to fit a regression with just one categorical variable and no intercept term. To see this we use the `tips` data in the **reshape2** package on which we will fit a regression.

```
> data(tips, package="reshape2")
> head(tips)

  total_bill  tip    sex smoker day   time size
1      16.99 1.01 Female     No Sun Dinner    2
2      10.34 1.66   Male     No Sun Dinner    3
3      21.01 3.50   Male     No Sun Dinner    3
4      23.68 3.31   Male     No Sun Dinner    2
5      24.59 3.61 Female     No Sun Dinner    4
6      25.29 4.71   Male     No Sun Dinner    4

> tipsAnova <- aov(tip ~ day - 1, data=tips)
> # putting -1 in the formula indicates that the intercept should not be
> # included in the model;
> # the categorical variable day is automatically setup to have a
> # coefficient for each level
> tipsLM <- lm(tip ~ day - 1, data=tips)
> summary(tipsAnova)

           Df Sum Sq Mean Sq F value Pr(>F)
day         4 2203.0   550.8   290.1 <2e-16 ***
Residuals 240  455.7     1.9
---
Signif. codes:  0 '***' 0.001 '**' 0.01 '*' 0.05 '.' 0.1 ' ' 1

> summary(tipsLM)

Call:
lm(formula = tip ~ day - 1, data = tips)

Residuals:
    Min      1Q  Median      3Q     Max
-2.2451 -0.9931 -0.2347  0.5382  7.0069

Coefficients:
        Estimate Std. Error t value Pr(>|t|)
dayFri    2.7347     0.3161   8.651 7.46e-16 ***
daySat    2.9931     0.1477  20.261  < 2e-16 ***
daySun    3.2551     0.1581  20.594  < 2e-16 ***
dayThur   2.7715     0.1750  15.837  < 2e-16 ***
---
Signif. codes:  0 '***' 0.001 '**' 0.01 '*' 0.05 '.' 0.1 ' ' 1
```

```
Residual standard error: 1.378 on 240 degrees of freedom
Multiple R-squared:  0.8286,Adjusted R-squared:  0.8257
F-statistic: 290.1 on 4 and 240 DF,  p-value: < 2.2e-16
```

Notice that the F-value or F-statistic is the same for both, as are the degrees of freedom. This is because the ANOVA and regression were derived along the same lines and can accomplish the same analysis. Visualizing the coefficients and standard errors should show the same results as computing them using the ANOVA formula. This is seen in Figure 19.2. The point estimates for the mean are identical and the confidence intervals are similar, the difference due to slightly different calculations.

```
> # first calculate the means and CI manually
> library(dplyr)
> tipsByDay <- tips %>%
+     group_by(day) %>%
+     dplyr::summarize(
+         tip.mean=mean(tip), tip.sd=sd(tip),
+         Length=NROW(tip),
+         tfrac=qt(p=.90, df=Length-1),
+         Lower=tip.mean - tfrac*tip.sd/sqrt(Length),
+         Upper=tip.mean + tfrac*tip.sd/sqrt(Length)
+     )
>
> # now extract them from the summary for tipsLM
> tipsInfo <- summary(tipsLM)
> tipsCoef <- as.data.frame(tipsInfo$coefficients[, 1:2])
> tipsCoef <- within(tipsCoef, {
+     Lower <- Estimate - qt(p=0.90, df=tipsInfo$df[2]) * `Std. Error`
+     Upper <- Estimate + qt(p=0.90, df=tipsInfo$df[2]) * `Std. Error`
+     day <- rownames(tipsCoef)
+ })
>
> # plot them both
> ggplot(tipsByDay, aes(x=tip.mean, y=day)) + geom_point() +
+     geom_errorbarh(aes(xmin=Lower, xmax=Upper), height=.3) +
+     ggtitle("Tips by day calculated manually")
>
> ggplot(tipsCoef, aes(x=Estimate, y=day)) + geom_point() +
+     geom_errorbarh(aes(xmin=Lower, xmax=Upper), height=.3) +
+     ggtitle("Tips by day calculated from regression model")
```

A new function and a new feature were used here. First, we introduced **within**, which is similar to **with** in that it lets us refer to columns in a `data.frame` by name but different in that we can create new columns within that `data.frame`, hence the name. This function has largely been superceded by **mutate** in **dplyr** but is still good to know. Second, one of the columns was named `Std. Error` with a space. In order to refer to a variable with spaces in its name, even as a column in a `data.frame`, we must enclose the name in back ticks (`).

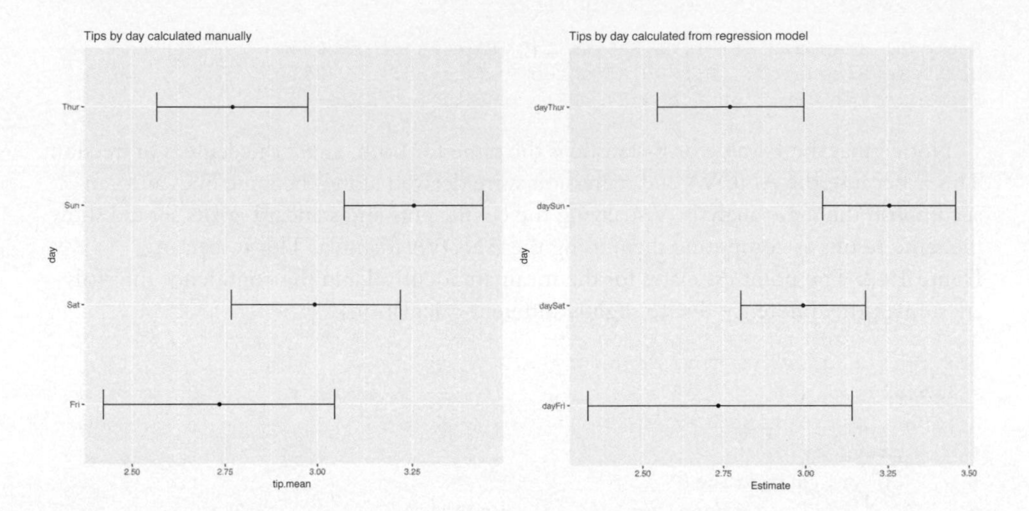

Figure 19.2 Regression coefficients and confidence intervals as taken from a regression model and calcualted manually. The point estimates for the mean are identical and the confidence intervals are very similar, the difference due to slightly different calculations. The y-axis labels are also different because when dealing with factors `lm` tacks on the name of the variable to the level value.

19.2 Multiple Regression

The logical extension of simple linear regression is multiple regression, which allows for multiple predictors. The idea is still the same; we are still making predictions or inferences[1] on the response, but we now have more information in the form of multiple predictors. The math requires some matrix algebra but fortunately the **lm** function is used with very little extra effort.

In this case the relationship between the response and the p predictors ($p - 1$ predictors and the intercept) is modeled as

$$\boldsymbol{Y} = \boldsymbol{X}\boldsymbol{\beta} + \epsilon \tag{19.5}$$

where $\boldsymbol{Y}$ is the nx1 response vector

$$\boldsymbol{Y} = \begin{bmatrix} Y_1 \\ Y_2 \\ Y_3 \\ \vdots \\ Y_n \end{bmatrix} \tag{19.6}$$

1. Prediction is the use of known predictors to predict an unknown response, while inference is figuring out how predictors affect a response.

$\boldsymbol{X}$ is the nxp matrix (n rows and $p-1$ predictors plus the intercept)

$$\boldsymbol{X} = \begin{bmatrix} 1 & X_{11} & X_{12} & \dots & X_{1,p-1} \\ 1 & X_{21} & X_{22} & \dots & X_{2,p-1} \\ \vdots & \vdots & \vdots & \ddots & \vdots \\ 1 & X_{n1} & X_{n2} & \dots & X_{n,p-1} \end{bmatrix} \tag{19.7}$$

$\boldsymbol{\beta}$ is the px1 vector of coefficients (one for each predictor and intercept)

$$\boldsymbol{\beta} = \begin{bmatrix} \beta_0 \\ \beta_1 \\ \beta_2 \\ \vdots \\ \beta_{p-1} \end{bmatrix} \tag{19.8}$$

and $\boldsymbol{\epsilon}$ is the nx1 vector of normally distributed errors

$$\boldsymbol{\epsilon} = \begin{bmatrix} \epsilon_1 \\ \epsilon_2 \\ \epsilon_3 \\ \vdots \\ \epsilon_n \end{bmatrix} \tag{19.9}$$

with

$$\epsilon_i \sim \mathcal{N}(0, \sigma^2 \boldsymbol{I}) \tag{19.10}$$

which seems more complicated than simple regression but the algebra actually gets easier.

The solution for the coefficients is simply written as in Equation 19.11.

$$\hat{\boldsymbol{\beta}} = (\boldsymbol{X}^T \boldsymbol{X})^{-1} \boldsymbol{X}^T \boldsymbol{Y} \tag{19.11}$$

To see this in action we use New York City condo evaluations for fiscal year 2011-2012, obtained through NYC Open Data. NYC Open Data is an initiative by New York City to make government more transparent and work better. It provides data on all manner of city services to the public for analysis, scrutiny and app building (through `http://nycbigapps.com/`). It has been surprisingly popular, spawning hundreds of mobile apps and being copied in other cities such as Chicago and Washington, DC. Its Web site is at `https://data.cityofnewyork.us/`.

The original data were separated by borough with one file each for Manhattan,[2] Brooklyn,[3] Queens,[4] the Bronx[5] and Staten Island,[6] and contained extra information we will not be using. So we combined the five files into one, cleaned up the column names and posted it at `http://www.jaredlander.com/data/housing.csv`. To access the data, either download it from that URL and use **read.table** on the now local file, or read it directly from the URL.

```
> housing <- read.table("http://www.jaredlander.com/data/housing.csv",
+                       sep = ",", header = TRUE,
+                       stringsAsFactors = FALSE)
```

A few reminders about what that code does: `sep` specifies that commas were used to separate columns; `header` means the first row contains the column names; and `stringsAsFactors` leaves `character` columns as they are and does not convert them to `factor`s, which speeds up loading time and also makes them easier to work with. Looking at the data, we see that we have a lot of columns and some bad names, so we should rename those.[7]

```
> names(housing) <- c("Neighborhood", "Class", "Units", "YearBuilt",
+                     "SqFt", "Income", "IncomePerSqFt", "Expense",
+                     "ExpensePerSqFt", "NetIncome", "Value",
+                     "ValuePerSqFt", "Boro")
> head(housing)

  Neighborhood           Class Units YearBuilt   SqFt   Income
1    FINANCIAL R9-CONDOMINIUM    42      1920  36500  1332615
2    FINANCIAL R4-CONDOMINIUM    78      1985 126420  6633257
3    FINANCIAL RR-CONDOMINIUM   500        NA 554174 17310000
4    FINANCIAL R4-CONDOMINIUM   282      1930 249076 11776313
5      TRIBECA R4-CONDOMINIUM   239      1985 219495 10004582
6      TRIBECA R4-CONDOMINIUM   133      1986 139719  5127687
  IncomePerSqFt Expense ExpensePerSqFt NetIncome    Value
1         36.51  342005           9.37    990610  7300000
2         52.47 1762295          13.94   4870962 30690000
3         31.24 3543000           6.39  13767000 90970000
4         47.28 2784670          11.18   8991643 67556006
```

2. `https://data.cityofnewyork.us/Finances/DOF-Condominium-Comparable-Rental-Income-Manhattan/dvzp h4k9`
3. `https://data.cityofnewyork.us/Finances/DOF-Condominium-Comparable-Rental-Income-Brooklyn-/bss9-579f`
4. `https://data.cityofnewyork.us/Finances/DOF-Condominium-Comparable-Rental-Income-Queens-FY/jcih-dj9q`
5. `https://data.cityofnewyork.us/Property/DOF-Condominium-Comparable-Rental-Income-Bronx-FY-/3qfc-4tta`
6. `https://data.cityofnewyork.us/Finances/DOF-Condominium-Comparable-Rental-Income-Staten-Is/tkdy-59zg`
7. A copy of this file that already has the fixed names is available at `http://www.jaredlander.com/data/housing1.csv`.

```
5        45.58 2783197          12.68    7221385 54320996
6        36.70 1497788          10.72    3629899 26737996
  ValuePerSqFt       Boro
1       200.00 Manhattan
2       242.76 Manhattan
3       164.15 Manhattan
4       271.23 Manhattan
5       247.48 Manhattan
6       191.37 Manhattan
```

For these data the response is the value per square foot and the predictors are everything else. However, we ignore the income and expense variables, as they are actually just estimates based on an arcane requirement that condos be compared to rentals for valuation purposes. The first step is to visualize the data in some exploratory data analysis. The natural place to start is with a histogram of `ValuePerSqFt`, which is shown in Figure 19.3.

```
> ggplot(housing, aes(x=ValuePerSqFt)) +
+     geom_histogram(binwidth=10) + labs(x="Value per Square Foot")
```

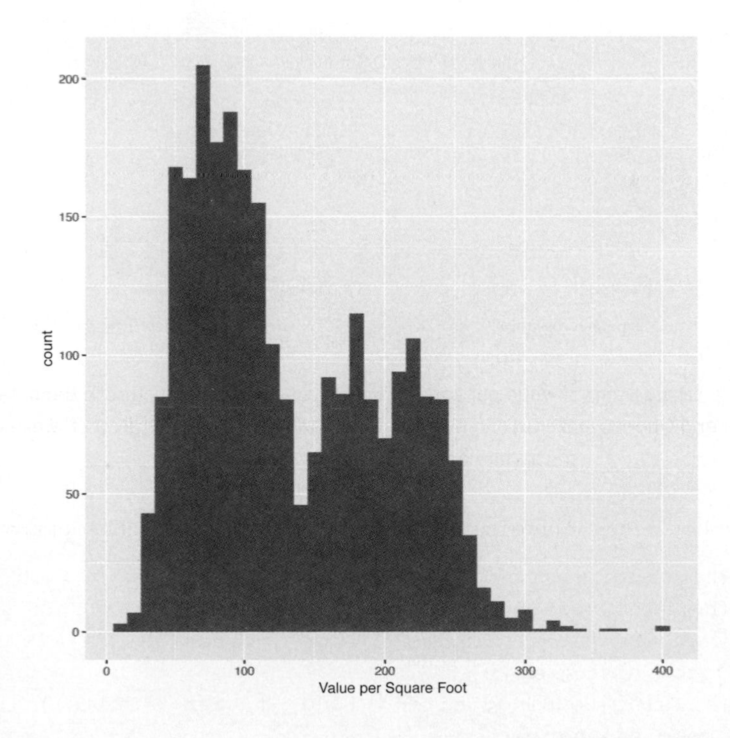

Figure 19.3 Histogram of value per square foot for NYC condos. It appears to be bimodal.

The bimodal nature of the histogram means there is something left to be explored. Mapping color to `Boro` in Figure 19.4a and faceting on `Boro` in Figure 19.4b reveal that Brooklyn and Queens make up one mode and Manhattan makes up the other, while there is not much data on the Bronx and Staten Island.

```
> ggplot(housing, aes(x=ValuePerSqFt, fill=Boro)) +
+     geom_histogram(binwidth=10) + labs
        (x="Value per Square Foot")
> ggplot(housing, aes(x=ValuePerSqFt, fill=Boro)) +
+     geom_histogram(binwidth=10) + labs
        (x="Value per Square Foot") +
+     facet_wrap(~Boro)
```

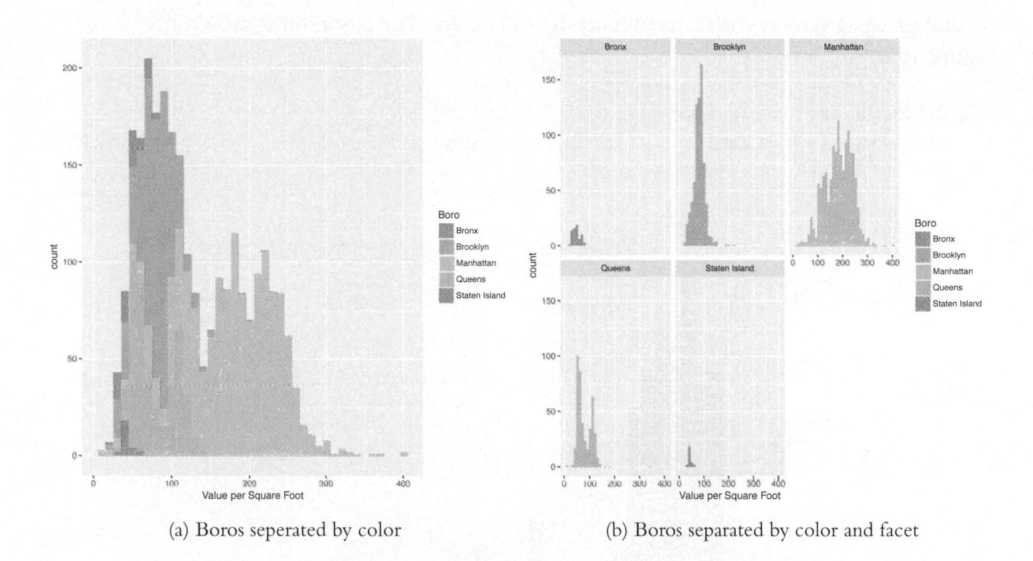

(a) Boros seperated by color

(b) Boros separated by color and facet

Figure 19.4 Histograms of value per square foot. These illustrate structure in the data, revealing that Brooklyn and Queens make up one mode and Manhattan makes up the other, while there is not much data on the Bronx and Staten Island.

Next we should look at histograms for square footage and the number of units.

```
> ggplot(housing, aes(x=SqFt)) + geom_histogram()
> ggplot(housing, aes(x=Units)) + geom_histogram()
> ggplot(housing[housing$Units < 1000, ], aes(x=SqFt)) +
+         geom_histogram()
> ggplot(housing[housing$Units < 1000, ], aes(x=Units)) +
+         geom_histogram()
```

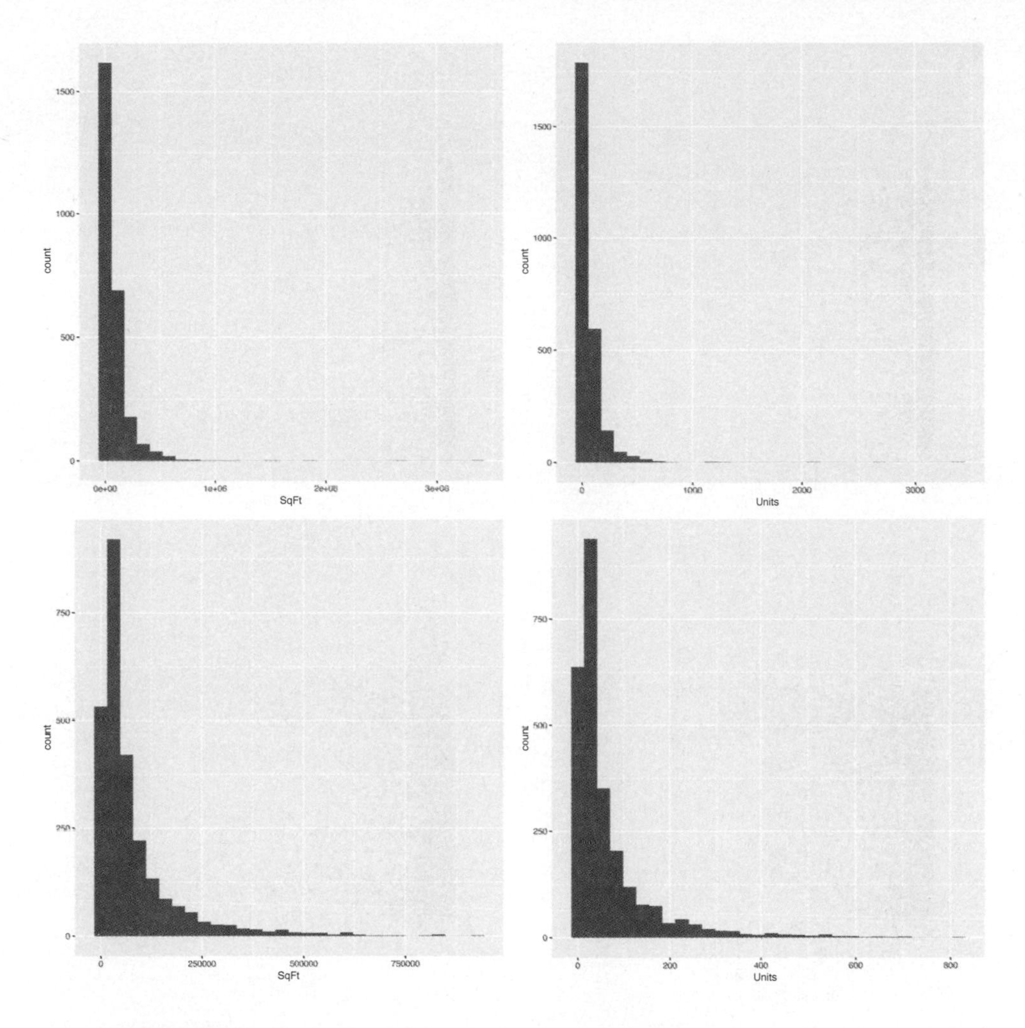

Figure 19.5 Histograms for total square feet and number of units. The distributions are highly right skewed in the top two graphs, so they were repeated after removing buildings with more than 1,000 units.

Figure 19.5 shows that there are quite a few buildings with an incredible number of units. Plotting scatterplots in Figure 19.6 of the value per square foot versus both number of units and square footage, with and without those outlying buildings, gives us an idea whether we can remove them from the analysis.

```
> ggplot(housing, aes(x=SqFt, y=ValuePerSqFt)) + geom_point()
> ggplot(housing, aes(x=Units, y=ValuePerSqFt)) + geom_point()
> ggplot(housing[housing$Units < 1000, ], aes(x=SqFt, y=ValuePerSqFt)) +
+     geom_point()
> ggplot(housing[housing$Units < 1000, ], aes(x=Units, y=ValuePerSqFt)) +
+     geom_point()
```

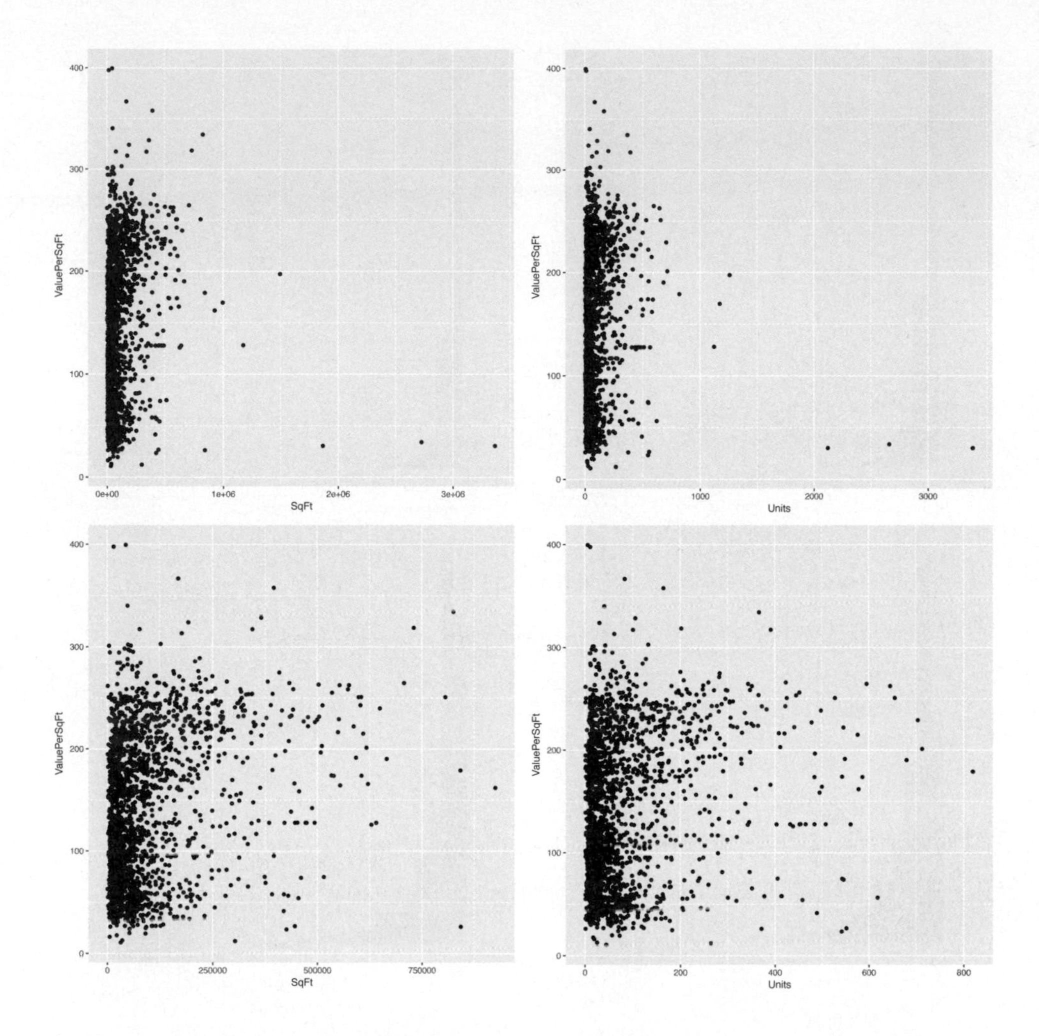

Figure 19.6 Scatterplots of value per square foot versus square footage and value versus number of units, both with and without the buildings that have over 1,000 units.

```
> # how many need to be removed?
> sum(housing$Units >= 1000)

[1] 6

> # remove them
> housing <- housing[housing$Units < 1000, ]
```

Even after we remove the outliers, it still seems like a log transformation of some data could be helpful. Figures 19.7 and 19.8 show that taking the log of square footage and number of units might prove helpful. It also shows what happens when taking the log of value.

```
> # plot ValuePerSqFt against SqFt
> ggplot(housing, aes(x=SqFt, y=ValuePerSqFt)) + geom_point()
> ggplot(housing, aes(x=log(SqFt), y=ValuePerSqFt)) + geom_point()
> ggplot(housing, aes(x=SqFt, y=log(ValuePerSqFt))) + geom_point()
> ggplot(housing, aes(x=log(SqFt), y=log(ValuePerSqFt))) +
+     geom_point()
```

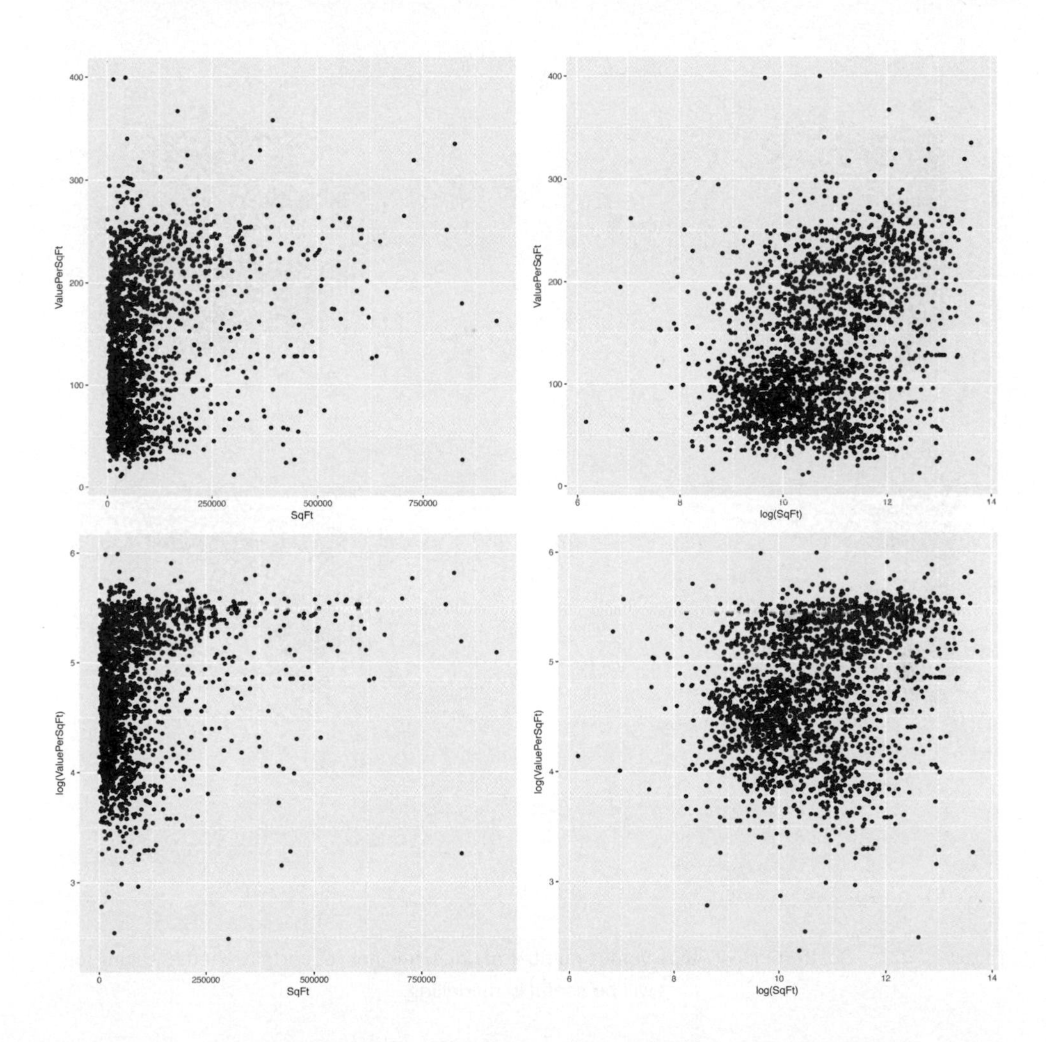

Figure 19.7 Scatterplots of value versus sqaure footage. The plots indicate that taking the log of `SqFt` might be useful in modelling.

```
> # plot ValuePerSqFt against Units
> ggplot(housing, aes(x=Units, y=ValuePerSqFt)) + geom_point()
> ggplot(housing, aes(x=log(Units), y=ValuePerSqFt)) + geom_point()
```

```
> ggplot(housing, aes(x=Units, y=log(ValuePerSqFt))) + geom_point()
> ggplot(housing, aes(x=log(Units), y=log(ValuePerSqFt))) +
+     geom_point()
```

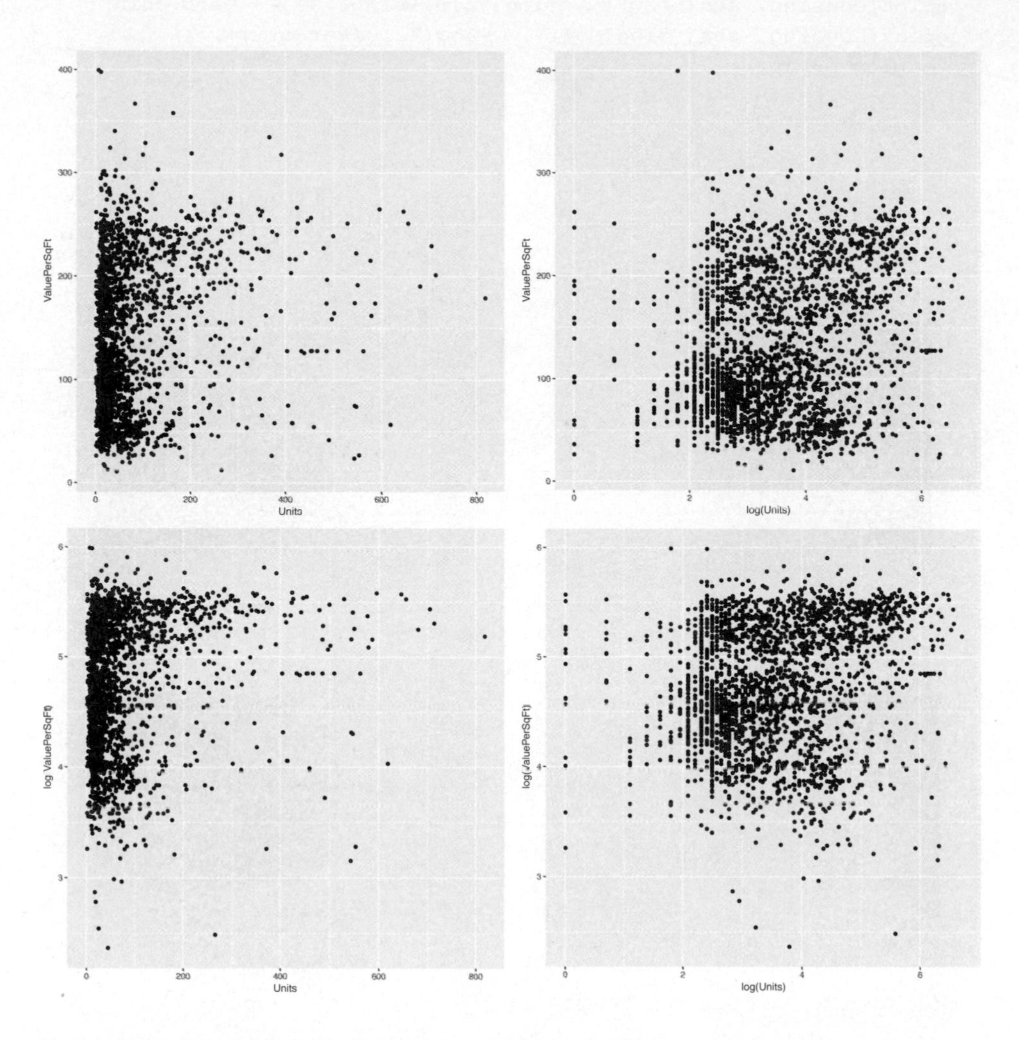

Figure 19.8 Scatterplots of value versus number of units. It is not yet certain whether taking logs will be useful in modelling.

Now that we have viewed our data a few different ways, it is time to start modelling. We already saw from Figure 19.4 that accounting for the different boroughs will be important and the various scatterplots indicated that `Units` and `SqFt` will be important as well.

Fitting the model uses the `formula` interface in **lm**. Now that there are multiple predictors, we separate them on the right side of the `formula` using plus signs (+).

```
> house1 <- lm(ValuePerSqFt ~ Units + SqFt + Boro, data=housing)
> summary(house1)

Call:
lm(formula = ValuePerSqFt ~ Units + SqFt + Boro, data = housing)

Residuals:
     Min       1Q   Median       3Q      Max
-168.458  -22.680    1.493   26.290  261.761

Coefficients:
                     Estimate Std. Error t value Pr(>|t|)
(Intercept)         4.430e+01  5.342e+00   8.293  < 2e-16 ***
Units              -1.532e-01  2.421e-02  -6.330 2.88e-10 ***
SqFt                2.070e-04  2.129e-05   9.723  < 2e-16 ***
BoroBrooklyn        3.258e+01  5.561e+00   5.858 5.28e-09 ***
BoroManhattan       1.274e+02  5.459e+00  23.343  < 2e-16 ***
BoroQueens          3.011e+01  5.711e+00   5.272 1.46e-07 ***
BoroStaten Island -7.114e+00  1.001e+01  -0.711    0.477
---
Signif. codes:  0 '***' 0.001 '**' 0.01 '*' 0.05 '.' 0.1 ' ' 1

Residual standard error: 43.2 on 2613 degrees of freedom
Multiple R-squared:  0.6034,Adjusted R-squared:  0.6025
F-statistic: 662.6 on 6 and 2613 DF,  p-value: < 2.2e-16
```

The first thing to notice is that in some versions of R there is a message warning us that `Boro` was converted to a `factor`. This is because `Boro` was stored as a `character`, and for modelling purposes `character` data must be represented using indicator variables, which is how `factor`s are treated inside modelling functions, as seen in Section 5.1.

The **summary** function prints out information about the model, including how the function was called, quantiles for the residuals, coefficient estimates, standard errors and p-values for each variable, and the degrees of freedom, p-value and F-statistic for the model. There is no coefficient for the Bronx because that is the baseline `level` of `Boro`, and all the other `Boro` coefficients are relative to that baseline.

The coefficients represent the effect of the predictors on the response and the standard errors are the uncertainty in the estimation of the coefficients. The t value (t-statistic) and p-value for the coefficients are numerical measures of statistical significance, though these should be viewed with caution, as most modern data scientists do not like to look at the statistical significance of individual coefficients but rather judge the model as a whole as covered in Chapter 21.

The model p-value and F-statistic are measures of its goodness of fit. The degrees of freedom for a regression are calculated as the number of observations minus the number of coefficients. In this example, there are `nrow(housing) − length(coef(house1))` = 2613 degrees of freedom.

A quick way to grab the coefficients from a model is to either use the **coef** function or get them from the model using the $ operator on the model object.

```
> house1$coefficients
        (Intercept)                Units                 SqFt
       4.430325e+01        -1.532405e-01         2.069727e-04
       BoroBrooklyn        BoroManhattan          BoroQueens
       3.257554e+01         1.274259e+02         3.011000e+01
BoroStaten Island
      -7.113688e+00

> coef(house1)
        (Intercept)                Units                 SqFt
       4.430325e+01        -1.532405e-01         2.069727e-04
       BoroBrooklyn        BoroManhattan          BoroQueens
       3.257554e+01         1.274259e+02         3.011000e+01
BoroStaten Island
      -7.113688e+00

> # works the same as coef
> coefficients(house1)
        (Intercept)                Units                 SqFt
       4.430325e+01        -1.532405e-01         2.069727e-04
       BoroBrooklyn        BoroManhattan          BoroQueens
       3.257554e+01         1.274259e+02         3.011000e+01
BoroStaten Island
      -7.113688e+00
```

As a repeated theme, we prefer visualizations over tables of information, and a great way of visualizing regression results is a coefficient plot, like the one shown in Figure 19.2. Rather than build it from scratch, we use the convenient **coefplot** package that we wrote. Figure 19.9 shows the result, where each coefficient is plotted as a point with a thick line representing the one standard error confidence interval and a thin line representing the two standard error confidence interval. There is a vertical line indicating 0. In general, a good rule of thumb is that if the two standard error confidence interval does not contain 0, it is statistically significant.

```
> library(coefplot)
> coefplot(house1)
```

Figure 19.9 shows that, as expected, being located in Manhattan has the largest effect on value per square foot. Surprisingly, the number of units or square feet in a building has little effect on value. This is a model with purely additive terms. Interactions between variables can be equally powerful. To enter them in a `formula`, separate the desired variables with a `*` instead of `+`. Doing so results in the individual variables plus the interaction term being included in the model. To include just the interaction term, and

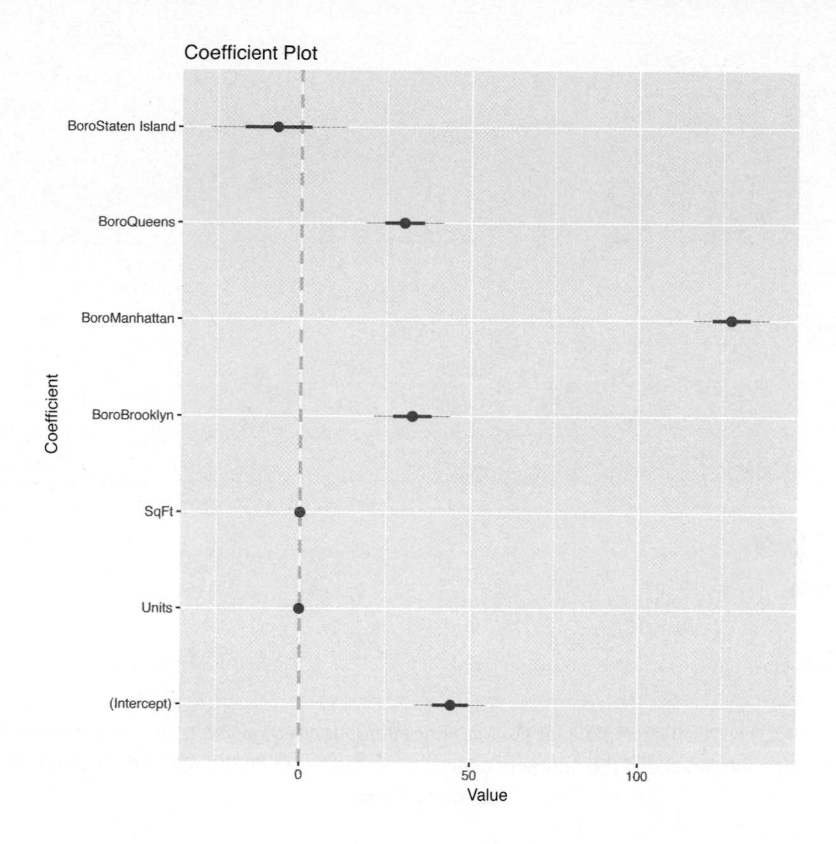

Figure 19.9 Coefficient plot for condo value regression.

not the individual variables, use : instead. The results of interacting Units and SqFt are shown in Figure 19.10.

```
> house2 <- lm(ValuePerSqFt ~ Units * SqFt + Boro, data=housing)
> house3 <- lm(ValuePerSqFt ~ Units : SqFt + Boro, data=housing)
> house2$coefficients

      (Intercept)            Units             SqFt
     4.093685e+01    -1.024579e-01     2.362293e-04
     BoroBrooklyn    BoroManhattan       BoroQueens
     3.394544e+01     1.272102e+02     3.040115e+01
BoroStaten Island       Units:SqFt
    -8.419682e+00    -1.809587e-07

> house3$coefficients

      (Intercept)      BoroBrooklyn    BoroManhattan
     4.804972e+01      3.141208e+01     1.302084e+02
       BoroQueens BoroStaten Island       Units:SqFt
     2.841669e+01     -7.199902e+00     1.088059e-07
```

```
> coefplot(house2)
> coefplot(house3)
```

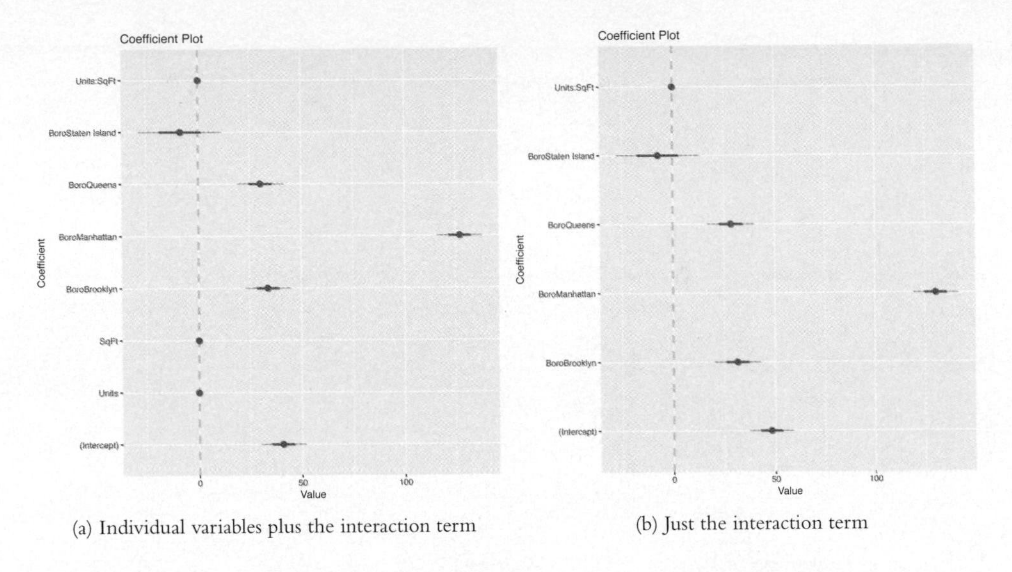

(a) Individual variables plus the interaction term

(b) Just the interaction term

Figure 19.10 Coefficient plots for models with interaction terms. The figure on the left includes individual variables and the interaction term, while the figure on the right only includes the interaction term.

If three variables all interact together, the resulting coefficients will be the three individual terms, three two-way interactions and one three-way interaction.

```
> house4 <- lm(ValuePerSqFt ~ SqFt*Units*Income, housing)
> house4$coefficients

      (Intercept)              SqFt             Units
     1.116433e+02     -1.694688e-03      7.142611e-03
           Income        SqFt:Units       SqFt:Income
     7.250830e-05      3.158094e-06     -5.129522e-11
     Units:Income SqFt:Units:Income
    -1.279236e-07      9.107312e-14
```

Interacting (from now on, unless otherwise specified, interacting will refer to the `*` operator) a continuous variable like `SqFt` with a `factor` like `Boro` results in individual terms for the continuous variable and each non-baseline `level` of the `factor` plus an interaction term between the continuous variable and each non-baseline `level` of the `factor`. Interacting two (or more) `factors` yields terms for all the individual non-baseline `levels` in both `factors` and an interaction term for every combination of non-baseline `levels` of the `factors`.

```
> house5 <- lm(ValuePerSqFt ~ Class*Boro, housing)
> house5$coefficients

                                  (Intercept)
                                    47.041481
                         ClassR4-CONDOMINIUM
                                     4.023852
                         ClassR9-CONDOMINIUM
                                    -2.838624
                         ClassRR-CONDOMINIUM
                                     3.688519
                                BoroBrooklyn
                                    27.627141
                               BoroManhattan
                                    89.598397
                                  BoroQueens
                                    19.144780
                           BoroStaten Island
                                    -9.203410
            ClassR4-CONDOMINIUM:BoroBrooklyn
                                     4.117977
            ClassR9-CONDOMINIUM:BoroBrooklyn
                                     2.660419
            ClassRR-CONDOMINIUM:BoroBrooklyn
                                   -25.607141
           ClassR4-CONDOMINIUM:BoroManhattan
                                    47.198900
           ClassR9-CONDOMINIUM:BoroManhattan
                                    33.479718
           ClassRR-CONDOMINIUM:BoroManhattan
                                    10.619231
              ClassR4-CONDOMINIUM:BoroQueens
                                    13.588293
              ClassR9-CONDOMINIUM:BoroQueens
                                    -9.830637
              ClassRR-CONDOMINIUM:BoroQueens
                                    34.675220
       ClassR4-CONDOMINIUM:BoroStaten Island
                                           NA
       ClassR9-CONDOMINIUM:BoroStaten Island
                                           NA
       ClassRR-CONDOMINIUM:BoroStaten Island
                                           NA
```

Neither SqFt nor Units appear to be significant in any model when viewed in a coefficient plot. However, zooming in on the plot shows that the coefficients for Units and SqFt are non-zero as seen in Figure 19.11.

```
> coefplot(house1, sort='mag') + scale_x_continuous(limits=c(-.25, .1))
> coefplot(house1, sort='mag') + scale_x_continuous(limits=c(-.0005, .0005))
```

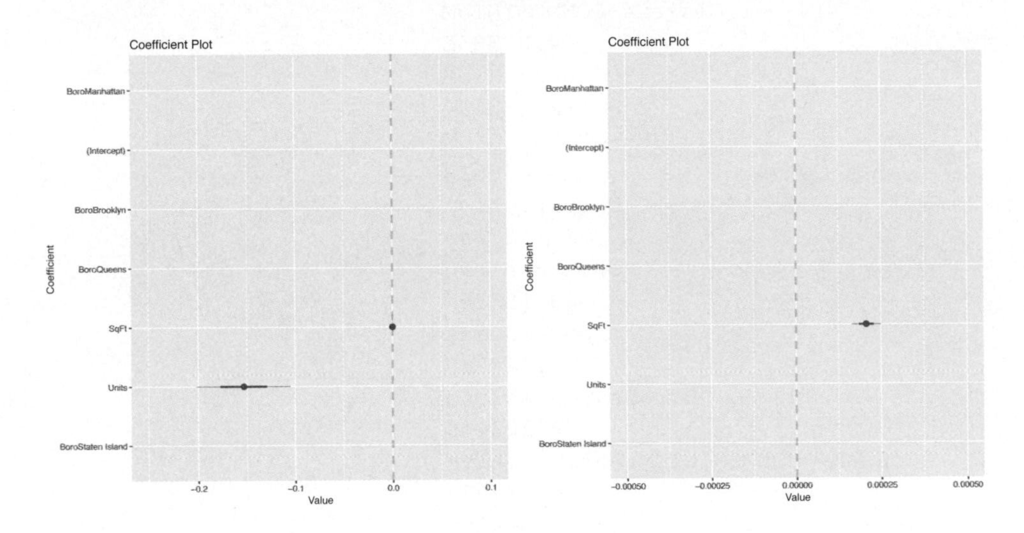

Figure 19.11 Coefficient plots for model house1 zoomed in to show the coefficients for Units and SqFt.

This is likely a scaling issue, as the indicator variables for Boro are on the scale of 0 and 1 while the range for Units is between 1 and 818 and SqFt is between 478 and 925,645. This can be resolved by standardizing, or scaling, the variables. This subtracts the mean and divides by the standard deviation. While the results of the model will mathematically be the same, the coefficients will have different values and different interpretations. Whereas before a coefficient was the change in the response corresponding to a one-unit increase in the predictor, the coefficientis now the change in the response corresponding to a one-standard-deviation increase in the predictor. Standardizing can be performed within the formula interface with the **scale** function.

```
> house1.b <- lm(ValuePerSqFt ~ scale(Units) + scale(SqFt) + Boro,
+                data=housing)
> coefplot(house1.b, sort='mag')
```

The coefficient plot in Figure 19.12 shows that for each change in the standard deviation of SqFt there is a change of 21.95 in ValuePerSqFt. We also see that Units

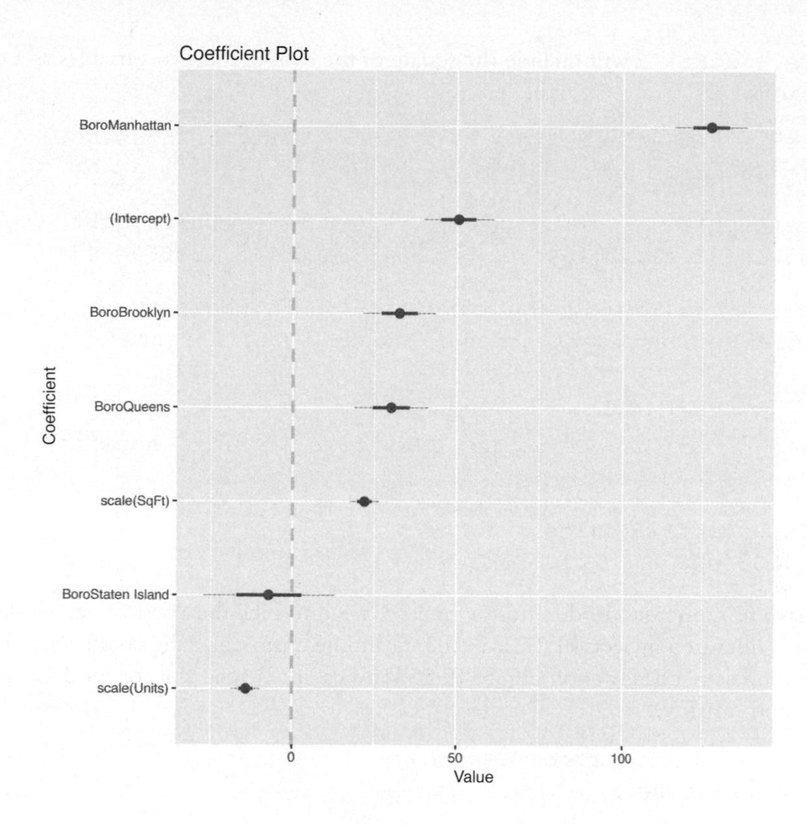

Figure 19.12 Coefficient plot for a model with standardized values for `Units` and `SqFt`. This shows that having fewer, but larger, units is beneficial to the value of a building.

has a negative impact. This implies that having fewer, but larger, units is beneficial to the value of a building.

Another good test is to include the ratio of `Units` and `SqFt` as a single variable. To simply divide one variable by another in a `formula`, the division must be wrapped in the **I** function.

```
> house6 <- lm(ValuePerSqFt ~ I(SqFt/Units) + Boro, housing)
> house6$coefficients

      (Intercept)     I(SqFt/Units)       BoroBrooklyn
     43.754838763       0.004017039       30.774343209
    BoroManhattan        BoroQueens BoroStaten Island
    130.769502685      29.767922792       -6.134446417
```

The **I** function is used to preserve a mathematical relationship in a `formula` and prevent it from being interpreted according to `formula` rules. For instance, using `(Units + SqFt)^2` in a `formula` is the same as using `Units * SqFt`, whereas

I(Units + SqFt)^2 will include the square of the sum of the two variables as a term in the formula.

```
> house7 <- lm(ValuePerSqFt ~ (Units + SqFt)^2, housing)
> house7$coefficients

  (Intercept)         Units          SqFt    Units:SqFt
 1.070301e+02 -1.125194e-01  4.964623e-04 -5.159669e-07

> house8 <- lm(ValuePerSqFt ~ Units * SqFt, housing)
> identical(house7$coefficients, house8$coefficients)

[1] TRUE

> house9 <- lm(ValuePerSqFt ~ I(Units + SqFt)^2, housing)
> house9$coefficients

   (Intercept) I(Units + SqFt)
  1.147034e+02    2.107231e-04
```

We have fit numerous models from which we need to pick the "best" one. Model selection is discussed in Section 21.2. In the meantime, visualizing the coefficients from multiple models is a handy tool. Figure 19.13 shows a coefficient plot for models house1, house2 and house3.

```
> # also from the coefplot package
> multiplot(house1, house2, house3)
```

Regression is often used for prediction, which in R is enabled by the **predict** function. For this example, new data are available at http://www.jaredlander.com/data/housingNew.csv.

```
> housingNew <- read.table("http://www.jaredlander.com/data/housingNew.csv",
+                          sep=",", header=TRUE, stringsAsFactors=FALSE)
```

Making the prediction can be as simple as calling **predict**, although caution must be used when dealing with factor predictors to ensure that they have the same levels as those used in building the model.

```
> # make prediction with new data and 95% confidence bounds
> housePredict <- predict(house1, newdata=housingNew, se.fit=TRUE,
+                        interval="prediction", level=.95)
> # view predictions with upper and lower bounds based on standard errors
> head(housePredict$fit)

        fit        lwr      upr
1  74.00645 -10.813887 158.8268
2  82.04988  -2.728506 166.8283
3 166.65975  81.808078 251.5114
4 169.00970  84.222648 253.7968
5  80.00129  -4.777303 164.7799
6  47.87795 -37.480170 133.2361

> # view the standard errors for the prediction
```

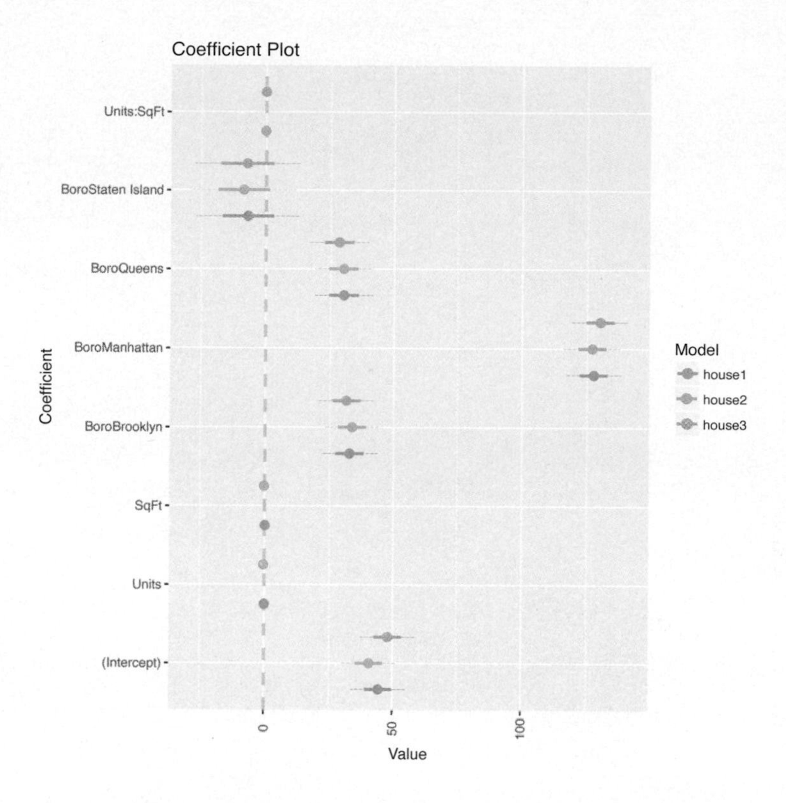

Figure 19.13 Coefficient plot for multiple condo models. The coefficients are plotted in the same spot on the y-axis for each model. If a model does not contain a particular coefficient, it is simply not plotted.

```
> head(housePredict$se.fit)

       1        2        3        4        5        6
2.118509 1.624063 2.423006 1.737799 1.626923 5.318813
```

19.3 Conclusion

Perhaps one of the most versatile tools in statistical analysis, regression is well handled using R's **lm** function. It takes the `formula` interface, where a response is modeled on a set of predictors. Other useful arguments to the function are `weights`, which specifies the weights attributed to observations (both probability and count weights), and `subset`, which will fit the model only on a subset of the data.

20

Generalized Linear Models

Not all data can be appropriately modeled with linear regression, because they are binomial (TRUE/FALSE), count data or some other form. To model these types of data, generalized linear models were developed. They are still modeled using a linear predictor, $\boldsymbol{X\beta}$, but they are transformed using some link function. To the R user, fitting a generalized linear model requires barely any more effort than running a linear regression.

20.1 Logistic Regression

A very powerful and common model—especially in fields such as marketing and medicine—is logistic regression. The examples in this section will use the a subset of data from the 2010 American Community Survey (ACS) for New York State.[1] ACS data contain a lot of information, so we have made a subset of it with 22,745 rows and 18 columns available at `http://jaredlander.com/data/acs_ny.csv`.

```
> acs <- read.table("http://jaredlander.com/data/acs_ny.csv",
+                   sep=",", header=TRUE, stringsAsFactors=FALSE)
```

Logistic regression models are formulated as

$$p(y_i = 1) = \text{logit}^{-1}(\boldsymbol{X}_i\boldsymbol{\beta}) \qquad (20.1)$$

where y_i is the ith response and $\boldsymbol{X}_i\boldsymbol{\beta}$ is the linear predictor. The inverse logit function

$$\text{logit}^{-1}(x) = \frac{e^x}{1 + e^x} = \frac{1}{1 + e^{-x}} \qquad (20.2)$$

transforms the continuous output from the linear predictor to fall between 0 and 1. This is the inverse of the link function.

We now formulate a question that asks whether a household has an income greater than \$150,000. To do this we need to create a new binary variable with TRUE for income above that mark and FALSE for income below.

1. The ACS is a large-scale survey very similar to the decennial census, except that is conducted on a more frequent basis.

```
> acs$Income <- with(acs, FamilyIncome >= 150000)
> library(ggplot2)
> library(useful)
> ggplot(acs, aes(x=FamilyIncome)) +
+     geom_density(fill="grey", color="grey") +
+     geom_vline(xintercept=150000) +
+     scale_x_continuous(label=multiple.dollar, limits=c(0, 1000000))
```

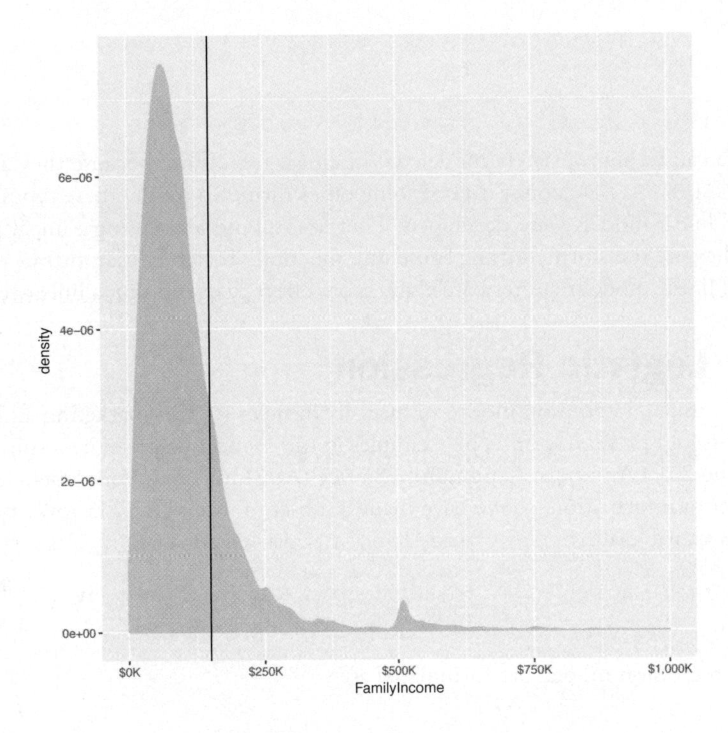

Figure 20.1 Density plot of family income, with a vertical line indicating the $150,000 mark.

```
> head(acs)

  Acres FamilyIncome  FamilyType NumBedrooms NumChildren NumPeople
1  1-10          150     Married           4           1         3
2  1-10          180 Female Head           3           2         4
3  1-10          280 Female Head           4           0         2
4  1-10          330 Female Head           2           1         2
5  1-10          330   Male Head           3           1         2
6  1-10          480   Male Head           0           3         4
  NumRooms         NumUnits NumVehicles NumWorkers  OwnRent
1        9 Single detached           1          0 Mortgage
2        6 Single detached           2          0   Rented
3        8 Single detached           3          1 Mortgage
```

```
4       4 Single detached          1         0   Rented
5       5 Single attached          1         0 Mortgage
6       1 Single detached          0         0   Rented
    YearBuilt HouseCosts ElectricBill FoodStamp HeatingFuel Insurance
1   1950-1959       1800           90        No         Gas      2500
2 Before 1939        850           90        No         Oil         0
3   2000-2004       2600          260        No         Oil      6600
4   1950-1959       1800          140        No         Oil         0
5 Before 1939        860          150        No         Gas       660
6 Before 1939        700          140        No         Gas         0
        Language Income
1        English  FALSE
2        English  FALSE
3 Other European  FALSE
4        English  FALSE
5        Spanish  FALSE
6        English  FALSE
```

Running a logistic regression is done very similarly to running a linear regression. It still uses the `formula` interface but the function is **glm** rather than **lm** (**glm** can actually fit linear regressions as well), and a few more options need to be set.

```
> income1 <- glm(Income ~ HouseCosts + NumWorkers + OwnRent  +
+                     NumBedrooms + FamilyType,
+                 data=acs, family=binomial(link="logit"))
> summary(income1)

Call:
glm(formula = Income ~ HouseCosts + NumWorkers + OwnRent + NumBedrooms +
    FamilyType, family = binomial(link = "logit"), data = acs)

Deviance Residuals:
    Min       1Q   Median       3Q      Max
-2.8452  -0.6246  -0.4231  -0.1743   2.9503

Coefficients:
                    Estimate Std. Error z value Pr(>|z|)
(Intercept)       -5.738e+00  1.185e-01 -48.421   <2e-16 ***
HouseCosts         7.398e-04  1.724e-05  42.908   <2e-16 ***
NumWorkers         5.611e-01  2.588e-02  21.684   <2e-16 ***
OwnRentOutright    1.772e+00  2.075e-01   8.541   <2e-16 ***
OwnRentRented     -8.886e-01  1.002e-01  -8.872   <2e-16 ***
NumBedrooms        2.339e-01  1.683e-02  13.895   <2e-16 ***
FamilyTypeMale Head 3.336e-01  1.472e-01   2.266   0.0235 *
FamilyTypeMarried  1.405e+00  8.704e-02  16.143   <2e-16 ***
---
Signif. codes:  0 '***' 0.001 '**' 0.01 '*' 0.05 '.' 0.1 ' ' 1

(Dispersion parameter for binomial family taken to be 1)
```

```
    Null deviance: 22808  on 22744  degrees of freedom
Residual deviance: 18073  on 22737  degrees of freedom
AIC: 18089

Number of Fisher Scoring iterations: 6

> library(coefplot)
> coefplot(income1)
```

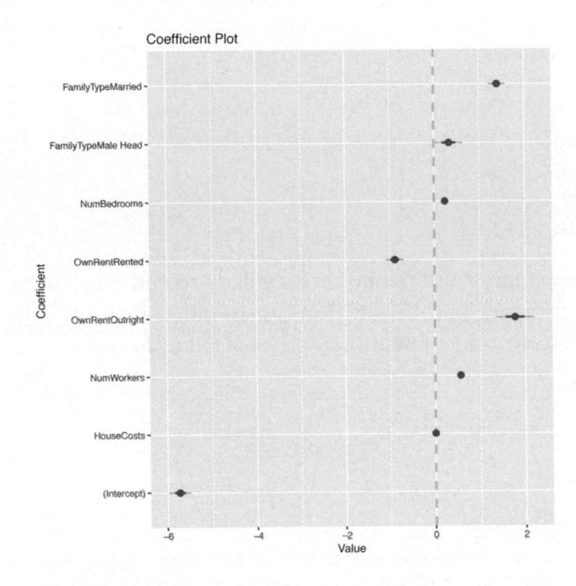

Figure 20.2 Coefficient plot for logistic regression on family income greater than $150,000, based on the American Community Survey.

The output from **summary** and **coefplot** for **glm** is similar to that of **lm**. There are coefficient estimates, standard errors, p-values—both overall and for the coefficients—and a measure of correctness, which in this case is the deviance and AIC. A general rule of thumb is that adding a variable (or a `level` of a `factor`) to a model should result in a drop in deviance of two; otherwise, the variable is not useful in the model. Interactions and all the other `formula` concepts work the same.

Interpreting the coefficients from a logistic regression necessitates taking the inverse logit.

```
> invlogit <- function(x)
+ {
+     1 / (1 + exp(-x))
+ }
> invlogit(income1$coefficients)

        (Intercept)           HouseCosts           NumWorkers
        0.003211572          0.500184950          0.636702036
```

```
   OwnRentOutright          OwnRentRented          NumBedrooms
       0.854753527            0.291408659          0.558200010
FamilyTypeMale Head    FamilyTypeMarried
       0.582624773            0.802983719
```

20.2 Poisson Regression

Another popular member of the generalized linear models is Poisson regression, which, much like the Poisson distribution, is used for count data. Like all other generalized linear models, it is called using **glm**. To illustrate we continue using the ACS data with the number of children (`NumChildren`) as the response.

The formulation for Poisson regression is

$$y_i \sim pois(\theta_i) \tag{20.3}$$

where y_i is the ith response and

$$\theta_i = e^{X_i\beta} \tag{20.4}$$

is the mean of the distribution for the ith observation.

Before fitting a model, we look at the histogram of the number of children in each household.

```
> ggplot(acs, aes(x=NumChildren)) + geom_histogram(binwidth=1)
```

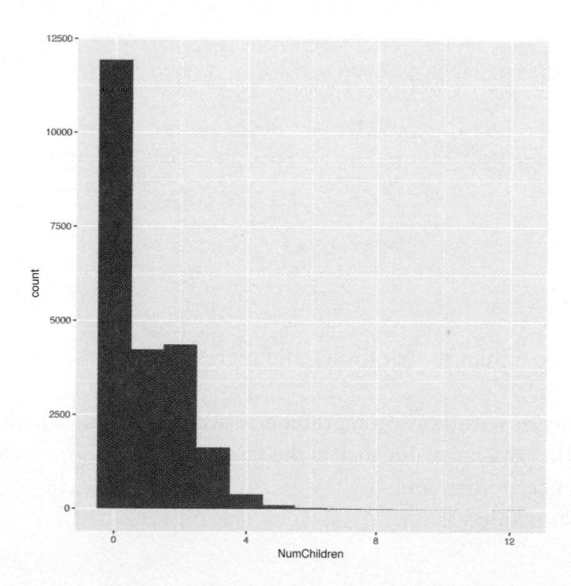

Figure 20.3 Histogram of the number of children per household from the American Community Survey. The distribution is not perfectly Poisson but it is sufficiently so for modelling with Poisson regression.

While Figure 20.3 does not show data that have a perfect Poisson distribution, it is close enough to fit a good model. The coeficient plot is shown in Figure 20.4.

```
> children1 <- glm(NumChildren ~ FamilyIncome + FamilyType + OwnRent,
+                  data=acs, family=poisson(link="log"))
> summary(children1)

Call:
glm(formula = NumChildren ~ FamilyIncome + FamilyType + OwnRent,
    family = poisson(link = "log"), data = acs)

Deviance Residuals:
    Min       1Q   Median       3Q      Max
-1.9950  -1.3235  -1.2045   0.9464   6.3781

Coefficients:
                     Estimate Std. Error z value Pr(>|z|)
(Intercept)        -3.257e-01  2.103e-02 -15.491  < 2e-16 ***
FamilyIncome        5.420e-07  6.572e-08   8.247  < 2e-16 ***
FamilyTypeMale Head -6.298e-02  3.847e-02  -1.637    0.102
FamilyTypeMarried   1.440e-01  2.147e-02   6.707 1.98e-11 ***
OwnRentOutright    -1.974e+00  2.292e-01  -8.611  < 2e-16 ***
OwnRentRented       4.086e-01  2.067e-02  19.773  < 2e-16 ***
---
Signif. codes:  0 '***' 0.001 '**' 0.01 '*' 0.05 '.' 0.1 ' ' 1

(Dispersion parameter for poisson family taken to be 1)

    Null deviance: 35240  on 22744  degrees of freedom
Residual deviance: 34643  on 22739  degrees of freedom
AIC: 61370

Number of Fisher Scoring iterations: 5

> coefplot(children1)
```

The output here is similar to that for logistic regression, and the same rule of thumb for deviance applies.

A particular concern with Poisson regression is overdispersion, which means that the variability seen in the data is greater than is theorized by the Poisson distribution where the mean and variance are the same.

Overdispersion is defined as

$$OD = \frac{1}{n-p}\sum_{i=1}^{n} z_i^2 \tag{20.5}$$

where

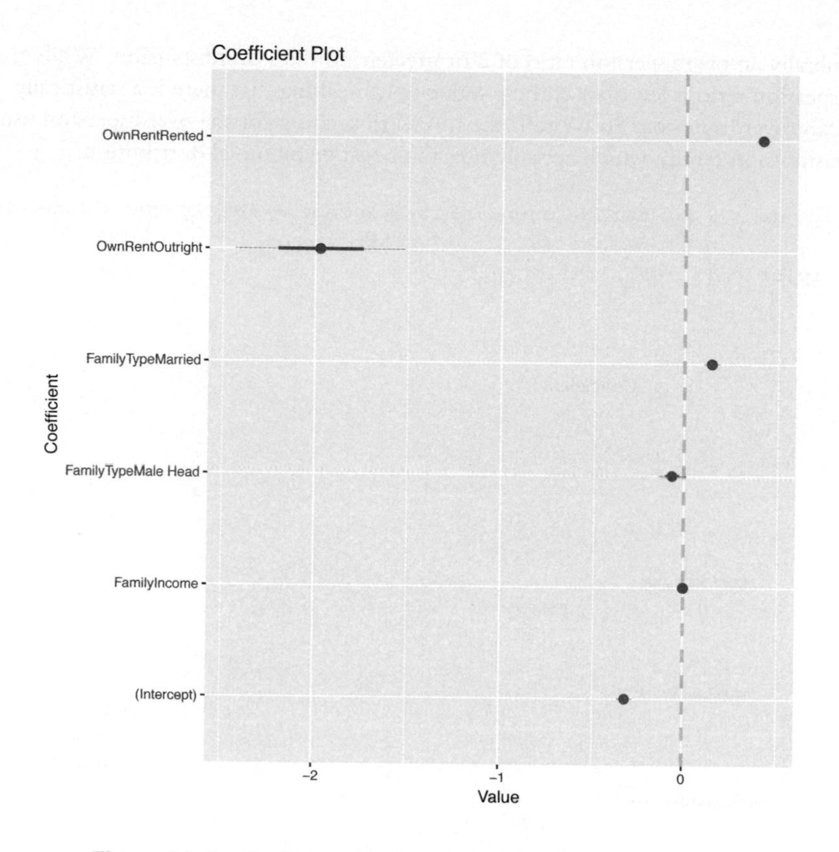

Figure 20.4 Coefficient plot for a logistic regression on ACS data.

$$z_i = \frac{y_i - \hat{y}_i}{sd(\hat{y}_i)} = \frac{y_i - u_i\hat{\theta}_i}{\sqrt{u_i\hat{\theta}_i}} \tag{20.6}$$

are the studentized residuals.

Calculating overdispersion in R is as follows.

```
> # the standardized residuals
> z <- (acs$NumChildren - children1$fitted.values) /
+     sqrt(children1$fitted.values)
> # Overdispersion Factor
> sum(z^2) / children1$df.residual

[1] 1.469747

> # Overdispersion p-value
> pchisq(sum(z^2), children1$df.residual)

[1] 1
```

Generally an overdispersion ratio of 2 or greater indicates overdispersion. While this overdispersion ratio is less than 2, the p-value is 1, meaning that there is a statistically significant overdispersion. So we refit the model to account for the overdispersion using the quasipoisson family, which actually uses the negative binomial distribution.

```
> children2 <- glm(NumChildren ~ FamilyIncome + FamilyType + OwnRent,
+                  data=acs, family=quasipoisson(link="log"))
> multiplot(children1, children2)
```

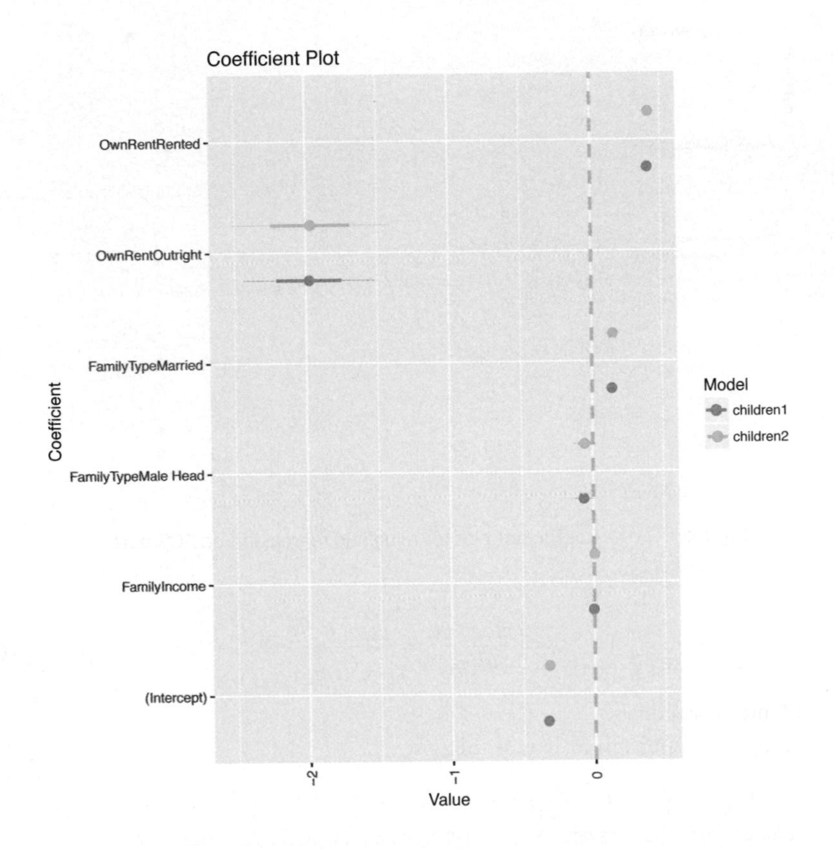

Figure 20.5 Coefficient plot for Poisson models. The first model, `children1`, does not account for overdispersion while `children2` does. Because the overdispersion was not too big, the coefficient estimates in the second model have just a bit more uncertainty.

Figure 20.5 shows a coefficient plot for a model with that accounts for overdispersion and one that does not. Since the overdispersion was not very large, the second model adds just a little uncertainty to the coefficient estimates.

20.3 Other Generalized Linear Models

Other common generalized linear models supported by the **glm** function are gamma, inverse gaussian and quasibinomial. Different link functions can be supplied, such as the following: logit, probit, cauchit, log and cloglog for binomial; inverse, identity and log for gamma; log, identity and sqrt for Poisson; and 1/mu^2, inverse, identity and log for inverse gaussian.

Multinomial regression, for classifying multiple categories, requires either running multiple logistic regressions (a tactic well supported in statistical literature) or using the **polr** function or the **multinom** function from the **nnet** package.

20.4 Survival Analysis

While not technically part of the family of generalized linear models, survival analysis is another important extension to regression. It has many applications, such as clinical medical trials, server failure times, number of accidents and time to death after a treatment or disease.

Data used for survival analysis are different from most other data in that they are censored, meaning there is unknown information, typically about what happens to a subject after a given amount of time. For an example, we look at the `bladder` data from the **survival** package.

```
> library(survival)
> head(bladder)

  id rx number size stop event enum
1  1  1      1    3    1     0    1
2  1  1      1    3    1     0    2
3  1  1      1    3    1     0    3
4  1  1      1    3    1     0    4
5  2  1      2    1    4     0    1
6  2  1      2    1    4     0    2
```

The columns of note are `stop` (when an event occurs or the patient leaves the study) and `event` (whether an event occurred at the time). Even if `event` is 0, we do not know if an event could have occurred later; this is why it is called censored. Making use of that structure requires the **Surv** function.

```
> # first look at a piece of the data
> bladder[100:105, ]

    id rx number size stop event enum
100 25  1      2    1   12     1    4
101 26  1      1    3   12     1    1
102 26  1      1    3   15     1    2
103 26  1      1    3   24     1    3
104 26  1      1    3   31     0    4
105 27  1      1    2   32     0    1
```

```
> # now look at the response variable built by build.y
> survObject <- with(bladder[100:105, ], Surv(stop, event))
> # nicely printed form
> survObject

[1] 12  12  15  24  31+ 32+

> # see its matrix form
> survObject[, 1:2]

     time status
[1,]   12      1
[2,]   12      1
[3,]   15      1
[4,]   24      1
[5,]   31      0
[6,]   32      0
```

This shows that for the first three rows where an event occurred, the time is known to be 12, whereas the bottom two rows had no event, so the time is censored because an event could have occurred afterward.

Perhaps the most common modelling technique in survival analysis is using a Cox proportional hazards model, which in R is done with **coxph**. The model is fitted using the familiar `formula` interface supplied to **coxph**. The **survfit** function builds the survival curve that can then be plotted as shown in Figure 20.6. The survival curve shows the percentage of participants surviving at a given time. The summary is similar to other summaries but tailored to survival analysis.

```
> cox1 <- coxph(Surv(stop, event) ~ rx + number + size + enum,
+               data=bladder)
> summary(cox1)

Call:
coxph(formula = Surv(stop, event) ~ rx + number + size + enum,
    data = bladder)

  n= 340, number of events= 112

           coef exp(coef) se(coef)      z Pr(>|z|)
rx     -0.59739   0.55024  0.20088 -2.974  0.00294 **
number  0.21754   1.24301  0.04653  4.675 2.93e-06 ***
size   -0.05677   0.94481  0.07091 -0.801  0.42333
enum   -0.60385   0.54670  0.09401 -6.423 1.34e-10 ***
---
Signif. codes:  0 '***' 0.001 '**' 0.01 '*' 0.05 '.' 0.1 ' ' 1

       exp(coef) exp(-coef) lower .95 upper .95
rx        0.5502     1.8174    0.3712    0.8157
number    1.2430     0.8045    1.1347    1.3617
```

```
size       0.9448     1.0584     0.8222     1.0857
enum       0.5467     1.8291     0.4547     0.6573

Concordance= 0.753  (se = 0.029 )
Rsquare= 0.179   (max possible= 0.971 )
Likelihood ratio test= 67.21  on 4 df,   p=8.804e-14
Wald test            = 64.73  on 4 df,   p=2.932e-13
Score (logrank) test = 69.42  on 4 df,   p=2.998e-14
```

```
> plot(survfit(cox1), xlab="Days", ylab="Survival Rate", conf.int=TRUE)
```

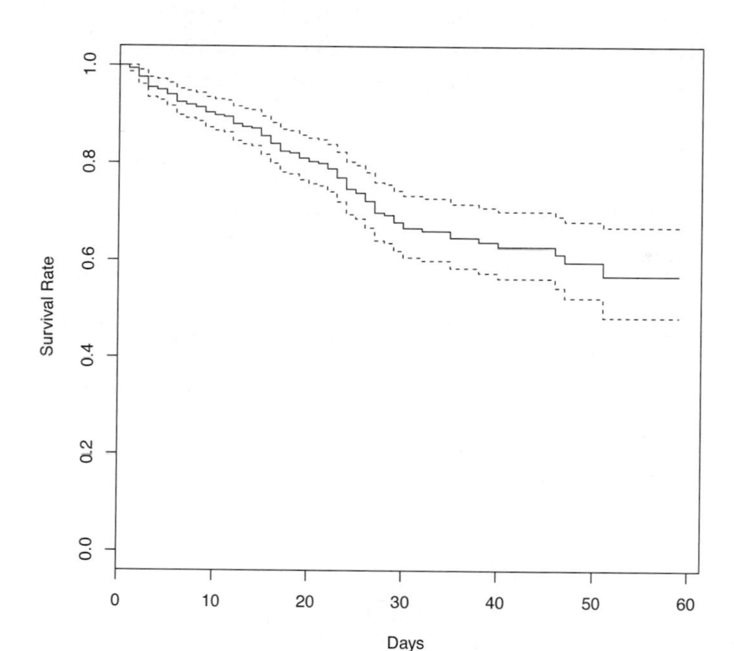

Figure 20.6 Survival curve for Cox proportional hazards model fitted on `bladder` data.

In this data, the `rx` variable indicates placebo versus treatment, which is a natural stratification of the patients. Passing `rx` to **strata** in the `formula` splits the data into two for analysis and will result in two survival curves like those in Figure 20.7.

```
> cox2 <- coxph(Surv(stop, event) ~ strata(rx) + number
+               + size + enum, data=bladder)
> summary(cox2)

Call:
coxph(formula = Surv(stop, event) ~ strata(rx) + number + size +
    enum, data = bladder)
```

```
  n= 340, number of events= 112

            coef exp(coef) se(coef)      z Pr(>|z|)
number  0.21371   1.23826  0.04648  4.598 4.27e-06 ***
size   -0.05485   0.94662  0.07097 -0.773     0.44
enum   -0.60695   0.54501  0.09408 -6.451 1.11e-10 ***
---
Signif. codes:  0 '***' 0.001 '**' 0.01 '*' 0.05 '.' 0.1 ' ' 1

       exp(coef) exp(-coef) lower .95 upper .95
number    1.2383     0.8076    1.1304    1.3564
size      0.9466     1.0564    0.8237    1.0879
enum      0.5450     1.8348    0.4532    0.6554

Concordance= 0.74  (se = 0.04 )
Rsquare= 0.166   (max possible= 0.954 )
Likelihood ratio test= 61.84  on 3 df,   p=2.379e-13
Wald test            = 60.04  on 3 df,   p=5.751e-13
Score (logrank) test = 65.05  on 3 df,   p=4.896e-14
```

```
> plot(survfit(cox2), xlab="Days", ylab="Survival Rate",
+      conf.int=TRUE, col=1:2)
> legend("bottomleft", legend=c(1, 2), lty=1, col=1:2,
+        text.col=1:2, title="rx")
```

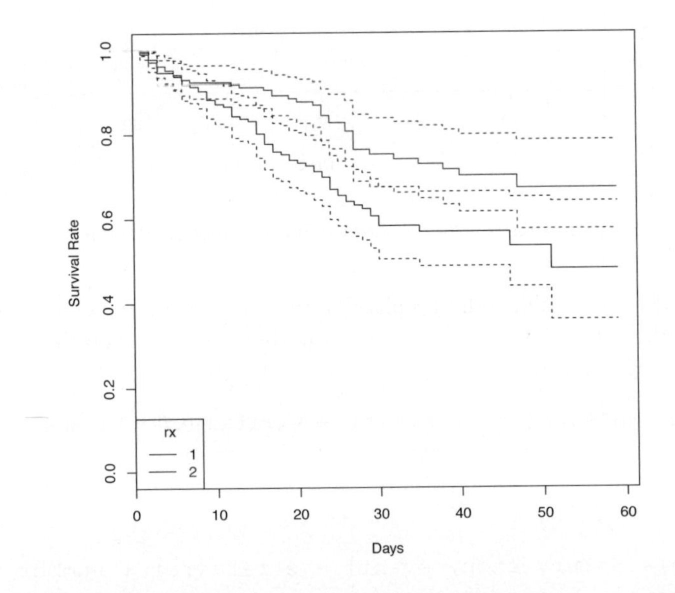

Figure 20.7 Survival curve for Cox proportional hazards model fitted on `bladder` data stratified on `rx`.

As an aside, this was a relatively simple legend to produce but it took a lot more effort than it would with **ggplot2**.

Testing the assumption of proportional hazards is done with **cox.zph**.

```
> cox.zph(cox1)

           rho   chisq        p
rx      0.0299  0.0957 7.57e-01
number  0.0900  0.6945 4.05e-01
size   -0.1383  2.3825 1.23e-01
enum    0.4934 27.2087 1.83e-07
GLOBAL      NA 32.2101 1.73e-06

> cox.zph(cox2)

           rho  chisq        p
number  0.0966  0.785 3.76e-01
size   -0.1331  2.197 1.38e-01
enum    0.4972 27.237 1.80e-07
GLOBAL      NA 32.101 4.98e-07
```

An Andersen-Gill analysis is similar to survival analysis, except it takes intervalized data and can handle multiple events, such as counting the number of emergency room visits as opposed to whether or not there is an emergency room visit. It is also performed using **coxph**, except an additional variable is passed to **Surv**, and the data must be clustered on an identification column (`id`) to keep track of multiple events. The corresponding survival curves are seen in Figure 20.8.

```
> head(bladder2)

  id rx number size start stop event enum
1  1  1      1    3     0    1     0    1
2  2  1      2    1     0    4     0    1
3  3  1      1    1     0    7     0    1
4  4  1      5    1     0   10     0    1
5  5  1      4    1     0    6     1    1
6  5  1      4    1     6   10     0    2

> ag1 <- coxph(Surv(start, stop, event) ~ rx + number + size + enum +
+                      cluster(id), data=bladder2)
> ag2 <- coxph(Surv(start, stop, event) ~ strata(rx) + number + size +
+                      enum + cluster(id), data=bladder2)
> plot(survfit(ag1), conf.int=TRUE)
> plot(survfit(ag2), conf.int=TRUE, col=1:2)
> legend("topright", legend=c(1, 2), lty=1, col=1:2,
+        text.col=1:2, title="rx")
```

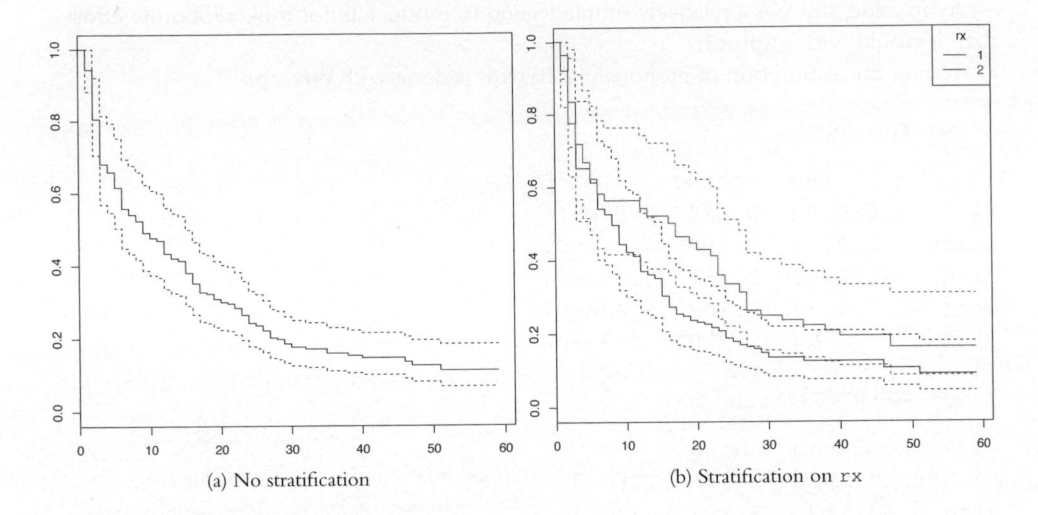

(a) No stratification

(b) Stratification on `rx`

Figure 20.8 Andersen-Gill survival curves for `bladder2` data.

20.5 Conclusion

Generalized linear models extend regression beyond linear relationships between the predictors and response. The most prominent types are logistic for binary data, Poisson for count data and survival analysis. Their uses go far beyond that, but those are by far the most common.

21

Model Diagnostics

Building a model can be a never ending process in which we constantly improve the model by adding interactions, taking away variables, doing transformations and so on. However, at some point we need to confirm that we have the best model at the time, or even a good model. That leads to the question: How do we judge the quality of a model? In almost all cases the answer has to be: in relation to other models. This could be an analysis of residuals, the results of an ANOVA test, a Wald test, drop-in deviance, the AIC or BIC score, cross-validation error or bootstrapping.

21.1 Residuals

One of the first-taught ways of assessing model quality is an analysis of the residuals, which is the difference between the actual response and the fitted values, values predicted by the model. This is a direct result of the formulation in Equation 19.1 where the errors, akin to residuals, are normally distributed. The basic idea is that if the model is appropriately fitted to the data, the residuals should be normally distributed as well. To see this, we start with the housing data to which we fit a regression and visualize with a coefficient plot, as shown in Figure 21.1.

```
> # read in the data
> housing <- read.table("data/housing.csv", sep=",", header=TRUE,
+                       stringsAsFactors=FALSE)
> # give the data good names
> names(housing) <- c("Neighborhood", "Class", "Units", "YearBuilt",
+                     "SqFt", "Income", "IncomePerSqFt", "Expense",
+                     "ExpensePerSqFt", "NetIncome", "Value",
+                     "ValuePerSqFt", "Boro")
> # eliminate some outliers
> housing <- housing[housing$Units < 1000, ]
> head(housing)

  Neighborhood           Class Units YearBuilt   SqFt   Income
1    FINANCIAL R9-CONDOMINIUM    42      1920  36500  1332615
2    FINANCIAL R4-CONDOMINIUM    78      1985 126420  6633257
3    FINANCIAL RR-CONDOMINIUM   500        NA 554174 17310000
4    FINANCIAL R4-CONDOMINIUM   282      1930 249076 11776313
```

```
5     TRIBECA R4-CONDOMINIUM   239       1985 219495 10004582
6     TRIBECA R4-CONDOMINIUM   133       1986 139719  5127687
  IncomePerSqFt Expense ExpensePerSqFt NetIncome    Value
1         36.51  342005           9.37    990610  7300000
2         52.47 1762295          13.94   4870962 30690000
3         31.24 3543000           6.39  13767000 90970000
4         47.28 2784670          11.18   8991643 67556006
5         45.58 2783197          12.68   7221385 54320996
6         36.70 1497788          10.72   3629899 26737996
  ValuePerSqFt      Boro
1       200.00 Manhattan
2       242.76 Manhattan
3       164.15 Manhattan
4       271.23 Manhattan
5       247.48 Manhattan
6       191.37 Manhattan

> # fit a model
> house1 <- lm(ValuePerSqFt ~ Units + SqFt + Boro, data=housing)
> summary(house1)

Call:
lm(formula = ValuePerSqFt ~ Units + SqFt + Boro, data = housing)

Residuals:
     Min       1Q   Median       3Q      Max
-168.458  -22.680    1.493   26.290  261.761

Coefficients:
                      Estimate Std. Error t value Pr(>|t|)
(Intercept)          4.430e+01  5.342e+00   8.293  < 2e-16 ***
Units               -1.532e-01  2.421e-02  -6.330 2.88e-10 ***
SqFt                 2.070e-04  2.129e-05   9.723  < 2e-16 ***
BoroBrooklyn         3.258e+01  5.561e+00   5.858 5.28e-09 ***
BoroManhattan        1.274e+02  5.459e+00  23.343  < 2e-16 ***
BoroQueens           3.011e+01  5.711e+00   5.272 1.46e-07 ***
BoroStaten Island -7.114e+00  1.001e+01  -0.711    0.477
---
Signif. codes:  0 '***' 0.001 '**' 0.01 '*' 0.05 '.' 0.1 ' ' 1

Residual standard error: 43.2 on 2613 degrees of freedom
Multiple R-squared:  0.6034,Adjusted R-squared:  0.6025
F-statistic: 662.6 on 6 and 2613 DF,  p-value: < 2.2e-16

> # visualize the model
> library(coefplot)
> coefplot(house1)
```

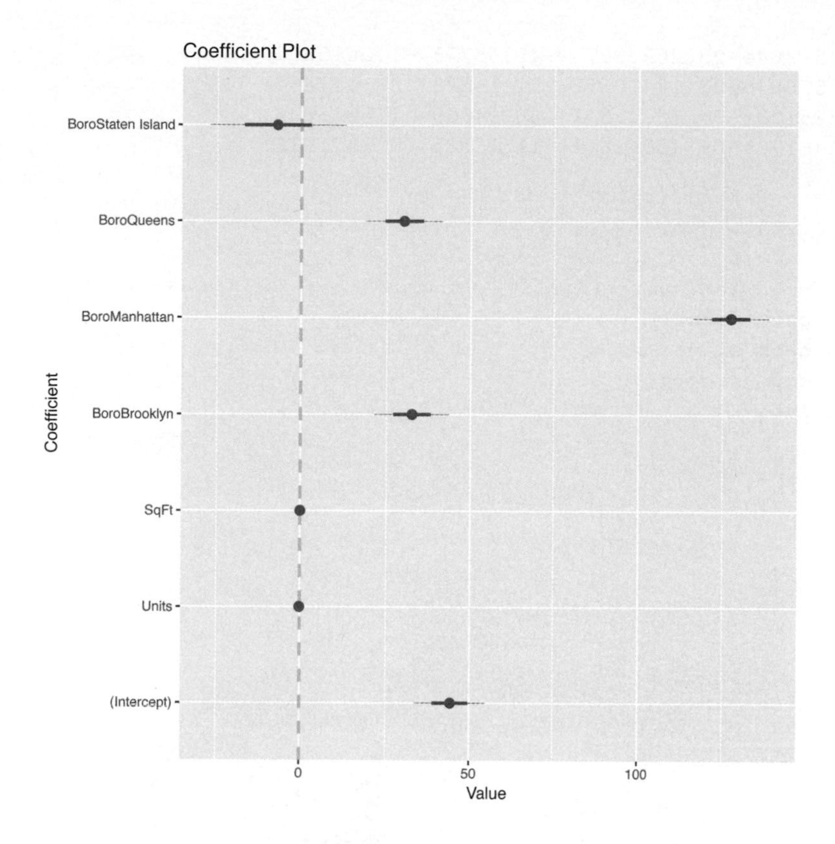

Figure 21.1 Coefficient plot for condo value data regression in `house1`.

For linear regression, three important residual plots are: fitted values against residuals, Q-Q plots and the histogram of the residuals. The first is easy enough with **ggplot2**. Fortunately, **ggplot2** has a handy trick for dealing with **lm** models. We can use the model as the data source and **ggplot2** "fortifies" it, creating new columns, for easy plotting.

```
> library(ggplot2)
> # see what a fortified lm model looks like
> head(fortify(house1))

  ValuePerSqFt Units   SqFt      Boro          .hat   .sigma
1       200.00    42  36500 Manhattan 0.0009594821 43.20952
2       242.76    78 126420 Manhattan 0.0009232393 43.19848
3       164.15   500 554174 Manhattan 0.0089836758 43.20347
4       271.23   282 249076 Manhattan 0.0035168641 43.17583
5       247.48   239 219495 Manhattan 0.0023865978 43.19289
6       191.37   133 139719 Manhattan 0.0008934957 43.21225
       .cooksd  .fitted    .resid  .stdresid
1 5.424169e-05 172.8475  27.15248  0.6287655
2 2.285253e-04 185.9418  56.81815  1.3157048
```

```
3 1.459368e-03 209.8077 -45.65775 -1.0615607
4 2.252653e-03 180.0672  91.16278  2.1137487
5 8.225193e-04 180.5341  66.94589  1.5513636
6 8.446170e-06 180.2661  11.10385  0.2571216

> # save a plot to an object
> # notice we are using the created columns for the x- and y-axes
> # they are .fitted and .resid
> h1 <- ggplot(aes(x=.fitted, y=.resid), data = house1) +
+     geom_point() +
+     geom_hline(yintercept = 0) +
+     geom_smooth(se = FALSE) +
+     labs(x="Fitted Values", y="Residuals")
>
> # print the plot
> h1
```

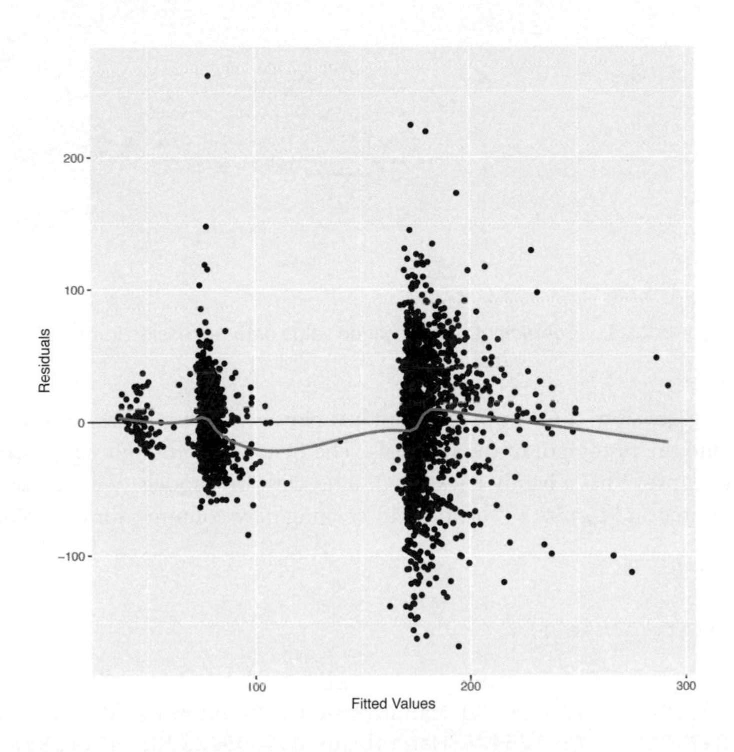

Figure 21.2 Plot of residuals versus fitted values for `house1`. This clearly shows a pattern in the data that does not appear to be random.

The plot of residuals versus fitted values shown in Figure 21.2 is at first glance disconcerting, because the pattern in the residuals shows that they are not as randomly dispersed as desired. However, further investigation reveals that this is due to the structure that `Boro` gives the data, as seen in Figure 21.3.

```
> h1 + geom_point(aes(color=Boro))
```

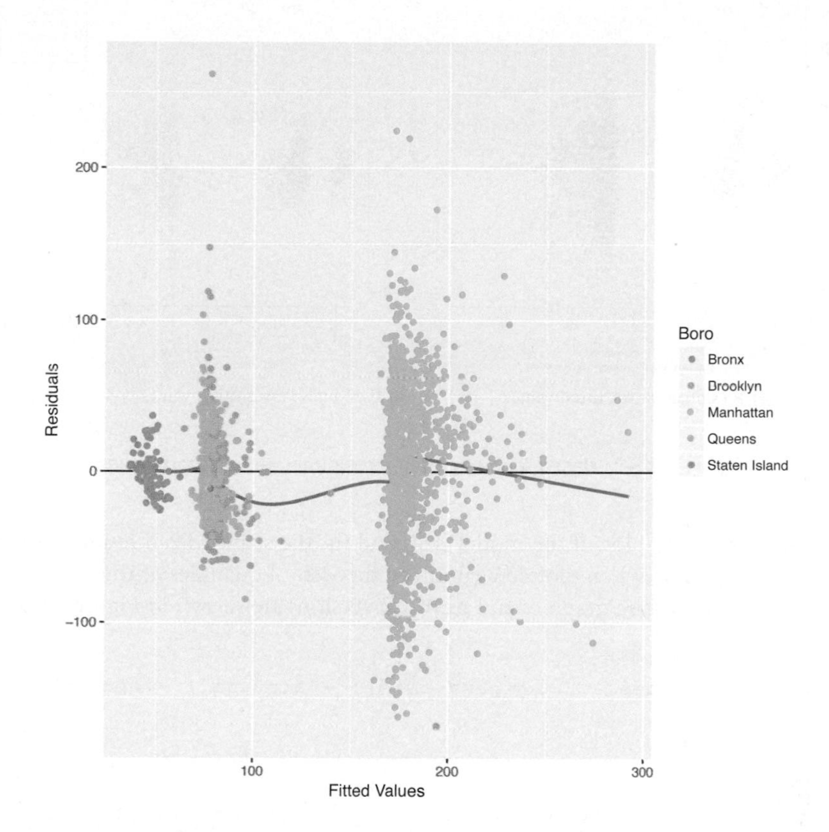

Figure 21.3 Plot of residuals versus fitted values for `house1` colored by `Boro`. The pattern in the residuals is revealed to be the result of the effect of `Boro` on the model. Notice that the points sit above the x-axis and the smoothing curve because `geom_point` was added after the other geoms, meaning it gets layered on top.

This plot could have been easily, although less attractively, plotted using the built-in plotting function, as shown in Figure 21.4.

```
> # basic plot
> plot(house1, which=1)

> # same plot but colored by Boro
> plot(house1, which=1, col=as.numeric(factor(house1$model$Boro)))
> # corresponding legend
> legend("topright", legend=levels(factor(house1$model$Boro)), pch=1,
+        col=as.numeric(factor(levels(factor(house1$model$Boro)))),
+        text.col=as.numeric(factor(levels(factor(house1$model$Boro)))),
+        title="Boro")
```

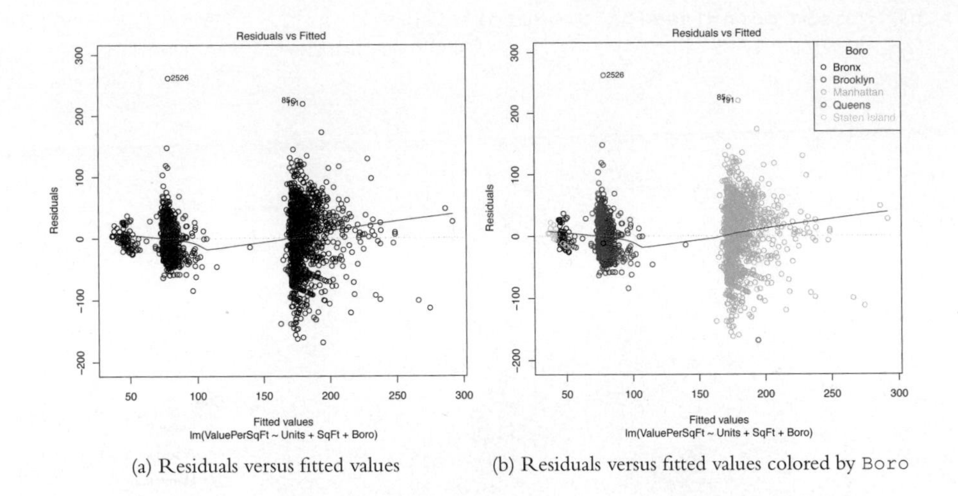

(a) Residuals versus fitted values

(b) Residuals versus fitted values colored by `Boro`

Figure 21.4 Base graphics plots for residuals versus fitted values.

Next up is the Q-Q plot. If the model is a good fit, the standardized residuals should all fall along a straight line when plotted against the theoretical quantiles of the normal distribution. Both the base graphics and **ggplot2** versions are shown in Figure 21.5.

```
> plot(house1, which=2)
> ggplot(house1, aes(sample=.stdresid)) + stat_qq() + geom_abline()
```

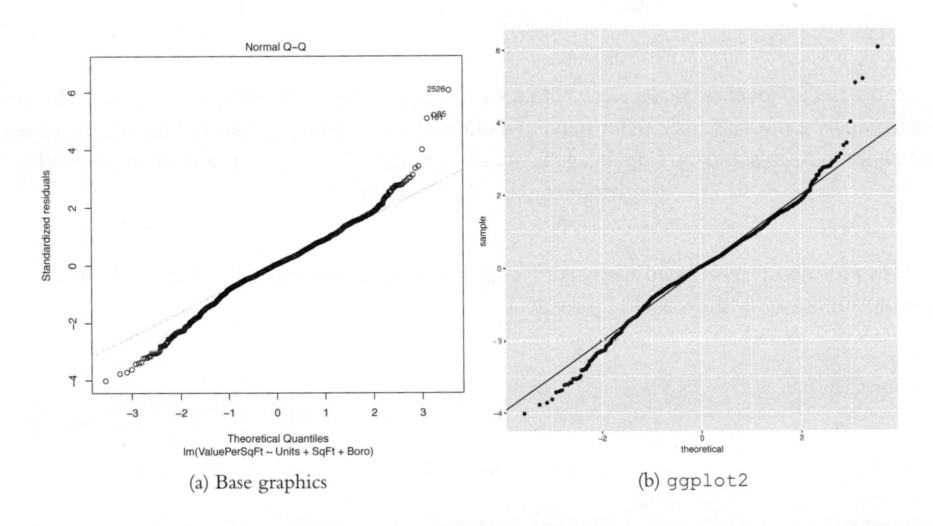

(a) Base graphics

(b) `ggplot2`

Figure 21.5 Q-Q plot for `house1`. The tails drift away from the ideal theoretical line, indicating that we do not have the best fit.

Another diagnostic is a histogram of the residuals. This time we will not be showing the base graphics alternative because a histogram is a standard plot that we have shown repeatedly. The histogram in Figure 21.6 is not normally distributed, meaning that our model is not an entirely correct specification.

```
> ggplot(house1, aes(x=.resid)) + geom_histogram()
```

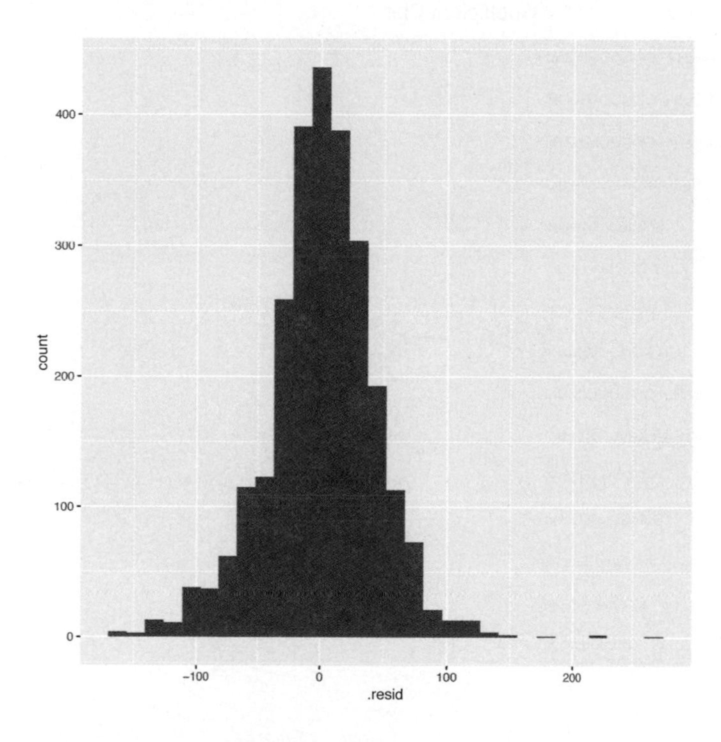

Figure 21.6 Histogram of residuals from `house1`. This does not look normally distributed, meaning our model is incomplete.

21.2 Comparing Models

All of this measuring of model fit only really makes sense when comparing multiple models, because all of these measures are relative. So we will fit a number of models in order to compare them to each other.

```
> house2 <- lm(ValuePerSqFt ~ Units * SqFt + Boro, data=housing)
> house3 <- lm(ValuePerSqFt ~ Units + SqFt * Boro + Class,
+              data=housing)
> house4 <- lm(ValuePerSqFt ~ Units + SqFt * Boro + SqFt*Class,
+              data=housing)
> house5 <- lm(ValuePerSqFt ~ Boro + Class, data=housing)
```

As usual, our first step is to visualize the models together using **multiplot** from the **coefplot** package. The result is in Figure 21.7 and shows that `Boro` is the only variable with a significant effect on `ValuePerSqFt` as do certain condominium types.

```
> multiplot(house1, house2, house3, house4, house5, pointSize=2)
```

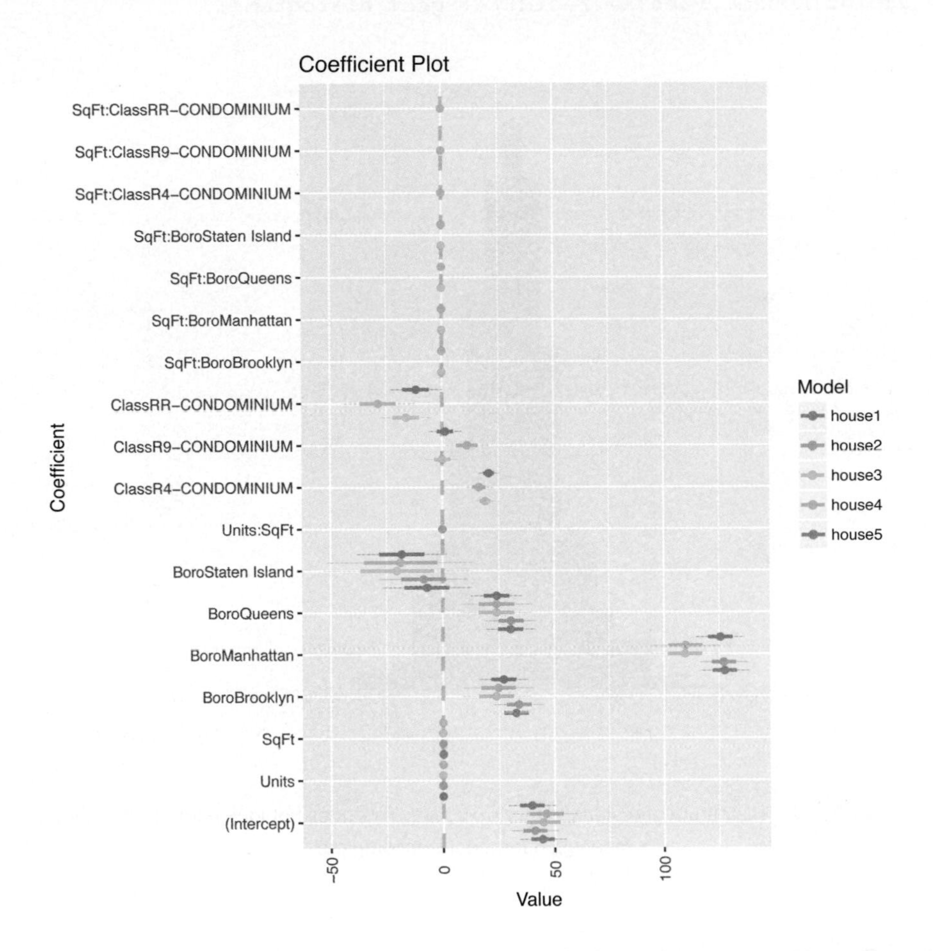

Figure 21.7 Coefficient plot of various models based on housing data. This shows that only `Boro` and some condominium types matter.

While we do not promote using ANOVA for a multisample test, we do believe it serves a useful purpose in testing the relative merits of different models. Simply passing multiple model objects to **anova** will return a table of results including the residual sum of squares (RSS), which is a measure of error, the lower the better.

```
> anova(house1, house2, house3, house4, house5)
```

```
Analysis of Variance Table

Model 1: ValuePerSqFt ~ Units + SqFt + Boro
Model 2: ValuePerSqFt ~ Units * SqFt + Boro
Model 3: ValuePerSqFt ~ Units + SqFt * Boro + Class
Model 4: ValuePerSqFt ~ Units + SqFt * Boro + SqFt * Class
Model 5: ValuePerSqFt ~ Boro + Class
  Res.Df     RSS Df Sum of Sq       F    Pr(>F)
1   2613 4877506
2   2612 4847886  1     29620 17.0360 3.783e-05 ***
3   2606 4576769  6    271117 25.9888 < 2.2e-16 ***
4   2603 4525783  3     50986  9.7749 2.066e-06 ***
5   2612 4895630 -9   -369847 23.6353 < 2.2e-16 ***
---
Signif. codes:  0 '***' 0.001 '**' 0.01 '*' 0.05 '.' 0.1 ' ' 1
```

This shows that the fourth model, `house4`, has the lowest RSS, meaning it is the best model of the bunch. The problem with RSS is that it always improves when an additional variable is added to the model. This can lead to excessive model complexity and overfitting. Another metric, which penalizes model complexity, is the Akaike Information Criterion (AIC). As with RSS, the model with the lowest AIC—even negative values—is considered optimal. The BIC (Bayesian Information Criterion) is a similar measure where, once again, lower is better.

The formula for AIC is

$$\text{AIC} = -2\ln(\mathcal{L}) + 2p \tag{21.1}$$

where $\ln(\mathcal{L})$ is the maximized log-likelihood and p is the number of coefficients in the model. As the model improves the log-likelihood gets bigger, and because that term is negated the AIC gets lower. However, adding coefficients increases the AIC; this penalizes model complexity. The formula for BIC is similar, except that instead of multiplying the number of coefficients by 2 it multiplies it by the natural log of the number of rows. This is seen in Equation 21.2.

$$\text{BIC} = -2\ln(\mathcal{L}) + \ln(n) \cdot p \tag{21.2}$$

The AIC and BIC for our models are calculated using the **AIC** and **BIC** functions, respectively.

```
> AIC(house1, house2, house3, house4, house5)

       df      AIC
house1  8 27177.78
house2  9 27163.82
house3 15 27025.04
house4 18 27001.69
house5  9 27189.50
```

```
> BIC(house1, house2, house3, house4, house5)

       df      BIC
house1  8 27224.75
house2  9 27216.66
house3 15 27113.11
house4 18 27107.37
house5  9 27242.34
```

When called on `glm` models, **anova** returns the deviance of the model, which is another measure of error. The general rule of thumb—according to Andrew Gelman—is that for every added variable in the model, the deviance should drop by two. For categorical (`factor`) variables, the deviance should drop by two for each `level`.

To illustrate we make a binary variable out of `ValuePerSqFt` and fit a few logistic regression models.

```
> # create the binary variable based on whether ValuePerSqFt is above 150
> housing$HighValue <- housing$ValuePerSqFt >= 150
>
> # fit a few models
> high1 <- glm(HighValue ~ Units + SqFt + Boro,
+              data=housing, family=binomial(link="logit"))
> high2 <- glm(HighValue ~ Units * SqFt + Boro,
+              data=housing, family=binomial(link="logit"))
> high3 <- glm(HighValue ~ Units + SqFt * Boro + Class,
+              data=housing, family=binomial(link="logit"))
> high4 <- glm(HighValue ~ Units + SqFt * Boro + SqFt*Class,
+              data=housing, family=binomial(link="logit"))
> high5 <- glm(HighValue ~ Boro + Class,
+              data=housing, family=binomial(link="logit"))
>
> # test the models using ANOVA (deviance), AIC and BIC
> anova(high1, high2, high3, high4, high5)

Analysis of Deviance Table

Model 1: HighValue ~ Units + SqFt + Boro
Model 2: HighValue ~ Units * SqFt + Boro
Model 3: HighValue ~ Units + SqFt * Boro + Class
Model 4: HighValue ~ Units + SqFt * Boro + SqFt * Class
Model 5: HighValue ~ Boro + Class
  Resid. Df Resid. Dev Df Deviance
1      2613     1687.5
2      2612     1678.8  1    8.648
3      2606     1627.5  6   51.331
4      2603     1606.1  3   21.420
5      2612     1662.3 -9  -56.205

> AIC(high1, high2, high3, high4, high5)

      df      AIC
high1  7 1701.484
high2  8 1694.835
high3 14 1655.504
```

```
high4 17 1640.084
high5  8 1678.290

> BIC(high1, high2, high3, high4, high5)

      df      BIC
high1  7 1742.580
high2  8 1741.803
high3 14 1737.697
high4 17 1739.890
high5  8 1725.257
```

Here, once again, the fourth model is the best. Notice that the fourth model added three variables (the three indicator variables for `Class` interacted with `SqFt`) and its deviance dropped by 21, which is greater than two for each additional variable.

21.3 Cross-Validation

Residual diagnostics and model tests such as ANOVA and AIC are a bit old fashioned and came along before modern computing horsepower. The preferred method to assess model quality—at least by most data scientists—is cross-validation, sometimes called k-fold cross-validation. The data is broken into k (usually five or ten) non-overlapping sections. Then a model is fitted on $k - 1$ sections of the data, which is then used to make predictions based on the kth section. This is repeated k times until every section has been held out for testing once and included in model fitting $k - 1$ times. Cross-validation provides a measure of the predictive accuracy of a model, which is largely considered a good means of assessing model quality.

There are a number of packages and functions that assist in performing cross validation. Each has its own limitations or quirks, so rather than going through a number of incomplete functions, we show one that works well for generalized linear models (including linear regression), and then build a generic framework that can be used generally for an arbitrary model type.

The **boot** package by Brian Ripley has **cv.glm** for performing cross-validation on. As the name implies, it works only for generalized linear models, which will suffice for a number of situations.

```
> library(boot)
> # refit house1 using glm instead of lm
> houseG1 <- glm(ValuePerSqFt ~ Units + SqFt + Boro,
+                data=housing, family=gaussian(link="identity"))
>
> # ensure it gives the same results as lm
> identical(coef(house1), coef(houseG1))

[1] TRUE

> # run the cross-validation with 5 folds
> houseCV1 <- cv.glm(housing, houseG1, K=5)
> # check the error
```

```
> houseCV1$delta

[1] 1870.317 1869.352
```

The results from **cv.glm** include `delta`, which has two numbers, the raw cross-validation error based on the cost function (in this case the mean squared error, which is a measure of correctness for an estimator and is defined in Equation 21.3) for all the folds and the adjusted cross-validation error. This second number compensates for not using leave-one-out cross-validation, which is like k-fold cross-validation except that each fold is the all but one data point with one point held out. This is very accurate but highly computationally intensive.

$$\text{MSE} = \frac{1}{n}\sum_{i=1}^{n}(\hat{y}_i - y_i)^2 \tag{21.3}$$

While we got a nice number for the error, it helps us only if we can compare it to other models, so we run the same process for the other models we built, rebuilding them with **glm** first.

```
> # refit the models using glm
> houseG2 <- glm(ValuePerSqFt ~ Units * SqFt + Boro, data=housing)
> houseG3 <- glm(ValuePerSqFt ~ Units + SqFt * Boro + Class,
+                data=housing)
> houseG4 <- glm(ValuePerSqFt ~ Units + SqFt * Boro + SqFt*Class,
+                data=housing)
> houseG5 <- glm(ValuePerSqFt ~ Boro + Class, data=housing)
>
> # run cross-validation
> houseCV2 <- cv.glm(housing, houseG2, K=5)
> houseCV3 <- cv.glm(housing, houseG3, K=5)
> houseCV4 <- cv.glm(housing, houseG4, K=5)
> houseCV5 <- cv.glm(housing, houseG5, K=5)
>
> ## check the error results
> # build a data.frame of the results
> cvResults <- as.data.frame(rbind(houseCV1$delta, houseCV2$delta,
+                                  houseCV3$delta, houseCV4$delta,
+                                  houseCV5$delta))
> ## do some cleaning up to make the results more presentable
> # give better column names
> names(cvResults) <- c("Error", "Adjusted.Error")
> # Add model name
> cvResults$Model <- sprintf("houseG%s", 1:5)
>
> # check the results
> cvResults

     Error Adjusted.Error   Model
1 1870.317       1869.352 houseG1
2 1866.730       1864.849 houseG2
```

```
3 1770.464         1767.784 houseG3
4 1758.651         1755.117 houseG4
5 1885.419         1883.539 houseG5
```

Once again, the fourth model, houseG4, is the superior model. Figure 21.8 shows how much ANOVA, AIC and cross-validation agree on the relative merits of the different models. The scales are all different but the shapes of the plots are identical.

```
> # visualize the results
> # test with ANOVA
> cvANOVA <-anova(houseG1, houseG2, houseG3, houseG4,  houseG5)
> cvResults$ANOVA <- cvANOVA$`Resid. Dev`
> # measure with AIC
> cvResults$AIC <- AIC(houseG1, houseG2, houseG3, houseG4, houseG5)$AIC
>
> # make the data.frame suitable for plotting
> library(reshape2)
> cvMelt <- melt(cvResults, id.vars="Model", variable.name="Measure",
+                 value.name="Value")
> cvMelt

     Model        Measure        Value
1  houseG1          Error     1870.317
2  houseG2          Error     1866.730
3  houseG3          Error     1770.464
4  houseG4          Error     1758.651
5  houseG5          Error     1885.419
6  houseG1 Adjusted.Error     1869.352
7  houseG2 Adjusted.Error     1864.849
8  houseG3 Adjusted.Error     1767.784
9  houseG4 Adjusted.Error     1755.117
10 houseG5 Adjusted.Error     1883.539
11 houseG1          ANOVA  4877506.411
12 houseG2          ANOVA  4847886.327
13 houseG3          ANOVA  4576768.981
14 houseG4          ANOVA  4525782.873
15 houseG5          ANOVA  4895630.307
16 houseG1            AIC    27177.781
17 houseG2            AIC    27163.822
18 houseG3            AIC    27025.042
19 houseG4            AIC    27001.691
20 houseG5            AIC    27189.499

> ggplot(cvMelt, aes(x=Model, y=Value)) +
+     geom_line(aes(group=Measure, color=Measure)) +
+     facet_wrap(~Measure, scales="free_y") +
+     theme(axis.text.x=element_text(angle=90, vjust=.5)) +
+     guides(color=FALSE)
```

We now present a general framework (loosely borrowed from **cv.glm**) for running our own cross-validation on models other than glm. This is not universal and will not work

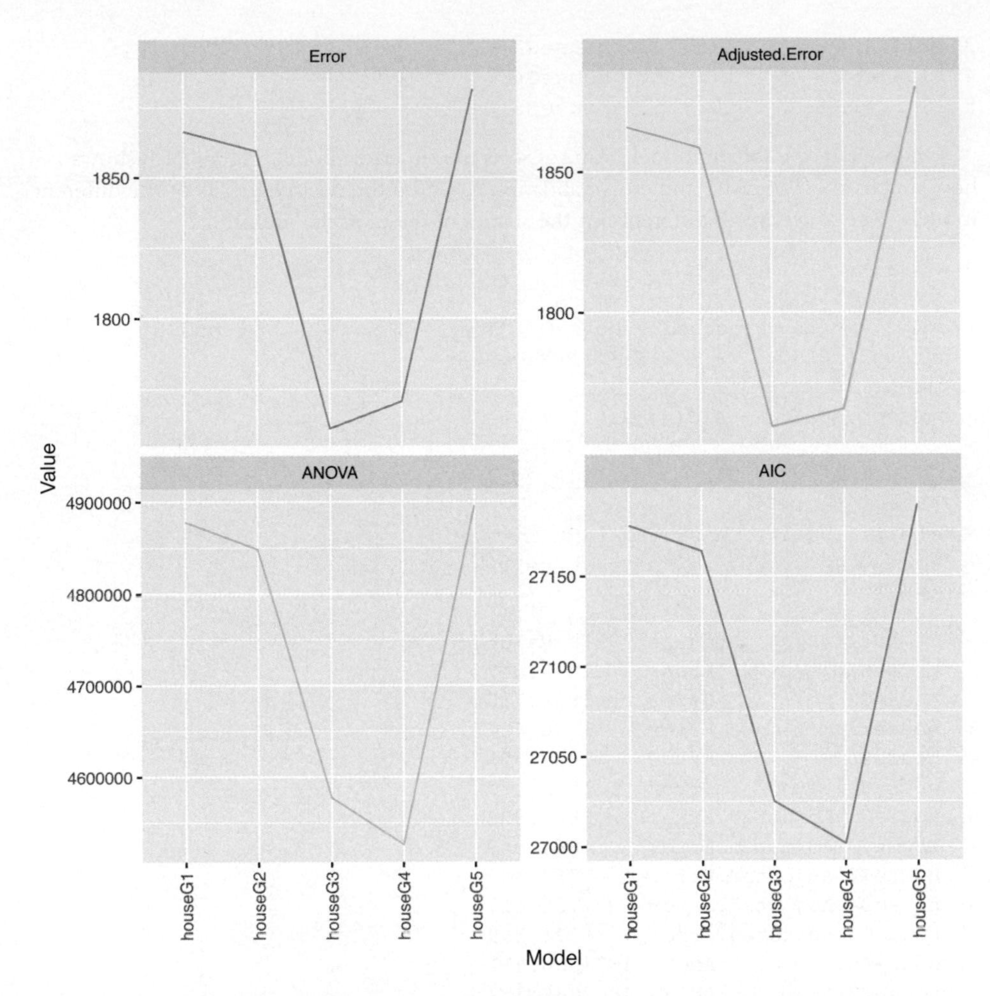

Figure 21.8 Plots for cross-validation error (raw and adjusted), ANOVA and AIC for housing models. The scales are different, as they should be, but the shapes are identical, indicating that `houseG4` truly is the best model.

for all models, but gives a general idea for how it should be done. In practice it should be abstracted into smaller parts and made more robust.

```
> cv.work <- function(fun, k = 5, data,
+                     cost = function(y, yhat) mean((y - yhat)^2),
+                     response="y", ...)
+ {
+     # generate folds
+     folds <- data.frame(Fold=sample(rep(x=1:k, length.out=nrow(data))),
+                         Row=1:nrow(data))
```

```
+     # start the error at 0
+     error <- 0

+     ## loop through each of the folds
+     ## for each fold:
+     ## fit the model on the training data
+     ## predict on the test data
+     ## compute the error and accumulate it
+     for(f in 1:max(folds$Fold))
+     {
+         # rows that are in test set
+         theRows <- folds$Row[folds$Fold == f]

+         ## call fun on data[-theRows, ]
+         ## predict on data[theRows, ]
+         mod <- fun(data=data[-theRows, ], ...)
+         pred <- predict(mod, data[theRows, ])

+         # add new error weighted by the number of rows in this fold
+         error <- error +
+             cost(data[theRows, response], pred) *
+             (length(theRows)/nrow(data))
+     }

+     return(error)
+ }
```

Applying that function to the various housing models we get their cross-validation errors.

```
> cv1 <- cv.work(fun=lm, k=5, data=housing, response="ValuePerSqFt",
+                formula=ValuePerSqFt ~ Units + SqFt + Boro)
> cv2 <- cv.work(fun=lm, k=5, data=housing, response="ValuePerSqFt",
+                formula=ValuePerSqFt ~ Units * SqFt + Boro)
> cv3 <- cv.work(fun=lm, k=5, data=housing, response="ValuePerSqFt",
+                formula=ValuePerSqFt ~ Units + SqFt * Boro + Class)
> cv4 <- cv.work(fun=lm, k=5, data=housing, response="ValuePerSqFt",
+                formula=ValuePerSqFt ~ Units + SqFt * Boro + SqFt*Class)
> cv5 <- cv.work(fun=lm, k=5, data=housing, response="ValuePerSqFt",
+                formula=ValuePerSqFt ~ Boro + Class)
> cvResults <- data.frame(Model=sprintf("house%s", 1:5),
+                         Error=c(cv1, cv2, cv3, cv4, cv5))
> cvResults

   Model    Error
1 house1 1867.451
2 house2 1868.941
3 house3 1769.159
4 house4 1751.637
5 house5 1871.996
```

This gives very similar results to **cv.glm** and again shows that the fourth parameterization is still the best. These measures do not always agree so nicely but it is great when they do.

21.4 Bootstrap

Sometimes, for one reason or another, there is not a good analytic solution to a problem and another tactic is needed. This is especially true for measuring uncertainty for confidence intervals. To overcome this, Bradley Efron introduced the bootstrap in 1979. Since then the bootstrap has grown to revolutionize modern statistics and is indispensable.

The idea is that we start with n rows of data. Some statistic (whether a mean, regression or some arbitrary function) is applied to the data. Then the data is sampled, creating a new dataset. This new set still has n rows except that there are repeats and other rows are entirely missing. The statistic is applied to this new dataset. The process is repeated R times (typically around 1,200), which generates an entire distribution for the statistic. This distribution can then be used to find the mean and confidence interval (typically 95%) for the statistic.

The **boot** package is a very robust set of tools for making the bootstrap easy to compute. Some care is needed when setting up the function call, but that can be handled easily enough.

Starting with a simple example, we analyze the batting average of Major League Baseball as a whole since 1990. The `baseball` data has information such as at bats (`ab`) and hits (`h`).

```
> library(plyr)
> baseball <- baseball[baseball$year >= 1990, ]
> head(baseball)

             id year stint team lg   g  ab  r   h X2b X3b hr rbi sb
67412 alomasa02 1990     1  CLE AL 132 445 60 129  26   2  9  66  4
67414 anderbr01 1990     1  BAL AL  89 234 24  54   5   2  3  24 15
67422 baergca01 1990     1  CLE AL 108 312 46  81  17   2  7  47  0
67424 baineha01 1990     1  TEX AL 103 321 41  93  10   1 13  44  0
67425 baineha01 1990     2  OAK AL  32  94 11  25   5   0  3  21  0
67442 bergmda01 1990     1  DET AL 100 205 21  57  10   1  2  26  3
      cs bb so ibb hbp sh sf gidp       OBP
67412  1 25 46   2   2  5  6   10 0.3263598
67414  2 31 46   2   5  4  5    4 0.3272727
67422  2 16 57   2   4  1  5    4 0.2997033
67424  1 47 63   9   0  0  3   13 0.3773585
67425  2 20 17   1   0  0  4    4 0.3813559
67442  2 33 17   3   0  1  2    7 0.3750000
```

The proper way to compute the batting average is to divide total hits by total at bats. This means we cannot simply run `mean(h/ab)` and `sd(h/ab)` to get the mean and standard deviation. Rather, the batting average is calculated as `sum(h)/sum(ab)` and its standard deviation is not easily calculated. This problem is a great candidate for using the bootstrap.

We calculate the overall batting average with the original data. Then we sample n rows with replacement and calculate the batting average again. We do this repeatedly until a distribution is formed. Rather that doing this manually, though, we use **boot**.

The first argument to **boot** is the data. The second argument is the function that is to be computed on the data. This function must take at least two arguments (unless `sim="parametric"` in which case only the first argument is necessary). The first is the original data and the second is a `vector` of indices, frequencies or weights. Additional named arguments can be passed into the function from **boot**.

```
> ## build a function for calculating batting average
> # data is the data
> # boot will pass varying sets of indices
> # some rows will be represented multiple times in a single pass
> # other rows will not be represented at all
> # on average about 63% of the rows will be present
> # this function is called repeatedly by boot

> bat.avg <- function(data, indices=1:NROW(data), hits="h", at.bats="ab")
+ {
+     sum(data[indices, hits], na.rm=TRUE) /
+         sum(data[indices, at.bats], na.rm=TRUE)
+ }
>
> # test it on the original data
> bat.avg(baseball)

[1] 0.2745988

> # bootstrap it
> # using the baseball data, call bat.avg 1,200 times
> # pass indices to the function
> avgBoot <- boot(data=baseball, statistic=bat.avg, R=1200, stype="i")
>
> # print original measure and estimates of bias and standard error
> avgBoot

ORDINARY NONPARAMETRIC BOOTSTRAP


Call:
boot(data = baseball, statistic = bat.avg, R = 1200, stype = "i")


Bootstrap Statistics :
     original          bias     std. error
t1* 0.2745988 -2.740059e-05 0.0006569477

> # print the confidence interval
> boot.ci(avgBoot, conf=.95, type="norm")

BOOTSTRAP CONFIDENCE INTERVAL CALCULATIONS
Based on 1200 bootstrap replicates
```

```
CALL :
boot.ci(boot.out = avgBoot, conf = 0.95, type = "norm")

Intervals :
Level      Normal
95%   ( 0.2733,  0.2759 )
Calculations and Intervals on Original Scale
```

Visualizing the distribution is as simple as plotting a histogram of the replicate results. Figure 21.9 shows the histogram for the batting average with vertical lines two standard errors on either side of the original estimate. These mark the (roughly) 95% confidence interval.

```
> ggplot() +
+     geom_histogram(aes(x=avgBoot$t), fill="grey", color="grey") +
+     geom_vline(xintercept=avgBoot$t0 + c(-1, 1)*2*sqrt(var(avgBoot$t)),
+                linetype=2)
```

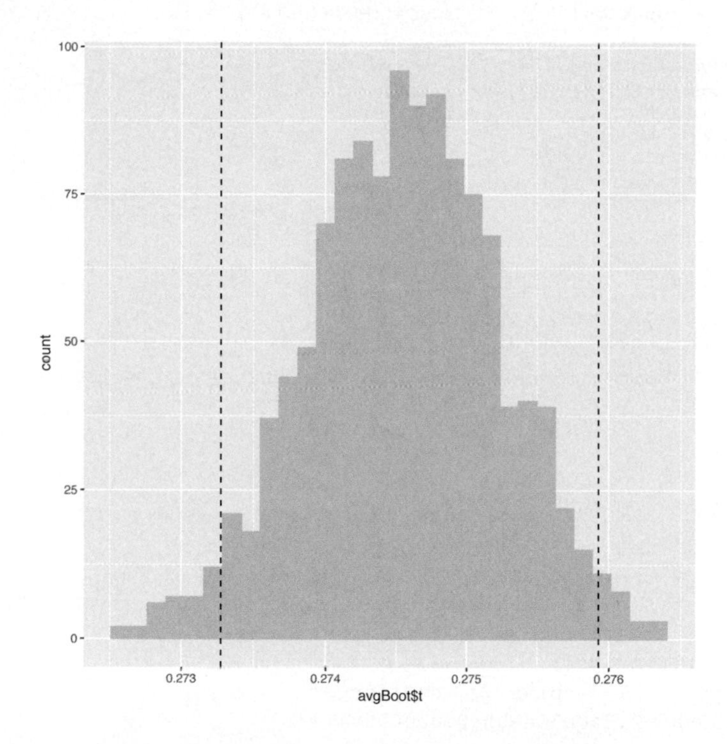

Figure 21.9 Histogram of the batting average bootstrap. The vertical lines are two standard errors from the original estimate in each direction. They make up the bootstrapped 95% confidence interval.

The bootstrap is an incredibly powerful tool that holds a great deal of promise. The **boot** package offers far more than what we have shown here, including the ability to bootstrap time series and censored data. The beautiful thing about the bootstrap is its near universal applicability. It can be used in just about any situation where an analytical solution is impractical or impossible. There are some instances where the bootstrap is inappropriate, such as for measuring uncertainty of biased estimators like those from the lasso, although such limitations are rare.

21.5 Stepwise Variable Selection

A common, though becoming increasingly discouraged, way to select variables for a model is stepwise selection. This is the process of iteratively adding and removing variables from a model and testing the model at each step, usually using AIC.

The **step** function iterates through possible models. The `scope` argument specifies a lower and upper bound on possible models. The `direction` argument specifies whether variables are just added into the model, just subtracted from the model or added and subtracted as necessary. When run, **step** prints out all the iterations it has taken to arrive at what it considers the optimal model.

```
> # the lowest model is the null model, basically the straight average
> nullModel <- lm(ValuePerSqFt ~ 1, data=housing)
> # the largest model we will accept
> fullModel <- lm(ValuePerSqFt ~ Units + SqFt*Boro + Boro*Class, data=housing)
> # try different models
> # start with nullModel
> # do not go above fullModel
> # work in both directions
> houseStep <- step(nullModel,
+                   scope=list(lower=nullModel, upper=fullModel),
+                   direction="both")

Start:  AIC=22151.56
ValuePerSqFt ~ 1

        Df Sum of Sq      RSS   AIC
+ Boro   4   7160206  5137931 19873
+ SqFt   1   1310379 10987758 21858
+ Class  3   1264662 11033475 21873
+ Units  1    778093 11520044 21982
<none>               12298137 22152

Step:  AIC=19872.83
ValuePerSqFt ~ Boro

        Df Sum of Sq      RSS   AIC
+ Class  3    242301  4895630 19752
+ SqFt   1    185635  4952296 19778
+ Units  1     83948  5053983 19832
<none>                5137931 19873
- Boro   4   7160206 12298137 22152
```

```
Step:  AIC=19752.26
ValuePerSqFt ~ Boro + Class

             Df Sum of Sq      RSS   AIC
+ SqFt        1    182170  4713460 19655
+ Units       1    100323  4795308 19700
+ Boro:Class  9    111838  4783792 19710
<none>                     4895630 19752
- Class       3    242301  5137931 19873
- Boro        4   6137845 11033475 21873

Step:  AIC=19654.91
ValuePerSqFt ~ Boro + Class + SqFt

             Df Sum of Sq      RSS   AIC
+ SqFt:Boro   4    113219  4600241 19599
+ Boro:Class  9     94590  4618870 19620
+ Units       1     37078  4676382 19636
<none>                     4713460 19655
- SqFt        1    182170  4895630 19752
- Class       3    238836  4952296 19778
- Boro        4   5480928 10194388 21668

Step:  AIC=19599.21
ValuePerSqFt ~ Boro + Class + SqFt + Boro:SqFt

             Df Sum of Sq     RSS   AIC
+ Boro:Class  9     68660 4531581 19578
+ Units       1     23472 4576769 19588
<none>                    4600241 19599
- Boro:SqFt   4    113219 4713460 19655
- Class       3    258642 4858883 19737

Step:  AIC=19577.81
ValuePerSqFt ~ Boro + Class + SqFt + Boro:SqFt + Boro:Class

             Df Sum of Sq     RSS   AIC
+ Units       1     20131 4511450 19568
<none>                    4531581 19578
- Boro:Class  9     68660 4600241 19599
- Boro:SqFt   4     87289 4618870 19620

Step:  AIC=19568.14
ValuePerSqFt ~ Boro + Class + SqFt + Units + Boro:SqFt + Boro:Class

             Df Sum of Sq     RSS   AIC
<none>                    4511450 19568
- Units       1     20131 4531581 19578
- Boro:Class  9     65319 4576769 19588
- Boro:SqFt   4     75955 4587405 19604
```

```
> # reveal the chosen model
> houseStep

Call:
lm(formula = ValuePerSqFt ~ Boro + Class + SqFt + Units + Boro:SqFt +
    Boro:Class, data = housing)

Coefficients:
                              (Intercept)
                                4.848e+01
                             BoroBrooklyn
                                2.655e+01
                            BoroManhattan
                                8.672e+01
                               BoroQueens
                                1.999e+01
                        BoroStaten Island
                               -1.132e+01
                      ClassR4-CONDOMINIUM
                                6.586e+00
                      ClassR9-CONDOMINIUM
                                4.553e+00
                      ClassRR-CONDOMINIUM
                                8.130e+00
                                     SqFt
                                1.373e-05
                                    Units
                               -8.296e-02
                        BoroBrooklyn:SqFt
                                3.798e-05
                       BoroManhattan:SqFt
                                1.594e-04
                          BoroQueens:SqFt
                                2.753e-06
                   BoroStaten Island:SqFt
                                4.362e-05
         BoroBrooklyn:ClassR4-CONDOMINIUM
                                1.933e+00
        BoroManhattan:ClassR4-CONDOMINIUM
                                3.436e+01
           BoroQueens:ClassR4-CONDOMINIUM
                                1.274e+01
    BoroStaten Island:ClassR4-CONDOMINIUM
                                       NA
         BoroBrooklyn:ClassR9-CONDOMINIUM
                               -3.440e+00
        BoroManhattan:ClassR9-CONDOMINIUM
                                1.497e+01
```

```
        BoroQueens:ClassR9-CONDOMINIUM
                            -9.967e+00
BoroStaten Island:ClassR9-CONDOMINIUM
                                    NA
      BoroBrooklyn:ClassRR-CONDOMINIUM
                            -2.901e+01
     BoroManhattan:ClassRR-CONDOMINIUM
                            -6.850e+00
        BoroQueens:ClassRR-CONDOMINIUM
                             2.989e+01
BoroStaten Island:ClassRR-CONDOMINIUM
                                    NA
```

Ultimately, **step** decided that `fullModel` was optimal with the lowest AIC. While this works, it is a bit of a brute force method and has its own theoretical problems. Lasso regression arguably does a better job of variable selection and is discussed in Section 22.1.

21.6 Conclusion

Determining the quality of a model is an important step in the model building process. This can take the form of traditional tests of fit such as ANOVA or more modern techniques like cross-validation. The bootstrap is another means of determining model uncertainty, especially for models where confidence intervals are impractical to calculate. These can all be shaped by helping select which variables are included in a model and which are excluded.

22

Regularization and Shrinkage

In today's era of high dimensional (many variables) data, methods are needed to prevent overfitting. Traditionally, this has been done with variable selection, as described in Chapter 21, although with a large number of variables that can become computationally prohibitive. These methods can take a number of forms; we focus on regularization and shrinkage. For these we will use **glmnet** from the **glmnet** package and **bayesglm** from the **arm** package.

22.1 Elastic Net

One of the most exciting algorithms to be developed in the past five years is the Elastic Net, which is a dynamic blending of lasso and ridge regression. The lasso uses an $L1$ penalty to perform variable selection and dimension reduction, while the ridge uses an $L2$ penalty to shrink the coefficients for more stable predictions. The formula for the Elastic Net is

$$\min_{\beta_0,\beta\in\mathbb{R}^{p+1}} \left[\frac{1}{2N}\sum_{i=1}^{N}\left(y_i - \beta_0 - x_i^T\beta\right)^2 + \lambda P_\alpha\left(\beta\right)\right] \tag{22.1}$$

where

$$P_\alpha\left(\beta\right) = \left(1-\alpha\right)\frac{1}{2}||\Gamma\beta||_{l_2}^2 + \alpha||\Gamma\beta||_{l_1} \tag{22.2}$$

where λ is a complexity parameter controlling the amount of shrinkage (0 is no penalty and ∞ is complete penalty) and α regulates how much of the solution is ridge versus lasso with $\alpha = 0$ being complete ridge and $\alpha = 1$ being complete lasso. Γ is a vector of penalty factors—one value per variable—that multiplies λ for fine tuning of the penalty applied to each variable; again 0 is no penalty and ∞ is complete penalty.

A fairly new package (this is a relatively new algorithm) is **glmnet**, which fits generalized linear models with the Elastic Net. It is written by Trevor Hastie, Robert Tibshirani and Jerome Friedman from Stanford University, who also published the landmark papers on the Elastic Net.

Because it is designed for speed and larger, sparser data, **glmnet** requires a little more effort to use than most other modelling functions in R. Where functions like **lm** and **glm** take a `formula` to specify the model, **glmnet** requires a `matrix` of predictors (including an intercept) and a response `matrix`.

Even though it is not incredibly high dimensional, we will look at the American Community Survey (ACS) data for New York State. We will throw every possible predictor into the model and see which are selected.

```
> acs <- read.table("http://jaredlander.com/data/acs_ny.csv", sep=",",
+                   header=TRUE, stringsAsFactors=FALSE)
```

Because **glmnet** requires a predictor `matrix`, it will be good to have a convenient way of building that `matrix`. This can be done simply enough using **model.matrix**, which at its most basic takes in a `formula` and a `data.frame` and returns a design `matrix`. As an example we create some fake data and run **model.matrix** on it.

```
> # build a data.frame where the first three columns are numeric
> testFrame <-
+     data.frame(First=sample(1:10, 20, replace=TRUE),
+                Second=sample(1:20, 20, replace=TRUE),
+                Third=sample(1:10, 20, replace=TRUE),
+                Fourth=factor(rep(c("Alice", "Bob", "Charlie", "David"), 5)),
+                Fifth=ordered(rep(c("Edward", "Frank", "Georgia",
+                                    "Hank", "Isaac"), 4)),
+                Sixth=rep(c("a", "b"), 10), stringsAsFactors=F)
> head(testFrame)

  First Second Third  Fourth   Fifth Sixth
1     3     11     2   Alice  Edward     a
2     4      1    10     Bob   Frank     b
3     5     19     2 Charlie Georgia     a
4     1      1     2   David    Hank     b
5     7     19     4   Alice   Isaac     a
6     6     10     8     Bob  Edward     b

> head(model.matrix(First ~ Second + Fourth + Fifth, testFrame))

  (Intercept) Second FourthBob FourthCharlie FourthDavid    Fifth.L
1           1     11         0             0           0 -0.6324555
2           1      1         1             0           0 -0.3162278
3           1     19         0             1           0  0.0000000
4           1      1         0             0           1  0.3162278
5           1     19         0             0           0  0.6324555
6           1     10         1             0           0 -0.6324555
     Fifth.Q       Fifth.C    Fifth^4
1  0.5345225 -3.162278e-01  0.1195229
2 -0.2672612  6.324555e-01 -0.4780914
3 -0.5345225 -4.095972e-16  0.7171372
4 -0.2672612 -6.324555e-01 -0.4780914
5  0.5345225  3.162278e-01  0.1195229
6  0.5345225 -3.162278e-01  0.1195229
```

This works very well and is simple, but first there are a few things to notice. As expected, `Fourth` gets converted into indicator variables with one less column than `levels` in `Fourth`. Initially, the parameterization of `Fifth` might seem odd, as there is one less column than there are `levels`, but their values are not just 1s and 0s. This is because Fifth is an ordered factor where one level is greater or less than another level.

Not creating an indicator variable for the base level of a factor is essential for most linear models to avoid multicollinearity.[1] However, it is generally considered undesirable for the predictor matrix to be designed this way for the Elastic Net. It is possible to have **model.matrix** return indicator variables for all levels of a factor, although doing so can take some creative coding.[2] To make the process easier we incorporated a solution in the **build.x** function in the **useful** package.

```
> library(useful)
> # always use all levels
> head(build.x(First ~ Second + Fourth + Fifth, testFrame,
+                 contrasts=FALSE))

  (Intercept) Second FourthAlice FourthBob FourthCharlie FourthDavid
1           1     11           1         0             0           0
2           1      1           0         1             0           0
3           1     19           0         0             1           0
4           1      1           0         0             0           1
5           1     19           1         0             0           0
6           1     10           0         1             0           0
  FifthEdward FifthFrank FifthGeorgia FifthHank FifthIsaac
1           1          0            0         0          0
2           0          1            0         0          0
3           0          0            1         0          0
4           0          0            0         1          0
5           0          0            0         0          1
6           1          0            0         0          0

> # just use all levels for Fourth
> head(build.x(First ~ Second + Fourth + Fifth, testFrame,
+               contrasts=c(Fourth=FALSE, Fifth=TRUE)))

  (Intercept) Second FourthAlice FourthBob FourthCharlie FourthDavid
1           1     11           1         0             0           0
2           1      1           0         1             0           0
3           1     19           0         0             1           0
4           1      1           0         0             0           1
5           1     19           1         0             0           0
6           1     10           0         1             0           0
     Fifth.L    Fifth.Q       Fifth.C    Fifth^4
1 -0.6324555  0.5345225 -3.162278e-01  0.1195229
2 -0.3162278 -0.2672612  6.324555e-01 -0.4780914
3  0.0000000 -0.5345225 -4.095972e-16  0.7171372
```

1. This is a characteristic of a matrix in linear algebra where the columns are not linearly independent. While this is an important concept, we do not need to concern ourselves with it much in the context of this book.
2. The difficulty is evidenced in this Stack Overflow question asked by us: http://stackoverflow.com/questions/4560459/all-levels-of-a-factor-in-a-model-matrix-in-r/15400119

```
4  0.3162278 -0.2672612 -6.324555e-01 -0.4780914
5  0.6324555  0.5345225  3.162278e-01  0.1195229
6 -0.6324555  0.5345225 -3.162278e-01  0.1195229
```

Using **build.x** appropriately on `acs` builds a nice predictor `matrix` for use in **glmnet**. We control the desired `matrix` by using a `formula` for our model specification just like we would in **lm**, interactions and all.

```
> # make a binary Income variable for building a logistic regression
> acs$Income <- with(acs, FamilyIncome >= 150000)
>
> head(acs)

  Acres FamilyIncome  FamilyType NumBedrooms NumChildren NumPeople
1  1-10          150     Married           4           1         3
2  1-10          180 Female Head           3           2         4
3  1-10          280 Female Head           4           0         2
4  1-10          330 Female Head           2           1         2
5  1-10          330   Male Head           3           1         2
6  1-10          480   Male Head           0           3         4
  NumRooms        NumUnits NumVehicles NumWorkers  OwnRent
1        9 Single detached           1          0 Mortgage
2        6 Single detached           2          0   Rented
3        8 Single detached           3          1 Mortgage
4        4 Single detached           1          0   Rented
5        5 Single attached           1          0 Mortgage
6        1 Single detached           0          0   Rented
    YearBuilt HouseCosts ElectricBill FoodStamp HeatingFuel Insurance
1   1950-1959       1800           90        No         Gas      2500
2 Before 1939        850           90        No         Oil         0
3   2000-2004       2600          260        No         Oil      6600
4   1950-1959       1800          140        No         Oil         0
5 Before 1939        860          150        No         Gas       660
6 Before 1939        700          140        No         Gas         0
        Language Income
1        English  FALSE
2        English  FALSE
3 Other European  FALSE
4        English  FALSE
5        Spanish  FALSE
6        English  FALSE

> # build predictor matrix
> # do not include the intercept as glmnet will add that automatically
> acsX <- build.x(Income ~ NumBedrooms + NumChildren + NumPeople +
+                     NumRooms + NumUnits + NumVehicles + NumWorkers +
+                     OwnRent + YearBuilt + ElectricBill + FoodStamp +
+                     HeatingFuel + Insurance + Language - 1,
+                 data=acs, contrasts=FALSE)
>
```

```
> # check class and dimensions
> class(acsX)

[1] "matrix"

> dim(acsX)

[1] 22745    44

> # view the top left and top right of the data
> topleft(acsX, c=6)

  NumBedrooms NumChildren NumPeople NumRooms NumUnitsMobile home
1           4           1         3        9                   0
2           3           2         4        6                   0
3           4           0         2        8                   0
4           2           1         2        4                   0
5           3           1         2        5                   0
  NumUnitsSingle attached
1                       0
2                       0
3                       0
4                       0
5                       1

> topright(acsX, c=6)

  Insurance LanguageAsian Pacific LanguageEnglish LanguageOther
1      2500                     0               1             0
2         0                     0               1             0
3      6600                     0               0             0
4         0                     0               1             0
5       660                     0               0             0
  LanguageOther European LanguageSpanish
1                      0               0
2                      0               0
3                      1               0
4                      0               0
5                      0               1

> # build response predictor
> acsY <- build.y(Income ~ NumBedrooms + NumChildren + NumPeople +
+                     NumRooms + NumUnits + NumVehicles + NumWorkers +
+                     OwnRent + YearBuilt + ElectricBill + FoodStamp +
+                     HeatingFuel + Insurance + Language - 1, data=acs)
>
> head(acsY)

[1] FALSE FALSE FALSE FALSE FALSE FALSE

> tail(acsY)

[1] TRUE TRUE TRUE TRUE TRUE TRUE
```

Now that the data is properly stored we can run **glmnet**. As seen in Equation 22.1, λ controls the amount of shrinkage. By default **glmnet** fits the regularization path on 100 different values of λ. The decision of which is best then falls upon the user with cross-validation being a good measure. Fortunately the **glmnet** package has a function, **cv.glmnet**, that computes the cross-validation automatically. By default $\alpha = 1$, meaning only the lasso is calculated. Selecting the best α requires an additional layer of cross-validation.

```
> library(glmnet)
> set.seed(1863561)
> # run the cross-validated glmnet
> acsCV1 <- cv.glmnet(x=acsX, y=acsY, family="binomial", nfold=5)
```

The most important information returned from **cv.glmnet** are the cross-validation and which value of λ minimizes the cross-validation error. Additionally, it also returns the largest value of λ with a cross-validation error that is within one standard error of the minimum. Theory suggests that the simpler model, even though it is slightly less accurate, should be preferred due to its parsimony. The cross-validation errors for differing values of λ are seen in Figure 22.1. The top row of numbers indicates how many variables (`factor levels` are counted as individual variables) are in the model for a given value of $\log(\lambda)$. The dots represent the cross-validation error at that point and the vertical lines are the confidence interval for the error. The leftmost vertical line indicates the value of λ where the error is minimized and the rightmost vertical line is the next largest value of λ error that is within one standard error of the minimum.

```
> acsCV1$lambda.min

[1] 0.0005258299

> acsCV1$lambda.1se

[1] 0.006482677

> plot(acsCV1)
```

Extracting the coefficients is done as with any other model, by using **coef**, except that a specific level of λ should be specified; otherwise, the entire path is returned. Dots represent variables that were not selected.

```
> coef(acsCV1, s="lambda.1se")

45 x 1 sparse Matrix of class "dgCMatrix"
                                    1
(Intercept)            -5.0552170103
NumBedrooms             0.0542621380
NumChildren             .
NumPeople               .
NumRooms                0.1102021934
NumUnitsMobile home    -0.8960712560
NumUnitsSingle attached .
```

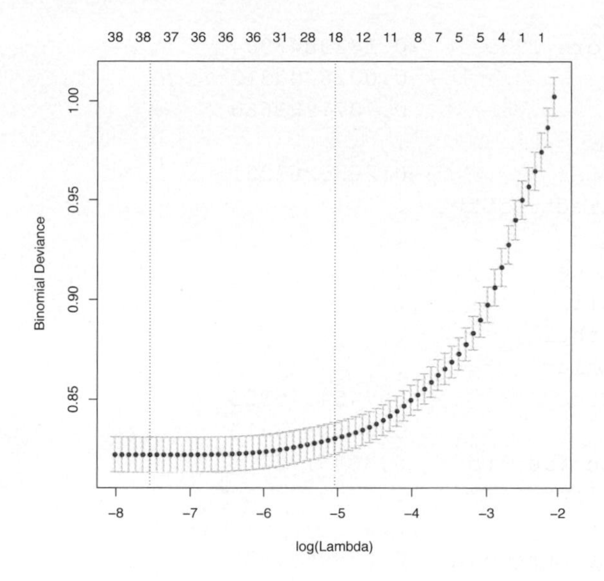

Figure 22.1 Cross-validation curve for the `glmnet` fitted on the American Community Survey data. The top row of numbers indicates how many variables (factor levels are counted as individual variables) are in the model for a given value of log(λ). The dots represent the cross-validation error at that point and the vertical lines are the confidence interval for the error. The leftmost vertical line indicates the value of λ where the error is minimized and the rightmost vertical line is the next largest value of λ error that is within one standard error of the minimum.

```
NumUnitsSingle detached  .
NumVehicles               0.1283171343
NumWorkers                0.4806697219
OwnRentMortgage           .
OwnRentOutright           0.2574766773
OwnRentRented            -0.1790627645
YearBuilt15               .
YearBuilt1940-1949       -0.0253908040
YearBuilt1950-1959        .
YearBuilt1960-1969        .
YearBuilt1970-1979       -0.0063336086
YearBuilt1980-1989        0.0147761442
YearBuilt1990-1999        .
YearBuilt2000-2004        .
YearBuilt2005             .
YearBuilt2006             .
YearBuilt2007             .
YearBuilt2008             .
YearBuilt2009             .
YearBuilt2010             .
```

```
YearBuiltBefore 1939      -0.1829643904
ElectricBill               0.0018200312
FoodStampNo                0.7071289660
FoodStampYes               .
HeatingFuelCoal           -0.2635263281
HeatingFuelElectricity     .
HeatingFuelGas             .
HeatingFuelNone            .
HeatingFuelOil             .
HeatingFuelOther           .
HeatingFuelSolar           .
HeatingFuelWood           -0.7454315355
Insurance                  0.0004973315
LanguageAsian Pacific      0.3606176925
LanguageEnglish            .
LanguageOther              .
LanguageOther European     0.0389641675
LanguageSpanish            .
```

It might seem weird that some `levels` of a `factor` were selected and others were not, but it ultimately makes sense because the lasso eliminates variables that are highly correlated with each other.

Another thing to notice is that there are no standard errors and hence no confidence intervals for the coefficients. The same is true of any predictions made from a **glmnet** model. This is due to the theoretical properties of the lasso and ridge, and is an open problem. Recent advancements have led to the ability to perform significance tests on lasso regressions, although the existing R package requires that the model be fitted using the **lars** package, not **glmnet**, at least until the research extends the testing ability to cover the Elastic Net as well.

Visualizing where variables enter the model along the λ path can be illuminating and is seen in Figure 22.2. Each line represents a coefficient's value at different values of λ. The leftmost vertical line indicates the value of λ where the error is minimized and the rightmost vertical line is the next largest value of λ error that is within one standard error of the minimum.

```
> # plot the path
> plot(acsCV1$glmnet.fit, xvar="lambda")
> # add in vertical lines for the optimal values of lambda
> abline(v=log(c(acsCV1$lambda.min, acsCV1$lambda.1se)), lty=2)
```

Setting α to 0 causes the results to be from the ridge. In this case, every variable is kept in the model but is just shrunk closer to 0. Notice in Figure 22.4 that for every value of λ there are still all the variables, just at different sizes. Figure 22.3 shows the cross-validation curve.

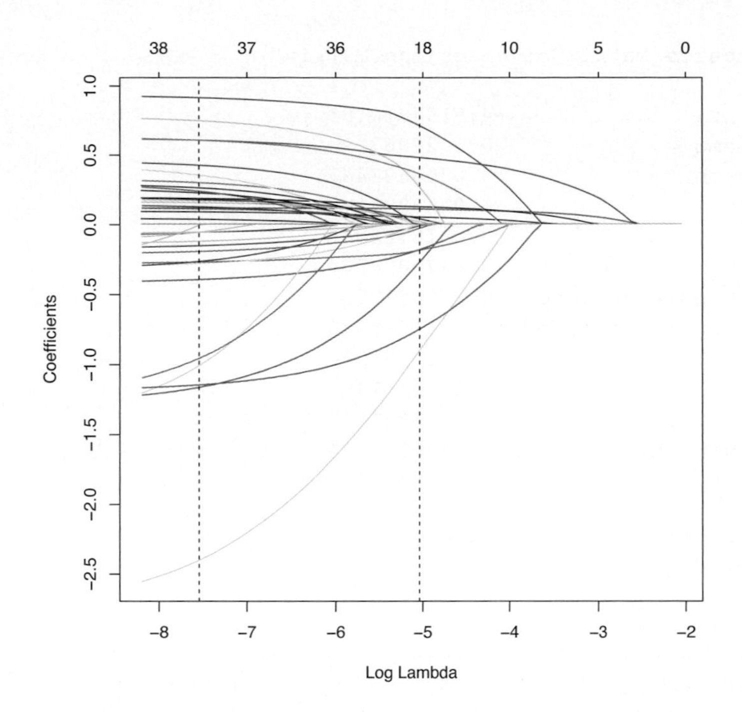

Figure 22.2 Coefficient profile plot of the `glmnet` model fitted on the ACS data. Each line represents a coefficient's value at different values of λ. The leftmost vertical line indicates the value of λ where the error is minimized and the rightmost vertical line is the next largest value of λ error that is within one standard error of the minimum.

```
> # fit the ridge model
> set.seed(71623)
> acsCV2 <- cv.glmnet(x=acsX, y=acsY, family="binomial",
+                     nfold=5, alpha=0)

> # look at the lambda values
> acsCV2$lambda.min

[1] 0.01396651

> acsCV2$lambda.1se

[1] 0.04681018

> # look at the coefficients
> coef(acsCV2, s="lambda.1se")
```

```
45 x 1 sparse Matrix of class "dgCMatrix"
                                       1
(Intercept)               -4.8197810188
NumBedrooms                0.1027963294
NumChildren                0.0308893447
NumPeople                 -0.0203037177
NumRooms                   0.0918136969
NumUnitsMobile home       -0.8470874369
NumUnitsSingle attached    0.1714879712
NumUnitsSingle detached    0.0841095530
NumVehicles                0.1583881396
NumWorkers                 0.3811651456
OwnRentMortgage            0.1985621193
OwnRentOutright            0.6480126218
OwnRentRented             -0.2548147427
YearBuilt15               -0.6828640400
YearBuilt1940-1949        -0.1082928305
YearBuilt1950-1959         0.0602009151
YearBuilt1960-1969         0.0081133932
YearBuilt1970-1979        -0.0816541923
YearBuilt1980-1989         0.1593567244
YearBuilt1990-1999         0.1218212609
YearBuilt2000-2004         0.1768690849
YearBuilt2005              0.2923210334
YearBuilt2006              0.2309044444
YearBuilt2007              0.3765019705
YearBuilt2008             -0.0648999685
YearBuilt2009              0.2382560699
YearBuilt2010              0.3804282473
YearBuiltBefore 1939      -0.1648659906
ElectricBill               0.0018576432
FoodStampNo                0.3886474609
FoodStampYes              -0.3886013004
HeatingFuelCoal           -0.7005075763
HeatingFuelElectricity    -0.1370927269
HeatingFuelGas             0.0873505398
HeatingFuelNone           -0.5983944720
HeatingFuelOil             0.1241958119
HeatingFuelOther          -0.1872564710
HeatingFuelSolar          -0.0870480957
HeatingFuelWood           -0.6699727752
Insurance                  0.0003881588
LanguageAsian Pacific      0.3982023046
LanguageEnglish           -0.0851389569
LanguageOther              0.1804675114
LanguageOther European     0.0964194255
LanguageSpanish           -0.1274688978
```

```
> # plot the cross-validation error path
> plot(acsCV2)
```

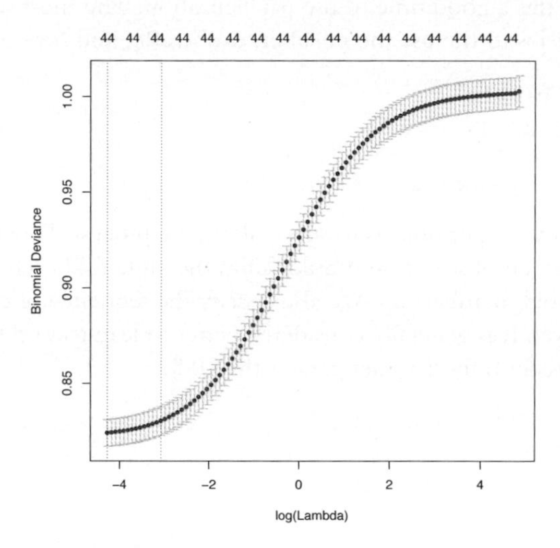

Figure 22.3 Cross-validation curve for ridge regression fitted on ACS data.

```
> # plot the coefficient path
> plot(acsCV2$glmnet.fit, xvar="lambda")
> abline(v=log(c(acsCV2$lambda.min, acsCV2$lambda.1se)), lty=2)
```

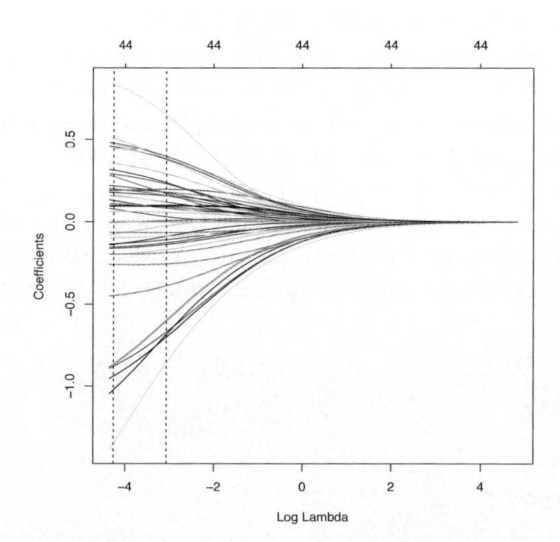

Figure 22.4 Coefficient profile plot for ridge regression fitted on ACS data.

Finding the optimal value of α requires an additional layer of cross-validation, and unfortunately **glmnet** does not do that automatically. This will require us to run **cv.glmnet** at various levels of α, which will take a fairly large chunk of time if performed sequentially, making this a good time to use parallelization. The most straightforward way to run code in parallel is to the use the **parallel**, **doParallel** and **foreach** packages.

```
> library(parallel)
> library(doParallel)

Loading required package: iterators
```

First, we build some helper objects to speed along the process. When a two-layered cross-validation is run, an observation should fall in the same fold each time, so we build a `vector` specifying fold membership. We also specify the sequence of α values that **foreach** will loop over. It is generally considered better to lean toward the lasso rather than the ridge, so we consider only α values greater than 0.5.

```
> # set the seed for repeatability of random results
> set.seed(2834673)
>
> # create folds
> # we want observations to be in the same fold each time it is run
> theFolds <- sample(rep(x=1:5, length.out=nrow(acsX)))
>
> # make sequence of alpha values
> alphas <- seq(from=.5, to=1, by=.05)
```

Before running a parallel job, a cluster (even on a single machine) must be started and registered with **makeCluster** and **registerDoParallel**. After the job is done the cluster should be stopped with **stopCluster**. Setting `.errorhandling` to `"remove"` means that if an error occurs, that iteration will be skipped. Setting `.inorder` to `FALSE` means that the order of combining the results does not matter and they can be combined whenever returned, which yields significant speed improvements. Because we are using the default combination function, **list**, which takes multiple arguments at once, we can speed up the process by setting `.multicombine` to `TRUE`. We specify in `.packages` that **glmnet** should be loaded on each of the workers, again leading to performance improvements. The operator **%dopar%** tells **foreach** to work in parallel. Parallel computing can be dependent on the `environment`, so we explicitly load some variables into the **foreach** `environment` using `.export`, namely, `acsX`, `acsY`, `alphas` and `theFolds`.

```
> # set the seed for repeatability of random results
> set.seed(5127151)
>
> # start a cluster with two workers
> cl <- makeCluster(2)
> # register the workers
> registerDoParallel(cl)
>
```

```
> # keep track of timing
> before <- Sys.time()
>
> # build foreach loop to run in parallel
> ## several arguments
> acsDouble <- foreach(i=1:length(alphas), .errorhandling="remove",
+                      .inorder=FALSE, .multicombine=TRUE,
+                      .export=c("acsX", "acsY", "alphas", "theFolds"),
+                      .packages="glmnet") %dopar%
+ {
+     print(alphas[i])
+     cv.glmnet(x=acsX, y=acsY, family="binomial", nfolds=5,
+               foldid=theFolds, alpha=alphas[i])
+ }
>
> # stop timing
> after <- Sys.time()
>
> # make sure to stop the cluster when done
> stopCluster(cl)
>
> # time difference
> # this will depend on speed, memory & number of cores of the machine
> after - before

Time difference of 1.182743 mins
```

The results in `acsDouble` should be a list with 11 instances of **cv.glmnet** objects. We can use **sapply** to check the `class` of each element of the `list`.

```
> sapply(acsDouble, class)

 [1] "cv.glmnet" "cv.glmnet" "cv.glmnet" "cv.glmnet" "cv.glmnet"
 [6] "cv.glmnet" "cv.glmnet" "cv.glmnet" "cv.glmnet" "cv.glmnet"
[11] "cv.glmnet"
```

The goal is to find the best combination of λ and α, so we need to build some code to extract the cross-validation error (including the confidence interval) and λ from each element of the list.

```
> # function for extracting info from cv.glmnet object
> extractGlmnetInfo <- function(object)
+ {
+     # find lambdas
+     lambdaMin <- object$lambda.min
+     lambda1se <- object$lambda.1se

+     # figure out where those lambdas fall in the path
+     whichMin <- which(object$lambda == lambdaMin)
+     which1se <- which(object$lambda == lambda1se)

+     # build a one line data.frame with each of the selected lambdas and
+     # its corresponding error figures
```

```
+     data.frame(lambda.min=lambdaMin, error.min=object$cvm[whichMin],
+                lambda.1se=lambda1se, error.1se=object$cvm[which1se])
+ }
>
> # apply that function to each element of the list
> # combine it all into a data.frame
> alphaInfo <- Reduce(rbind, lapply(acsDouble, extractGlmnetInfo))
>
> # could also be done with ldply from plyr
> alphaInfo2 <- plyr::ldply(acsDouble, extractGlmnetInfo)
> identical(alphaInfo, alphaInfo2)

[1] TRUE

> # make a column listing the alphas
> alphaInfo$Alpha <- alphas
> alphaInfo

     lambda.min error.min  lambda.1se error.1se Alpha
1  0.0009582333 0.8220268 0.008142621 0.8275240  0.50
2  0.0009560545 0.8220229 0.007402382 0.8273831  0.55
3  0.0008763832 0.8220198 0.006785517 0.8272666  0.60
4  0.0008089692 0.8220180 0.006263554 0.8271680  0.65
5  0.0008244253 0.8220170 0.005816158 0.8270837  0.70
6  0.0007694636 0.8220153 0.005428414 0.8270087  0.75
7  0.0007213721 0.8220140 0.005585323 0.8276055  0.80
8  0.0006789385 0.8220131 0.005256774 0.8275457  0.85
9  0.0006412197 0.8220125 0.004964731 0.8274930  0.90
10 0.0006074713 0.8220120 0.004703430 0.8274462  0.95
11 0.0005770977 0.8220121 0.004468258 0.8274054  1.00
```

Now that we have this nice unintelligible set of numbers, we should plot it to easily pick out the best combination of α and λ, which is where the plot shows minimum error. Figure 22.5 indicates that by using the one standard error methodology, the optimal α and λ are 0.75 and 0.0054284, respectively.

```
> ## prepare the data.frame for plotting multiple pieces of information
> library(reshape2)
> library(stringr)
>
> # melt the data into long format
> alphaMelt <- melt(alphaInfo, id.vars="Alpha", value.name="Value",
+                   variable.name="Measure")
> alphaMelt$Type <- str_extract(string=alphaMelt$Measure,
+                               pattern="(min)|(1se)")
>
> # some housekeeping
> alphaMelt$Measure <- str_replace(string=alphaMelt$Measure,
+                                  pattern="\\.(min|1se)",
+                                  replacement="")
> alphaCast <- dcast(alphaMelt, Alpha + Type ~ Measure,
+                    value.var="Value")
>
> ggplot(alphaCast, aes(x=Alpha, y=error)) +
```

```
+        geom_line(aes(group=Type)) +
+        facet_wrap(~Type, scales="free_y", ncol=1) +
+        geom_point(aes(size=lambda))
```

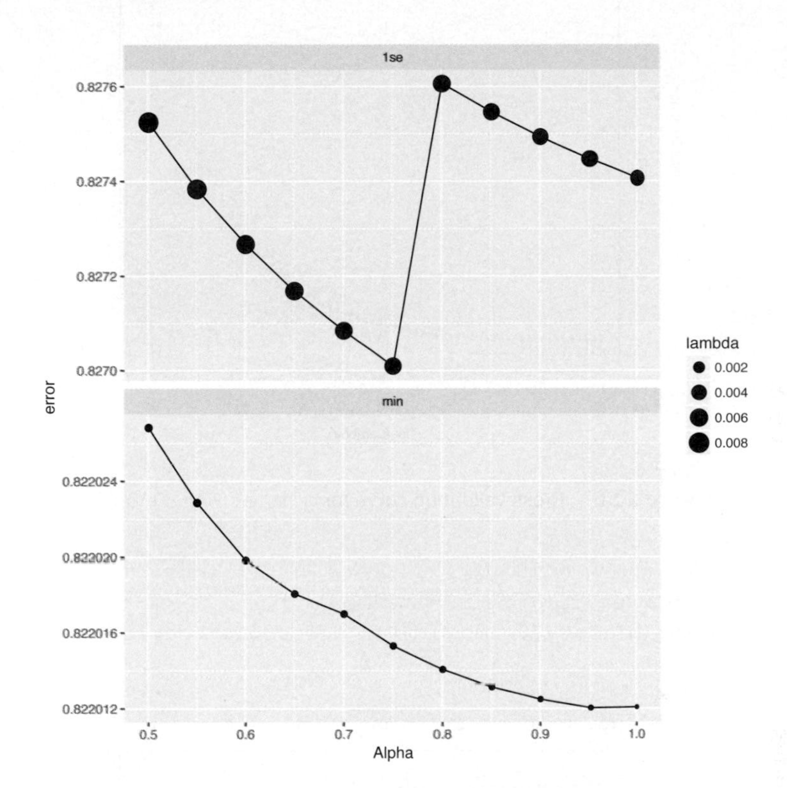

Figure 22.5 Plot of α versus error for `glmnet` cross-validation on the ACS data. The lower the error the better. The size of the dot represents the value of lambda. The top pane shows the error using the one standard error methodology (0.0054) and the bottom pane shows the error by selecting the λ (6e-04) that minimizes the error. In the top pane the error is minimized for an α of 0.75 and in the bottom pane the optimal α is 0.95.

Now that we have found the optimal value of α (0.75), we refit the model and check the results.

```
> set.seed(5127151)
> acsCV3 <- cv.glmnet(x=acsX, y=acsY, family="binomial", nfold=5,
+                     alpha=alphaInfo$Alpha[which.min(
+                         alphaInfo$error.1se
+                     )])
```

After fitting the model we check the diagnostic plots shown in Figures 22.6 and 22.7.

```
> plot(acsCV3)
```

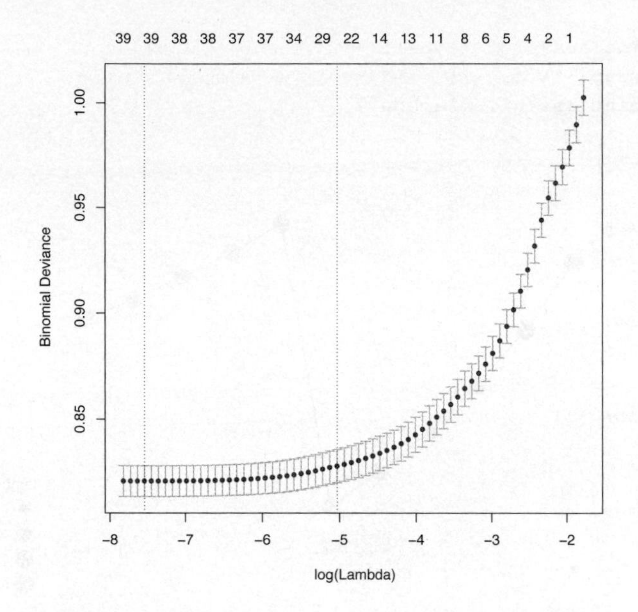

Figure 22.6 Cross-validation curve for `glmnet` with α = 0.75.

```
> plot(acsCV3$glmnet.fit, xvar = "lambda")
> abline(v = log(c(acsCV3$lambda.min, acsCV3$lambda.1se)), lty = 2)
```

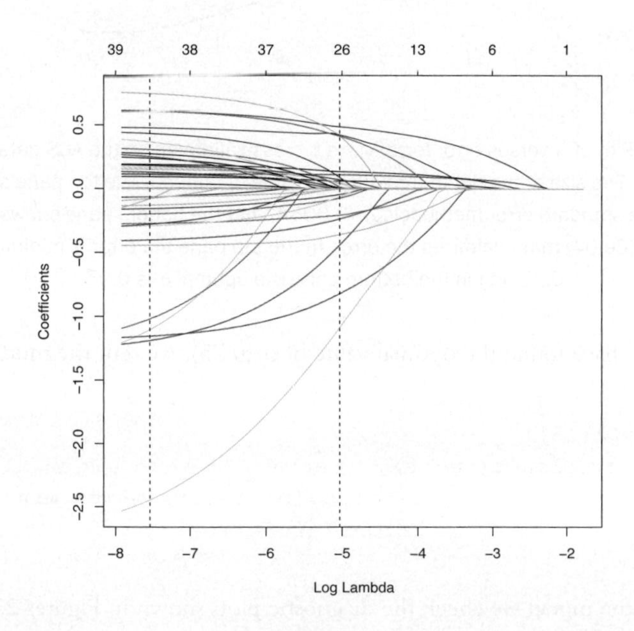

Figure 22.7 Coefficient path for `glmnet` with α = 0.75.

Viewing the coefficient plot for a `glmnet` object is not yet implemented in **coefplot**, so we build it manually. Figure 22.8 shows that the number of workers in the family and not being on foodstamps are the strongest indicators of having high income, and using coal heat and living in a mobile home are the strongest indicators of having low income. There are no standard errors because **glmnet** does not calculate them.

```
> theCoef <- as.matrix(coef(acsCV3, s="lambda.1se"))
> coefDF <- data.frame(Value=theCoef, Coefficient=rownames(theCoef))
> coefDF <- coefDF[nonzeroCoef(coef(acsCV3, s="lambda.1se")), ]
> ggplot(coefDF, aes(x=X1, y=reorder(Coefficient, X1))) +
+     geom_vline(xintercept=0, color="grey", linetype=2) +
+     geom_point(color="blue") +
+     labs(x="Value", y="Coefficient", title="Coefficient Plot")
```

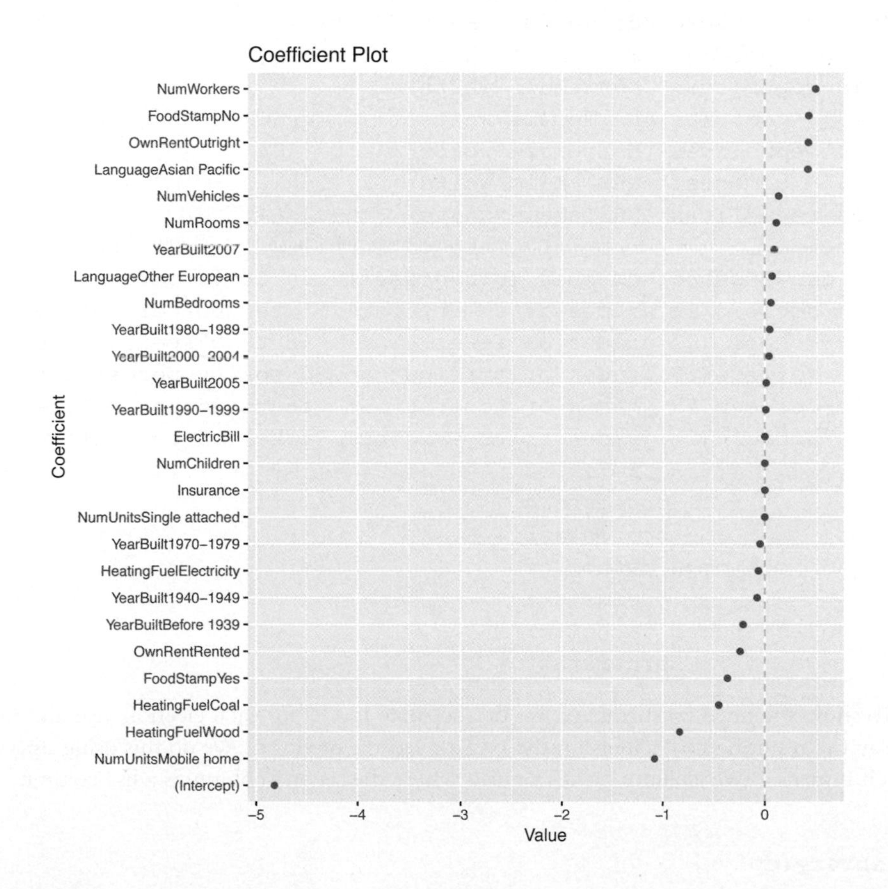

Figure 22.8 Coefficient plot for `glmnet` on ACS data. This shows that the number of workers in the family and not being on foodstamps are the strongest indicators of having high income, and using coal heat and living in a mobile home are the strongest indicators of having low income. There are no standard errors because `glmnet` does not calculate them.

22.2 Bayesian Shrinkage

For Bayesians, shrinkage can come in the form of weakly informative priors.[3] This can be particularly useful when a model is built on data that does not have a large enough number of rows for some combinations of the variables. To illustrate this, we blatantly steal an example from Andrew Gelman's and Jennifer Hill's book, *Data Analysis Using Regression and Multilevel/Hierarchical Models*, examining voter preference. The data have been cleaned up and posted at `http://jaredlander.com/data/ideo.rdata`.

```
> download.data('http://jaredlander.com/data/ideo.rdata',
                'data/ideo.rdata')

> load("data/ideo.rdata")
> head(ideo)

  Year       Vote Age Gender  Race
1 1948   democrat  NA   male white
2 1948 republican  NA female white
3 1948   democrat  NA female white
4 1948 republican  NA female white
5 1948   democrat  NA   male white
6 1948 republican  NA female white
                                     Education                Income
1     grade school of less (0-8 grades)  34 to 67 percentile
2 high school (12 grades or fewer, incl 96 to 100 percentile
3 high school (12 grades or fewer, incl  68 to 95 percentile
4 some college(13 grades or more,but no 96 to 100 percentile
5 some college(13 grades or more,but no  68 to 95 percentile
6 high school (12 grades or fewer, incl 96 to 100 percentile
                  Religion
1               protestant
2               protestant
3 catholic (roman catholic)
4               protestant
5 catholic (roman catholic)
6               protestant
```

To show the need for shrinkage, we fit a separate model for each election year and then display the resulting coefficients for the `black level` of `Race`. We do this using **dplyr**, which returns a two-column `data.frame` where the second column is a list-column.

```
> ## fit a bunch of models
> library(dplyr)
> results <- ideo %>%
+     # group the data by year
```

3. From a Bayesian point of view, the penalty terms in the Elastic Net could be considered log-priors as well.

```
+       group_by(Year) %>%
+       # fit a model to each grouping of data
+       do(Model=glm(Vote ~ Race + Income + Gender + Education,
+                data=.,
+                family=binomial(link="logit")))
> # Model is a list-column so we treat it as a column
> # give the list good names
> names(results$Model) <- as.character(results$Year)
>
> results

Source: local data frame [14 x 2]
Groups: <by row>

# A tibble: 14 × 2
     Year      Model
*   <dbl>     <list>
1    1948 <S3: glm>
2    1952 <S3: glm>
3    1956 <S3: glm>
4    1960 <S3: glm>
5    1964 <S3: glm>
6    1968 <S3: glm>
7    1972 <S3: glm>
8    1976 <S3: glm>
9    1980 <S3: glm>
10   1984 <S3: glm>
11   1988 <S3: glm>
12   1992 <S3: glm>
13   1996 <S3: glm>
14   2000 <S3: glm>
```

Now that we have all of these models, we can plot the coefficients with **multiplot**. Figure 22.9 shows the coefficient for the `black` `level` of `Race` for each model. The result for the model from 1964 is clearly far different from the other models. Figure 22.9 shows standard errors, which threw off the scale so much that we had to restrict the plot window to still see variation in the other points. Fitting a series of models like this and then plotting the coefficients over time has been termed the "secret weapon" by Gelman due to its usefulness and simplicity.

```
> library(coefplot)
> # get the coefficient information
> voteInfo <- multiplot(results$Model,
+                       coefficients="Raceblack", plot=FALSE)
> head(voteInfo)
```

```
        Value Coefficient   HighInner      LowInner   HighOuter
1   0.07119541   Raceblack   0.6297813    -0.4873905   1.1883673
2  -1.68490828   Raceblack  -1.3175506    -2.0522659  -0.9501930
3  -0.89178359   Raceblack  -0.5857195    -1.1978476  -0.2796555
4  -1.07674848   Raceblack  -0.7099648    -1.4435322  -0.3431811
5 -16.85751152   Raceblack 382.1171424 -415.8321655 781.0917963
6  -3.65505395   Raceblack  -3.0580572    -4.2520507  -2.4610605
    LowOuter Model
1   -1.045976  1948
2   -2.419624  1952
3   -1.503912  1956
4   -1.810316  1960
5 -814.806819  1964
6   -4.849047  1968

> # plot it restricting the window to (-20, 10)
> multiplot(results$Model,
+           coefficients="Raceblack", secret.weapon=TRUE) +
+     coord_flip(xlim=c(-20, 10))
```

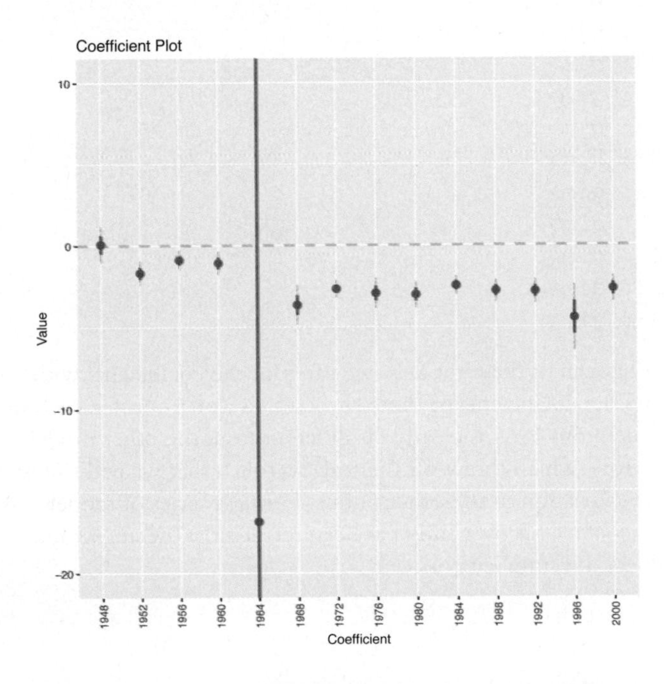

Figure 22.9 Plot showing the coefficient for the black level of Race for each of the models. The coefficient for 1964 has a standard error that is orders of magnitude bigger than for the other years. It is so out of proportion that the plot had to be truncated to still see variation in the other data points.

By comparing the model for 1964 to the other models, we can see that something is clearly wrong with the estimate. To fix this we put a prior on the coefficients in the model. The simplest way to do this is to use Gelman's **bayesglm** function in the **arm** package. By default it sets a Cauchy prior with scale 2.5. Because the **arm** package `namespace` interferes with the **coefplot** `namespace`, we do not load the package but rather just call the function using the `::` operator.

```
> resultsB <- ideo %>%
+     # group the data by year
+     group_by(Year) %>%
+     # fit a model to each grouping of data
+     do(Model=arm::bayesglm(Vote ~ Race + Income + Gender + Education,
+                            data=.,
+                            family=binomial(link="logit"),
+                            prior.scale=2.5, prior.df=1))
> # give the list good names
> names(resultsB$Model) <- as.character(resultsB$Year)
>
> # build the coefficient plot
> multiplot(resultsB$Model, coefficients="Raceblack", secret.weapon=TRUE)
```

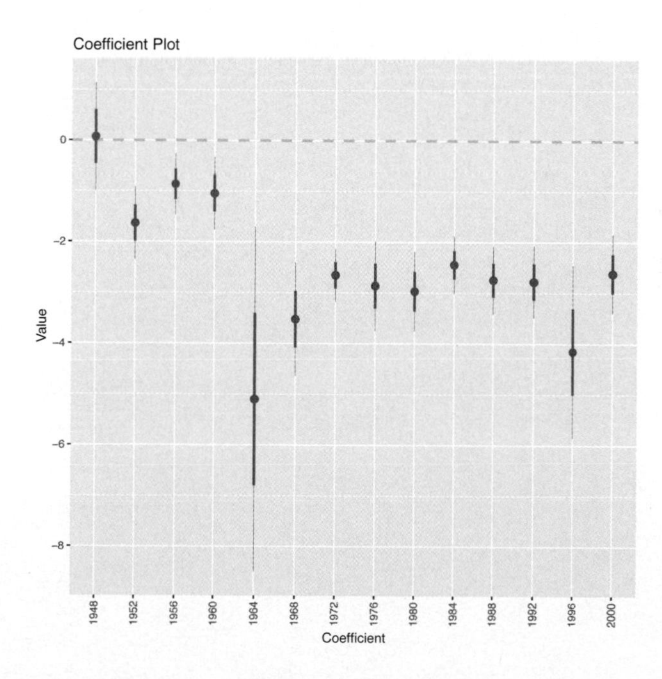

Figure 22.10 Coefficient plot (the secret weapon) for the `black` `level` of `Race` for each of the models with a Cauchy prior. A simple change like adding a prior dramatically changed the point estimate and standard error.

Simply adding Cauchy priors dramatically shrinks both the estimate and the standard error of the coefficient, as seen in Figure 22.10. Remember, the models were fitted independently, meaning that it was simply the prior that did the fix and not information from the other years. It turns out that the survey conducted in 1964 underrepresented black respondents, which led to a highly inaccurate measure.

The default prior is a Cauchy with scale 2.5, which is the same as a t distribution with 1 degree of freedom. These arguments, `prior.scale` and `prior.df`, can be changed to represent a t distribution with any degrees of freedom. Setting both to infinity (`Inf`) makes them normal priors, which is identical to running an ordinary **glm**.

22.3 Conclusion

Regularization and shrinkage play important roles in modern statistics. They help fit models to poorly designed data, and prevent overfitting of complex models. The former is done using Bayesian methods, in this case the simple **bayesglm**; the latter is done with the lasso, ridge or Elastic Net using **glmnet**. Both are useful tools to have.

23

Nonlinear Models

A key tenet of linear models is a linear relationship, which is actually reflected in the coefficients, not the predictors. While this is a nice simplifying assumption, in reality nonlinearity often holds. Fortunately, modern computing makes fitting nonlinear models not much more difficult than fitting linear models. Typical implementations are nonlinear least squares, splines, decision trees and random forests and generalized additive models (GAMs).

23.1 Nonlinear Least Squares

The nonlinear least squares model uses squared error loss to find the optimal parameters of a generic (nonlinear) function of the predictors.

$$y_i = f(x_i, \beta) \tag{23.1}$$

A common application for a nonlinear model is using the location of WiFi-connected devices to determine the location of the WiFi hotspot. In a problem like this, the locations of the devices in a two-dimensional grid are known, and they report their distance to the hotspot but with some random noise due to the fluctuation of the signal strength. A sample dataset is available at `http://jaredlander.com/data/wifi.rdata`.

```
> load("data/wifi.rdata")
> head(wifi)

  Distance        x         y
1 21.87559 28.60461 68.429628
2 67.68198 90.29680 29.155945
3 79.25427 83.48934  0.371902
4 44.73767 61.39133 80.258138
5 39.71233 19.55080 83.805855
6 56.65595 71.93928 65.551340
```

This dataset is easy to plot with **ggplot2**. The x- and y-axes are the devices' positions in the grid, and the color represents how far the device is from the hotspot, blue being closer and red being farther.

```
> library(ggplot2)
> ggplot(wifi, aes(x=x, y=y, color=Distance)) + geom_point() +
+     scale_color_gradient2(low="blue", mid="white", high="red",
+                           midpoint=mean(wifi$Distance))
```

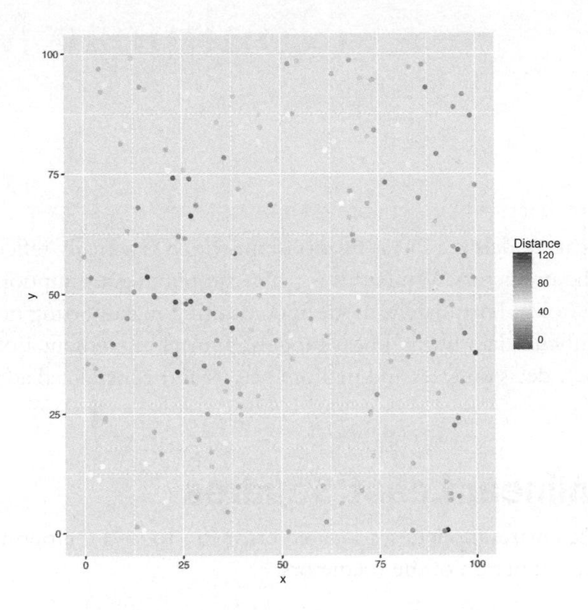

Figure 23.1 Plot of WiFi device position colored by distance from the hotspot. Blue points are closer and red points are farther.

The distance between a device i and the hotspot is

$$d_i = \sqrt{(\beta_x - x_i)^2 + (\beta_y - y_i)^2} \tag{23.2}$$

where β_x and β_y are the unknown x- and y-coordinates of the hotspot.

A standard function in R for computing nonlinear least squares is **nls**. Since these problems are usually intractable, numerical methods are used, which can be sensitive to starting values, so best guesses need to be specified. The function takes a formula—just like **lm**—except the equation and coefficients are explicitly specified. The starting values for the coefficients are given in a named `list`.

```
> # specify the square root model
> # starting values are at the center of the grid
> wifiMod1 <- nls(Distance ~ sqrt((betaX - x)^2 + (betaY - y)^2),
+                 data=wifi, start=list(betaX=50, betaY=50))
> summary(wifiMod1)
```

```
Formula: Distance ~ sqrt((betaX - x)^2 + (betaY - y)^2)

Parameters:
      Estimate Std. Error t value Pr(>|t|)
betaX   17.851      1.289   13.85   <2e-16 ***
betaY   52.906      1.476   35.85   <2e-16 ***
---
Signif. codes:  0 '***' 0.001 '**' 0.01 '*' 0.05 '.' 0.1 ' ' 1

Residual standard error: 13.73 on 198 degrees of freedom

Number of iterations to convergence: 6
Achieved convergence tolerance: 3.846e-06
```

This estimates that the hotspot is located at (17.8506668, 52.9056438). Plotting this in Figure 23.2, we see that the hotspot is located amidst the "close" blue points, indicating a good fit.

```
> ggplot(wifi, aes(x=x, y=y, color=Distance)) + geom_point() +
+     scale_color_gradient2(low="blue", mid="white", high="red",
+                           midpoint=mean(wifi$Distance)) +
+     geom_point(data=as.data.frame(t(coef(wifiMod1))),
+                aes(x=betaX, y=betaY), size=5, color="green")
```

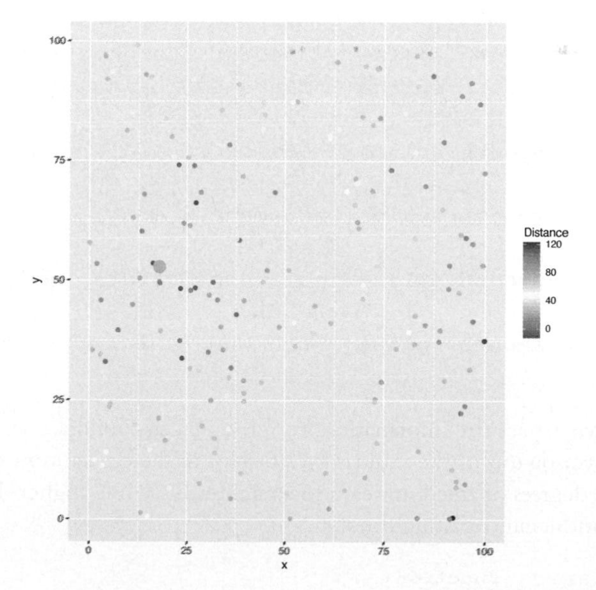

Figure 23.2 Plot of WiFi devices. The hotspot is the large green dot. Its position in the middle of the blue dots indicates a good fit.

23.2 Splines

A smoothing spline can be used to fit a smooth to data that exhibit nonlinear behavior and even make predictions on new data. A spline is a function f that is a linear combination of N functions (one for each unique data point) that are transformations of the variable x.

$$f(x) = \sum_{j=1}^{N} N_J(x)\theta_j \tag{23.3}$$

The goal is to find the function f that minimizes

$$RSS(f, \lambda) = \sum_{i=1}^{N} \{y_i - f(x_i)\}^2 + \lambda \int \{f''(t)\}^2 \, dt \tag{23.4}$$

where λ is the smoothing parameter. Small λs make for a rough smooth and large λs make for a smooth smooth.

This is accomplished in R using **smooth.spline**. It returns a list of items where x holds the unique values of the data, y are the corresponding fitted values and `df` is the degrees of freedom used. We demonstrate with the `diamonds` data.

```
> data(diamonds)
> # fit with a few different degrees of freedom
> # the degrees of freedom must be greater than 1
> # but less than the number of unique x values in the data
> diaSpline1 <- smooth.spline(x=diamonds$carat, y=diamonds$price)
> diaSpline2 <- smooth.spline(x=diamonds$carat, y=diamonds$price,
+                             df=2)
> diaSpline3 <- smooth.spline(x=diamonds$carat, y=diamonds$price,
+                             df=10)
> diaSpline4 <- smooth.spline(x=diamonds$carat, y=diamonds$price,
+                             df=20)
> diaSpline5 <- smooth.spline(x=diamonds$carat, y=diamonds$price,
+                             df=50)
> diaSpline6 <- smooth.spline(x=diamonds$carat, y=diamonds$price,
+                             df=100)
```

To plot these we extract the information from the objects, build a `data.frame`, and then add a new layer on top of the standard scatterplot of the `diamonds` data. Figure 23.3 shows this. Fewer degrees of freedom leads to straighter fits while higher degrees of freedom leads to more interpolating lines.

```
> get.spline.info <- function(object)
+ {
+     data.frame(x=object$x, y=object$y, df=object$df)
+ }
>
```

```
> library(plyr)
> # combine results into one data.frame
> splineDF <- ldply(list(diaSpline1, diaSpline2, diaSpline3, diaSpline4,
+                        diaSpline5, diaSpline6), get.spline.info)
> head(splineDF)

     x        y       df
1 0.20 361.9112 101.9053
2 0.21 397.1761 101.9053
3 0.22 437.9095 101.9053
4 0.23 479.9756 101.9053
5 0.24 517.0467 101.9053
6 0.25 542.2470 101.9053

> g <- ggplot(diamonds, aes(x=carat, y=price)) + geom_point()
> g + geom_line(data=splineDF,
+               aes(x=x, y=y, color=factor(round(df, 0)), group=df)) +
+     scale_color_discrete("Degrees of \nFreedom")
```

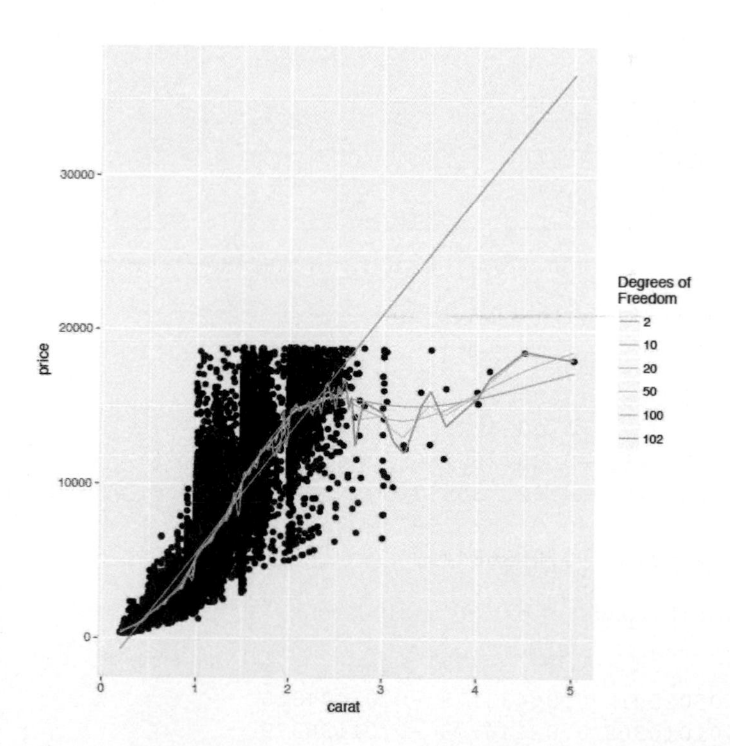

Figure 23.3 Diamonds data with a number of different smoothing splines.

Making predictions on new data is done, as usual, with **predict**.

Another type of spline is the basis spline, which creates new predictors based on transformations of the original predictors. The best basis spline is the natural cubic spline because it creates smooth transitions at interior breakpoints and forces linear behavior beyond the endpoints of the input data. A natural cubic spline with K breakpoints (knots) is made of K basis functions

$$N_1(X) = 1, N_2(X) = X, N_{k+2} = d_k(X) - d_{K-1}(X) \tag{23.5}$$

where

$$d_k(X) = \frac{(X - \xi_k)^3_+ - (X - \xi_K)^3_+}{\xi_K - \xi_k} \tag{23.6}$$

and ξ is the location of a knot and t_+ denotes the positive part of t.

While the math may seem complicated, natural cubic splines are easily fitted using **ns** from the **splines** package. It takes a predictor variable and the number of new variables to return.

```
> library(splines)
> head(ns(diamonds$carat, df=1))

              1
[1,] 0.00500073
[2,] 0.00166691
[3,] 0.00500073
[4,] 0.01500219
[5,] 0.01833601
[6,] 0.00666764

> head(ns(diamonds$carat, df=2))

               1            2
[1,] 0.013777685 -0.007265289
[2,] 0.004593275 -0.002422504
[3,] 0.013777685 -0.007265289
[4,] 0.041275287 -0.021735857
[5,] 0.050408348 -0.026525299
[6,] 0.018367750 -0.009684459

> head(ns(diamonds$carat, df=3))

               1          2           3
[1,] -0.03025012 0.06432178 -0.03404826
[2,] -0.01010308 0.02146773 -0.01136379
[3,] -0.03025012 0.06432178 -0.03404826
[4,] -0.08915435 0.19076693 -0.10098109
[5,] -0.10788271 0.23166685 -0.12263116
[6,] -0.04026453 0.08566738 -0.04534740

> head(ns(diamonds$carat, df=4))
```

```
                 1           2          3           4
[1,] 3.214286e-04 -0.04811737 0.10035562 -0.05223825
[2,] 1.190476e-05 -0.01611797 0.03361632 -0.01749835
[3,] 3.214286e-04 -0.04811737 0.10035562 -0.05223825
[4,] 8.678571e-03 -0.13796549 0.28774667 -0.14978118
[5,] 1.584524e-02 -0.16428790 0.34264579 -0.17835789
[6,] 7.619048e-04 -0.06388053 0.13323194 -0.06935141
```

These new predictors can then be used in any model just like any other predictor. More knots means a more interpolating fit. Plotting the result of a natural cubic spline overlaid on data is easy with **ggplot2**. Figure 23.4a shows this for the `diamonds` data and six knots, and Figure 23.4b shows it with three knots. Notice that having six knots fits the data more smoothly.

```
> g + stat_smooth(method="lm", formula=y ~ ns(x, 6), color="blue")
> g + stat_smooth(method="lm", formula=y ~ ns(x, 3), color="red")
```

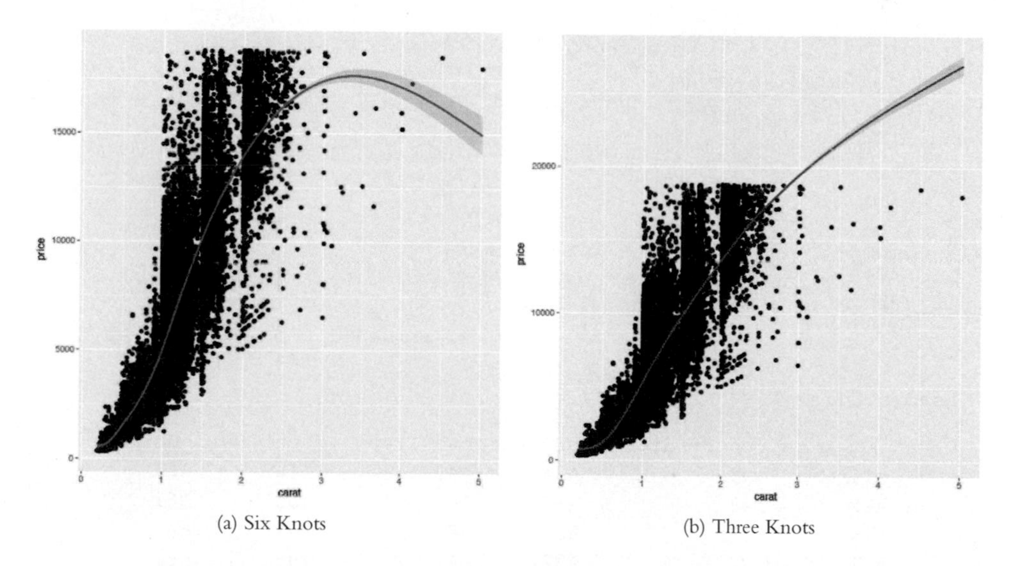

(a) Six Knots (b) Three Knots

Figure 23.4 Scatterplot of price versus carat with a regression fitted on a natural cubic spline.

23.3 Generalized Additive Models

Another method for fitting nonlinear models is generalized additive models (GAMs), which fit a separate smoothing function on each predictor independently. As the name implies, these are general and work in a number of regression contexts, meaning the response can be continuous, binary, count and other types. Like many of the best modern techniques in machine learning, this is the brainchild of Trevor Hastie and Robert Tibshirani based on work from John Chambers, the creator of `S`, the precursor of R.

They are specified as

$$E(Y|X_1, X_2, \ldots, X_p) = \alpha + f_1(X_1) + f_2(X_2) + \cdots + f_p(X_p) \tag{23.7}$$

where $X_1, X_2, \ldots, X_p$ are ordinary predictors and the f_j's are any smoothing functions.

The **mgcv** package fits GAMs with a syntax very similar to **glm**. To illustrate we use data on credit scores from the University of California–Irvine Machine Learning Repository at `http://archive.ics.uci.edu/ml/datasets/Statlog+(German+Credit+Data)`. The data is stored in a space-separated text file with no headers where categorical data have been labeled with non-obvious codes. This arcane file format goes back to a time when data storage was more limited but has, for some reason, persisted.

The first step is reading the data like any other file except that the column names need to be specified.

```
> # make vector of column names
> creditNames <- c("Checking", "Duration", "CreditHistory", "Purpose",
+                  "CreditAmount", "Savings", "Employment",
+                  "InstallmentRate", "GenderMarital", "OtherDebtors",
+                  "YearsAtResidence", "RealEstate", "Age",
+                  "OtherInstallment", "Housing", "ExistingCredits",
+                  "Job", "NumLiable", "Phone", "Foreign", "Credit")
>
> # use read.table to read the file
> # specify that headers are not included
> # the col.names are from creditNames
> theURL <- "http://archive.ics.uci.edu/ml/
              machine-learning-databases/statlog/german/german.data"
> credit <- read.table(theURL, sep=" ", header=FALSE,
+                      col.names=creditNames, stringsAsFactors=FALSE)
>
> head(credit)

  Checking Duration CreditHistory Purpose CreditAmount Savings
1      A11        6           A34     A43         1169     A65
2      A12       48           A32     A43         5951     A61
3      A14       12           A34     A46         2096     A61
4      A11       42           A32     A42         7882     A61
5      A11       24           A33     A40         4870     A61
6      A14       36           A32     A46         9055     A65
  Employment InstallmentRate GenderMarital OtherDebtors
1        A75               4           A93         A101
2        A73               2           A92         A101
3        A74               2           A93         A101
4        A74               2           A93         A103
5        A73               3           A93         A101
6        A73               2           A93         A101
  YearsAtResidence RealEstate Age OtherInstallment Housing
1                4       A121  67             A143    A152
2                2       A121  22             A143    A152
```

```
3               3    A121  49              A143     A152
4               4    A122  45              A143     A153
5               4    A124  53              A143     A153
6               4    A124  35              A143     A153
  ExistingCredits  Job NumLiable Phone Foreign Credit
1               2 A173         1  A192    A201      1
2               1 A173         1  A191    A201      2
3               1 A172         2  A191    A201      1
4               1 A173         2  A191    A201      1
5               2 A173         2  A191    A201      2
6               1 A172         2  A192    A201      1
```

Now comes the unpleasant task of translating the codes to meaningful data. To save time and effort we decode only the variables we care about for a simple model. The simplest way of decoding is to create named vectors where the name is the code and the value is the new data.

```
> # before
> head(credit[, c("CreditHistory", "Purpose", "Employment", "Credit")])

  CreditHistory Purpose Employment Credit
1           A34     A43        A75      1
2           A32     A43        A73      2
3           A34     A46        A74      1
4           A32     A42        A74      1
5           A33     A40        A73      2
6           A32     A46        A73      1

> creditHistory <- c(A30="All Paid", A31="All Paid This Bank",
+                    A32="Up To Date", A33="Late Payment",
+                    A34="Critical Account")
>
> purpose <- c(A40="car (new)", A41="car (used)",
+              A42="furniture/equipment", A43="radio/television",
+              A44="domestic appliances", A45="repairs", A46="education",
+              A47="(vacation - does not exist?)", A48="retraining",
+              A49="business", A410="others")
>
> employment <- c(A71="unemployed", A72="< 1 year", A73="1 - 4 years",
+                 A74="4 - 7 years", A75=">= 7 years")
>
> credit$CreditHistory <- creditHistory[credit$CreditHistory]
> credit$Purpose <- purpose[credit$Purpose]
> credit$Employment <- employment[credit$Employment]
>
> # code credit as good/bad
> credit$Credit <- ifelse(credit$Credit == 1, "Good", "Bad")
> # make good the base levels
> credit$Credit <- factor(credit$Credit, levels=c("Good", "Bad"))
>
> # after
> head(credit[, c("CreditHistory", "Purpose", "Employment", "Credit")])

     CreditHistory           Purpose  Employment Credit
1 Critical Account  radio/television  >= 7 years   Good
2       Up To Date  radio/television 1 - 4 years    Bad
```

```
3 Critical Account              education 4 - 7 years   Good
4      Up To Date furniture/equipment 4 - 7 years   Good
5    Late Payment               car (new) 1 - 4 years    Bad
6      Up To Date              education 1 - 4 years   Good
```

Viewing the data will help give a sense of the relationship between the variables. Figures 23.5 and 23.6 show that there is not a clear linear relationship, so a GAM may be appropriate.

```
> library(useful)
> ggplot(credit, aes(x=CreditAmount, y=Credit)) +
+     geom_jitter(position = position_jitter(height = .2)) +
+     facet_grid(CreditHistory ~ Employment) +
+     xlab("Credit Amount") +
+     theme(axis.text.x=element_text(angle=90, hjust=1, vjust=.5)) +
+     scale_x_continuous(labels=multiple)
```

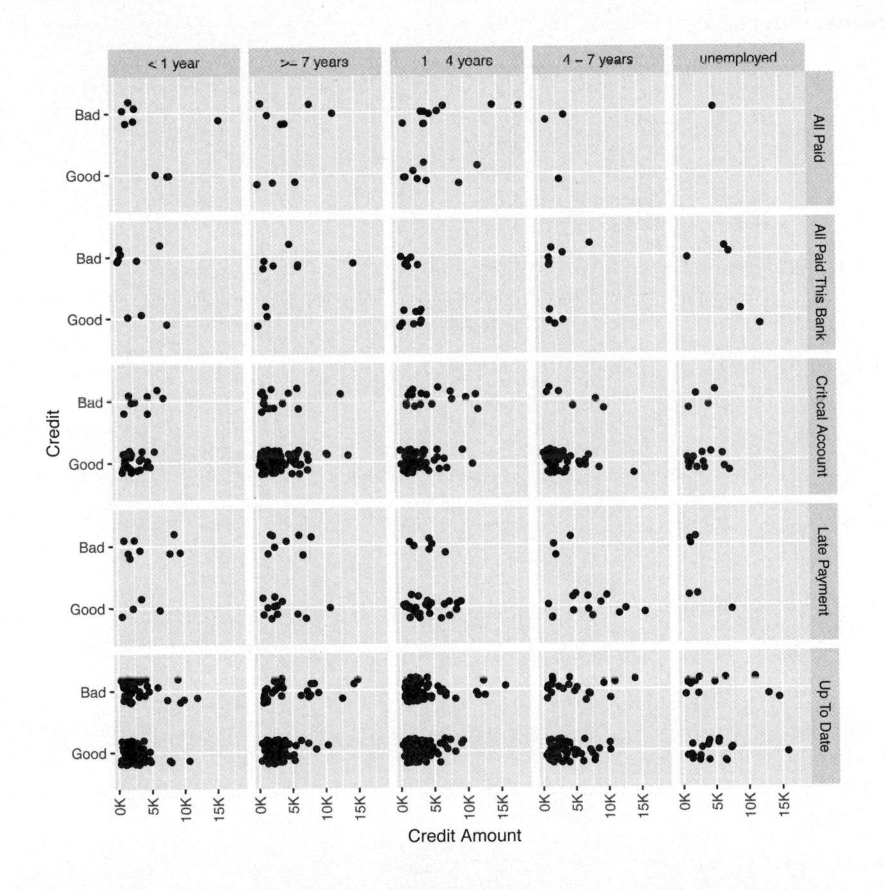

Figure 23.5 Plot of good credit versus bad based on credit amount, credit history and employment status.

```
> ggplot(credit, aes(x=CreditAmount, y=Age)) +
+     geom_point(aes(color=Credit)) +
+     facet_grid(CreditHistory ~ Employment) +
+     xlab("Credit Amount") +
+     theme(axis.text.x=element_text(angle=90, hjust=1, vjust=.5)) +
+     scale_x_continuous(labels=multiple)
```

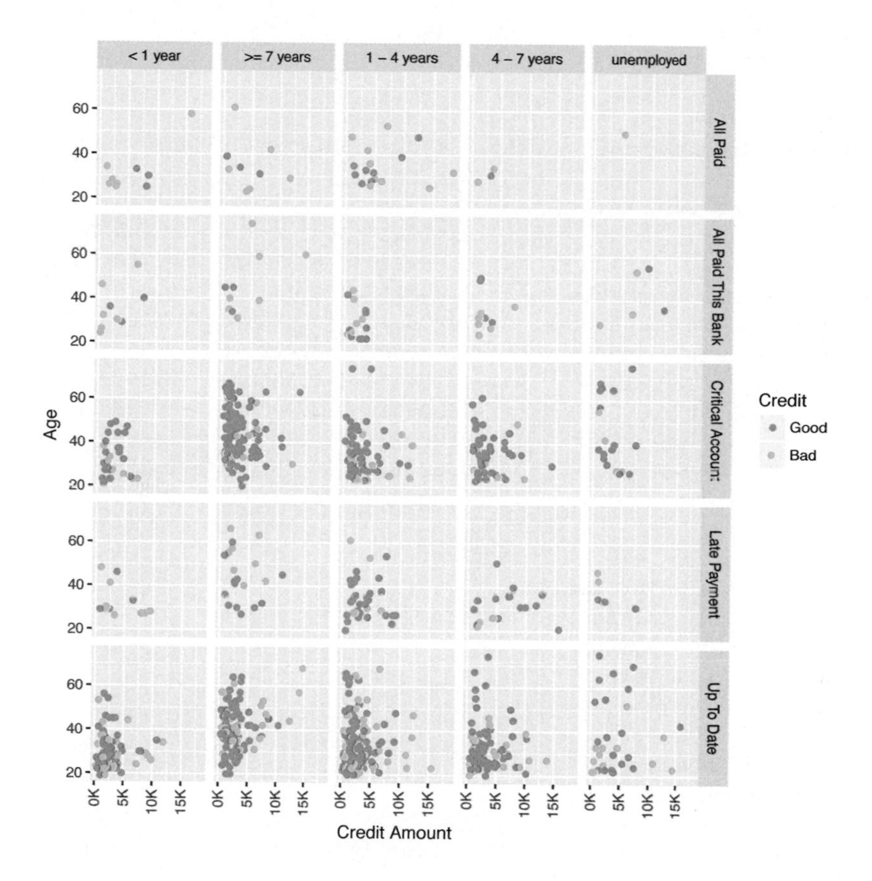

Figure 23.6 Plot of age versus credit amount faceted by credit history and employment status, color coded by credit.

Using **gam** is very similar to using other modelling functions like **lm** and **glm** that take a `formula` argument. The difference is that continuous variables, such as `CreditAmount` and `Age`, can be transformed using a nonparametric smoothing function such as a spline or tensor product.[1]

1. Tensor products are a way of representing transformation functions of predictors, possibly measured on different units.

```
> library(mgcv)
> # fit a logistic GAM
> # apply a tensor product on CreditAmount and a spline on Age
> creditGam <- gam(Credit ~ te(CreditAmount) + s(Age) + CreditHistory +
+                      Employment,
+                  data=credit, family=binomial(link="logit"))
> summary(creditGam)

Family: binomial
Link function: logit

Formula:
Credit ~ te(CreditAmount) + s(Age) + CreditHistory + Employment

Parametric coefficients:
                                Estimate Std. Error z value Pr(>|z|)
(Intercept)                     0.662840   0.372377   1.780  0.07507
CreditHistoryAll Paid This Bank 0.008412   0.453267   0.019  0.98519
CreditHistoryCritical Account  -1.809046   0.376326  -4.807 1.53e-06
CreditHistoryLate Payment      -1.136008   0.412776  -2.752  0.00592
CreditHistoryUp To Date        -1.104274   0.355208  -3.109  0.00188
Employment>= 7 years           -0.388518   0.240343  -1.617  0.10598
Employment1 - 4 years          -0.380981   0.204292  -1.865  0.06220
Employment4 - 7 years          -0.820943   0.252069  -3.257  0.00113
Employmentunemployed           -0.092727   0.334975  -0.277  0.78192

(Intercept)                     .
CreditHistoryAll Paid This Bank
CreditHistoryCritical Account   ***
CreditHistoryLate Payment       **
CreditHistoryUp To Date         **
Employment>= 7 years
Employment1 - 4 years           .
Employment4 - 7 years           **
Employmentunemployed
---
Signif. codes:  0 '***' 0.001 '**' 0.01 '*' 0.05 '.' 0.1 ' ' 1

Approximate significance of smooth terms:
                   edf Ref.df Chi.sq  p-value
te(CreditAmount) 2.415  2.783 20.896 7.26e-05 ***
s(Age)           1.932  2.435  7.383   0.0495 *
---
Signif. codes:  0 '***' 0.001 '**' 0.01 '*' 0.05 '.' 0.1 ' ' 1

R-sq.(adj) =  0.0922   Deviance explained = 8.57%
UBRE = 0.1437  Scale est. = 1         n = 1000
```

The smoother is fitted automatically in the fitting process and can be viewed after the fact. Figure 23.7 shows `CreditAmount` and `Age` with their applied smoothers, a tensor product and a spline, respectively. The gray, shaded area represents the confidence interval for the smooths.

```
> plot(creditGam, select=1, se=TRUE, shade=TRUE)
> plot(creditGam, select=2, se=TRUE, shade=TRUE)
```

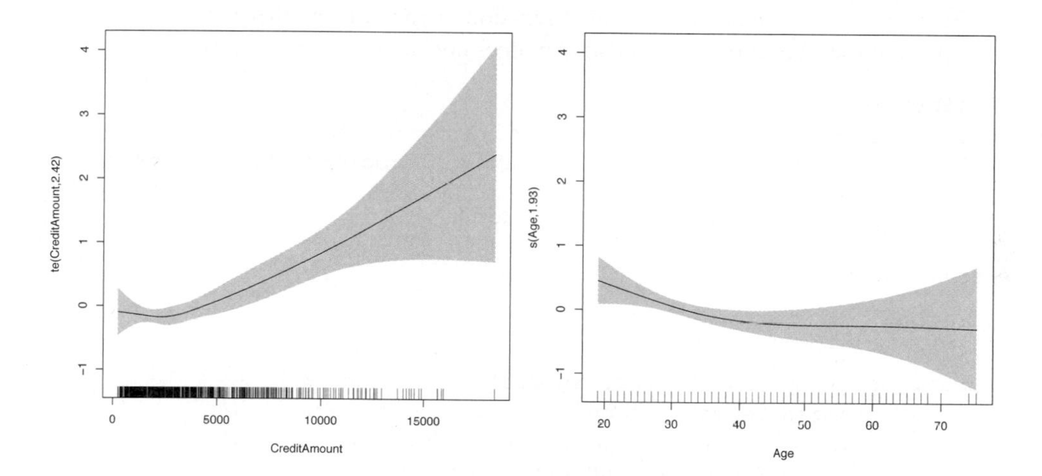

Figure 23.7 The smoother result for fitting a GAM on credit data. The shaded region represents two pointwise standard deviations.

23.4 Decision Trees

A relatively modern technique for fitting nonlinear models is the decision tree. Decision trees work for both regression and classification by performing binary splits on the recursive predictors.

For regression trees, the predictors are partitioned into M regions $R_1, R_2, \ldots, R_M$ and the response y is modeled as the average for a region with

$$\hat{f}(x) = \sum_{m=1}^{M} \hat{c}_m I(x \in R_m) \tag{23.8}$$

where

$$\hat{c}_m = \text{avg}(y_i | x_i \in R_m) \tag{23.9}$$

is the average y value for the region.

The method for classification trees is similar. The predictors are partitioned into M regions and the proportion of each class in each of the regions, $\hat{p}_{mk}$, is calculated as

$$\hat{p}_{mk} = \frac{1}{N_m} \sum_{x_i \in R_m} I(y_i = k) \tag{23.10}$$

where N_m is the number of items in region m and the summation counts the number of observations of class k in region m.

Trees can be calculated with the **rpart** function in **rpart**. Like other modelling functions, it uses the `formula` interface but does not work with interactions.

```
> library(rpart)
> creditTree <- rpart(Credit ~ CreditAmount + Age +
+                          CreditHistory + Employment, data=credit)
```

Printing the object displays the tree in text form.

```
> creditTree

n= 1000

node), split, n, loss, yval, (yprob)
      * denotes terminal node

 1) root 1000 300 Good (0.7000000 0.3000000)
   2) CreditHistory=Critical Account,Late Payment,Up To
      Date 911 247 Good (0.7288694 0.2711306)
     4) CreditAmount< 7760.5 846 211 Good (0.7505910 0.2494090) *
     5) CreditAmount>=7760.5 65  29 Bad (0.4461538 0.5538462)
      10) Age>=29.5 40  17 Good (0.5750000 0.4250000)
        20) Age< 38.5 19   4 Good (0.7894737 0.2105263) *
        21) Age>=38.5 21   8 Bad (0.3809524 0.6190476) *
      11) Age< 29.5 25   6 Bad (0.2400000 0.7600000) *
   3) CreditHistory=All Paid,All Paid This Bank 89  36
      Bad (0.4044944 0.5955056) *
```

The printed tree has one line per node. The first node is the `root` for all the data and shows that there are 1,000 observations of which 300 are considered "Bad." The next level of indentation is the first split, which is on `CreditHistory`. One direction—where `CreditHistory` equals either "Critical Account," "Late Payment" or "Up To Date"—contains 911 observations, of which 247 are considered "Bad." This has a 73% probability of having good credit. The other direction—where `CreditHistory` equals either "All Paid" or "All Paid This Bank"—has a 60% probability of having bad credit. The next level of indentation represents the next split.

Continuing to read the results this way could be laborious; plotting will be easier. Figure 23.8 shows the splits. Nodes split to the left meet the criteria while nodes to the right do not. Each terminal node is labelled by the predicted class, either "Good" or "Bad." The percentage is read from left to right, with the probability of being "Good" on the left.

```
> library(rpart.plot)
> rpart.plot(creditTree, extra=4)
```

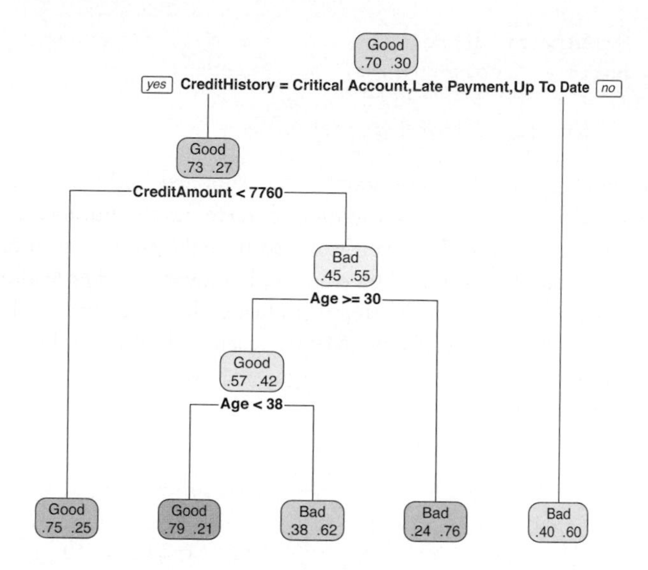

Figure 23.8 Display of decision tree based on credit data. Nodes split to the left meet the criteria while nodes to the right do not. Each terminal node is labelled by the predicted class, either "Good" or "Bad." The percentage is read from left to right, with the probability of being "Good" on the left.

While trees are easy to interpret and fit data nicely, they tend to be unstable with high variance due to overfitting. A slight change in the training data can cause a significant difference in the model.

23.5 Boosted Trees

Boosting is a popular way to improve predictions, particularly for decision trees. The main idea is that the model, or rather models, learn slowly through sequential fitting. First, a model is fit on the data with all observations having equal weight. Then the observations for which the model performed poorly are upweighted and the observations for which the model performed well are downweighted and a new model is fit. This process is repeated a set number of times and the final model is the accumulation of these little models.

The two most common functions for fitting boosted trees are **gbm** from the **gbm** package and **xgboost** from the **xgboost** package. In recent years **xgboost** has proven to be the more popular of the two. To see it in action we look at the credit data. Unlike with **rpart** we cannot use the `formula` interface and must build a predictor `matrix` and response `vector`. Unlike with **glm**, the response must be 0 and 1 and not a `logical vector`.

```
> library(useful)
> # the formula that describes the model
```

```
> # we do not need an intercept since it is a tree
> creditFormula <- Credit ~ CreditHistory + Purpose + Employment +
+     Duration + Age + CreditAmount - 1
> # we use all levels of the categorical variables since it is a tree
> creditX <- build.x(creditFormula, data=credit, contrasts=FALSE)
> creditY <- build.y(creditFormula, data=credit)
> # convert the logical vector to [0,1]
> creditY <- as.integer(relevel(creditY, ref='Bad')) - 1
```

The predictor `matrix` and response `vector` are supplied to the `data` and `label` arguments, respectively. The `nrounds` argument determines the number of passes on the data. Too many passes can lead to overfitting, so thought must go into this number. The learning rate is controlled by `eta`, with a lower number leading to less overfitting. The maximum depth of the trees is indicated by `max.depth`. Parallel processing is automatically enabled if OpenMP is present, and the number of parallel threads is controlled by the `nthread` argument. We specify the type of model with the `objective` argument.

```
> library(xgboost)
> creditBoost <- xgboost(data=creditX, label=creditY, max.depth=3,
+                        eta=.3, nthread=4, nrounds=3,
+                        objective="binary:logistic")

[1] train-error:0.261000
[2] train-error:0.262000
[3] train-error:0.255000
```

By default **xgboost** prints the evaluation metric result for each round. As the number of rounds increases the metric gets better as well.

```
> creditBoost20 <- xgboost(data=creditX, label=creditY, max.depth=3,
+                          eta=.3, nthread=4, nrounds=20,
+                          objective="binary:logistic")

[1] train-error:0.261000
[2] train-error:0.262000
[3] train-error:0.255000
[4] train-error:0.258000
[5] train-error:0.260000
[6] train-error:0.257000
[7] train-error:0.256000
[8] train-error:0.248000
[9] train-error:0.246000
[10] train-error:0.227000
[11] train-error:0.230000
[12] train-error:0.230000
[13] train-error:0.227000
[14] train-error:0.223000
[15] train-error:0.223000
```

```
[16] train-error:0.218000
[17] train-error:0.217000
[18] train-error:0.216000
[19] train-error:0.211000
[20] train-error:0.211000
```

The model generated by **xgboost** is saved to disk as a binary file, with `xgboost.model` as the default name. The file name can be set with the `save_name` argument.

Visualizing the boosted tree is achieved using the **htmlwidgets**-based **DiagrammeR** package through the **xgb.plot.multi.trees**. This function attempts to amalgamate the numerous trees into one cohesive visualization. The `feature_names` argument provides labels for the nodes. Figure 23.9 shows that for each node one or more questions are asked, depending on how each tree was fit.

```
> xgb.plot.multi.trees(creditBoost, feature_names=colnames(creditX))
```

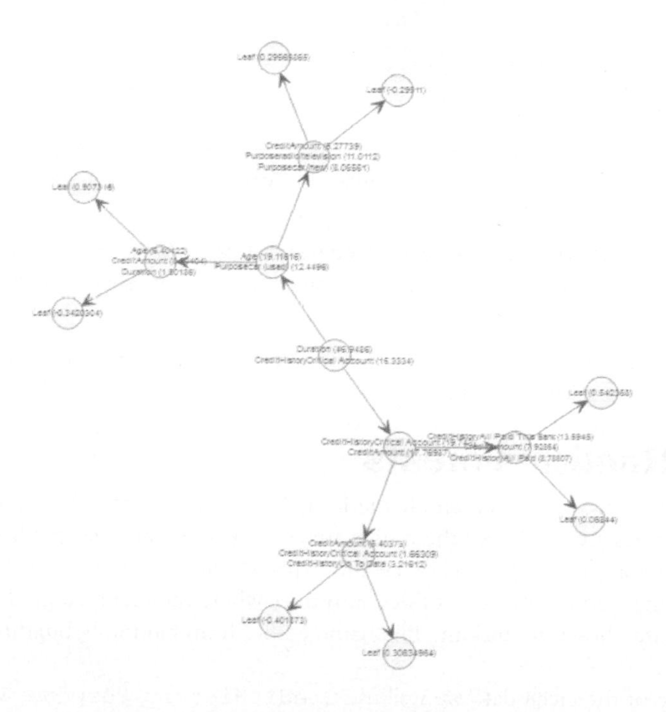

Figure 23.9 Projection of boosted trees onto one tree.

The plot can be a bit hard to make sense of, so another good visualization is a variable importance plot, which shows how much each feature contributes to the model. Figure 23.10 shows that `Duration` and `CreditAmount` are the most important variables to the model.

```
> xgb.plot.importance(xgb.importance(creditBoost,
+                                    feature_names=colnames(creditX)))
```

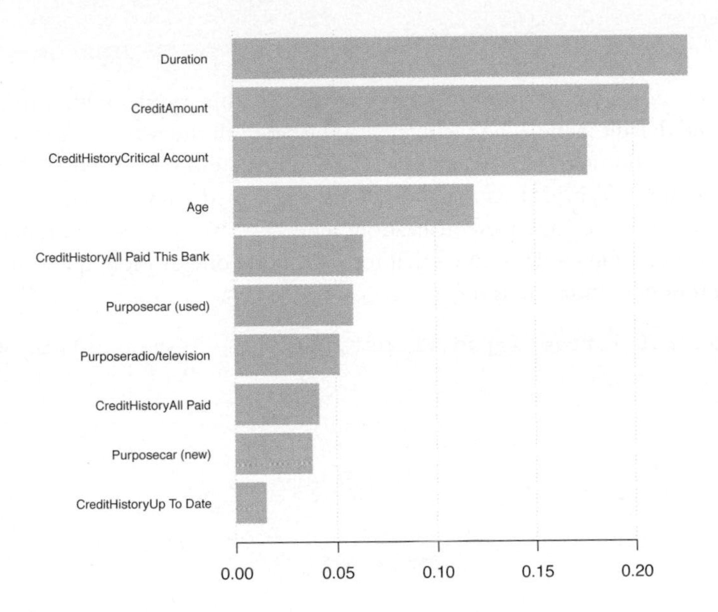

Figure 23.10 Variable importance plot for a boosted tree fit to the credit data. It shows that `Duration` and `CreditAmount` are the most important variables to the model.

Boosted trees provide a fast way to get better results than a regular decision tree, and **xgboost** is a fast implementation.

23.6 Random Forests

Random forests are a type of ensemble method. An ensemble method is a process in which numerous models are fitted, and the results are combined for stronger predictions. While this provides great predictions, inference and explainability are often limited. Random forests are composed of a number of decision trees where the included predictors and observations are chosen at random. The name comes from randomly building trees to make a forest.

In the case of the credit data we will use `CreditHistory`, `Purpose`, `Employment`, `Duration`, `Age` and `CreditAmount`. Some trees will have just `CreditHistory` and `Employment`, another will have `Purpose`, `Employment` and `Age`, while another will have `CreditHistory`, `Purpose`, `Employment` and `Age`. All of these different trees cover all the bases and make for a random forest that should have strong predictive power.

Fitting the random forest is done with **randomForest** from the **randomForest** package. Normally, **randomForest** can be used with a `formula`, but categorical variables must be stored as factors. To avoid having to convert the variables, we provide individual predictor and response matrices. This requirement for factor variables is due to the author's (Andy Liaw) frustration with the formula interface. He even warned users "I will take the formula interface away." We have seen, for this function, that using matrices is generally faster than formulas.

```
> library(randomForest)
> creditFormula <- Credit ~ CreditHistory + Purpose + Employment +
+     Duration + Age + CreditAmount - 1
> # we use all levels of the categorical variables since it is a tree
> creditX <- build.x(creditFormula, data=credit, contrasts=FALSE)
> creditY <- build.y(creditFormula, data=credit)
>
> # fit the random forest
> creditForest <- randomForest(x=creditX, y=creditY)
>
> creditForest

Call:
 randomForest(x = creditX, y = creditY)
               Type of random forest: classification
                     Number of trees: 500
No. of variables tried at each split: 4

        OOB estimate of  error rate: 27.4%
Confusion matrix:
     Good Bad class.error
Good  644  56   0.0800000
Bad   218  82   0.7266667
```

The displayed information shows that 500 trees were built and four variables were assessed at each split; the confusion matrix shows that this is not exactly the best fit, and that there is room for improvement.

Due to the similarity between boosted trees and random forests, it is possible to use **xgboost** to build a random forest by tweaking a few arguments. We fit 1000 trees in parallel (num_parallel_tree=1000) and set the row (subsample=0.5) and column (colsample_bytree=0.5) sampling to be done at random.

```
> # build the response matrix
> creditY2 <- as.integer(relevel(creditY, ref='Bad')) - 1
> # Fit the random forest
> boostedForest <- xgboost(data=creditX, label=creditY2, max_depth=4,
+                          num_parallel_tree=1000,
+                          subsample=0.5, colsample_bytree=0.5,
+                          nrounds=3, objective="binary:logistic")

[1] train-error:0.282000
[2] train-error:0.283000
[3] train-error:0.279000
```

In this case the error rate for the boosted-derived random forest is about the same as for the one fit by **randomForest**. Increasing the `nrounds` argument will improve the error rate, although it could also lead to overfitting. A nice benefit of using **xgboost** is that we can visualize the resulting random forest as a single tree as shown in Figure 23.11.

```
> xgb.plot.multi.trees(boostedForest, feature_names=colnames(creditX))
```

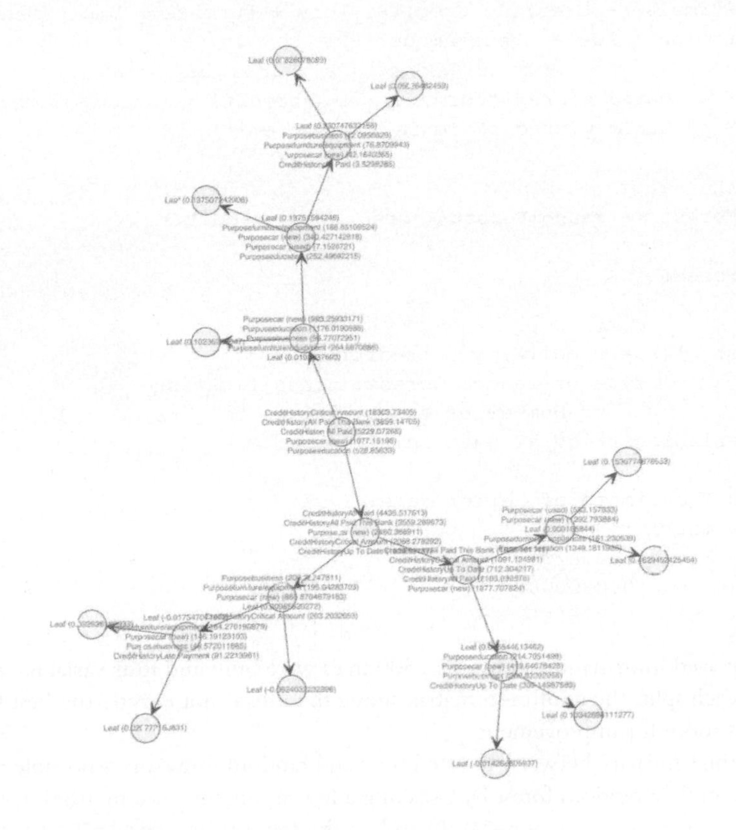

Figure 23.11 Projection of boosted random forest trees onto one tree.

23.7 Conclusion

With modern computing power, the previously necessary simplifying assumptions of linearity and normality are starting to give way to nonparametric techniques. Popular implementations are nonlinear least squares, splines, generalized additive models, decision trees and random forests. As with every other method, these all haves their benefits and costs.

24

Time Series and Autocorrelation

A big part of statistics, particularly for financial and econometric data, is analyzing time series, data that are autocorrelated over time. That is, one observation depends on previous observations and the order matters. Special care needs to be taken to account for this dependency. R has a number of built-in functions and packages to make working with time series easier.

24.1 Autoregressive Moving Average

One of the most common ways of fitting time series models is to use either autoregressive (AR), moving average (MA) or both (ARMA). These models are well represented in R and are fairly easy to work with. The formula for an ARMA(p, q) is

$$X_t - \Phi_1 X_{t-1} - \cdots - \Phi_p X_{t-p} = Z_t + \theta_1 Z_{t-1} + \cdots + \theta_q Z_{t-q} \tag{24.1}$$

where

$$Z_t \sim \mathrm{WN}(0, \sigma^2) \tag{24.2}$$

is white noise, which is essentially random data.

AR models can be thought of as linear regressions of the current value of the time series against previous values. MA models are, similarly, linear regressions of the current value of the time series against current and previous residuals.

For an illustration, we will make use of the World Bank API to download gross domestic product (GDP) for a number of countries from 1960 through 2011.

```
> # load the World Bank API package
> library(WDI)
> # pull the data
> gdp <- WDI(country=c("US", "CA", "GB", "DE", "CN", "JP", "SG", "IL"),
+            indicator=c("NY.GDP.PCAP.CD", "NY.GDP.MKTP.CD"),
+            start=1960, end=2011)
> # give it good names
> names(gdp) <- c("iso2c", "Country", "Year", "PerCapGDP", "GDP")
```

After downloading, we can inspect the data, which are stored in long country-year format with a plot of per capita GDP shown in Figure 24.1a. Figure 24.1b shows absolute

GDP, illustrating that while China's GDP has jumped significantly in the past ten years, its per capita GDP has only marginally increased.

```
> head(gdp)

  iso2c Country Year PerCapGDP         GDP
1    CA  Canada 1960  2294.569 41093453545
2    CA  Canada 1961  2231.294 40767969454
3    CA  Canada 1962  2255.230 41978852041
4    CA  Canada 1963  2354.839 44657169109
5    CA  Canada 1964  2529.518 48882938810
6    CA  Canada 1965  2739.586 53909570342

> library(ggplot2)
> library(scales)
> # per capita GDP
> ggplot(gdp, aes(Year, PerCapGDP, color=Country, linetype=Country)) +
+     geom_line() + scale_y_continuous(label=dollar)
>
> library(useful)
> # absolute GDP
> ggplot(gdp, aes(Year, GDP, color=Country, linetype=Country)) +
+     geom_line() +
+     scale_y_continuous(label=multiple_format(extra=dollar,
+                                              multiple="M"))
```

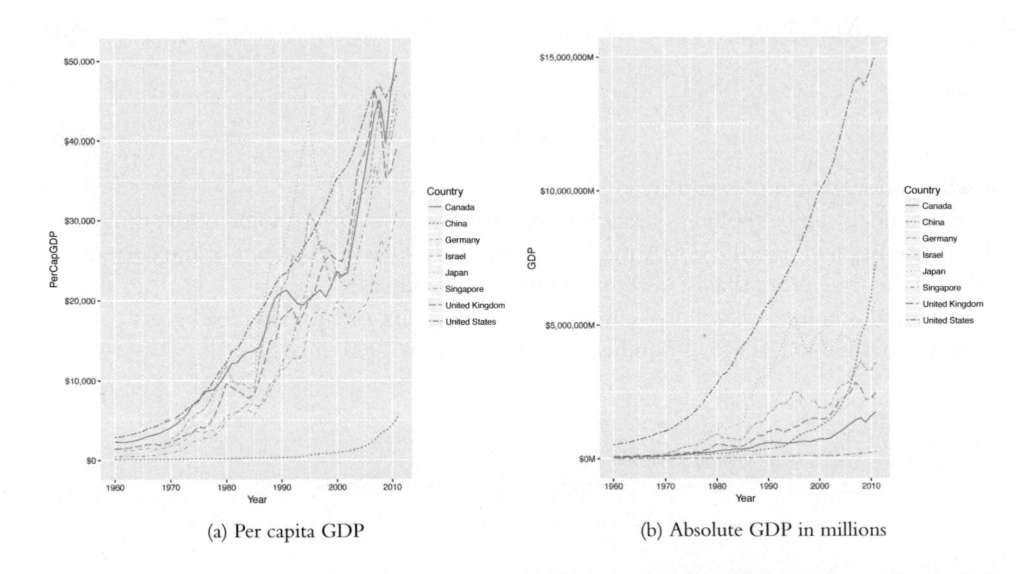

(a) Per capita GDP

(b) Absolute GDP in millions

Figure 24.1 GDP for a number of nations from 1960 to 2011.

First we will only look at only one time series, so we extract the data for the United States.

```
> # get US data
> us <- gdp$PerCapGDP[gdp$Country == "United States"]
> # convert it to a time series
> us <- ts(us, start=min(gdp$Year), end=max(gdp$Year))
> us

Time Series:
Start = 1960
End = 2011
Frequency = 1
 [1]  2881.100  2934.553  3107.937  3232.208  3423.396  3664.802
 [7]  3972.123  4152.020  4491.424  4802.642  4997.757  5360.178
[13]  5836.224  6461.736  6948.198  7516.680  8297.292  9142.795
[19] 10225.307 11301.682 12179.558 13526.187 13932.678 15000.086
[25] 16539.383 17588.810 18427.288 19393.782 20703.152 22039.227
[31] 23037.941 23443.263 24411.143 25326.736 26577.761 27559.167
[37] 28772.356 30281.636 31687.052 33332.139 35081.923 35912.333
[43] 36819.445 38224.739 40292.304 42516.393 44622.642 46349.115
[49] 46759.560 45305.052 46611.975 48111.967

> plot(us, ylab="Per Capita GDP", xlab="Year")
```

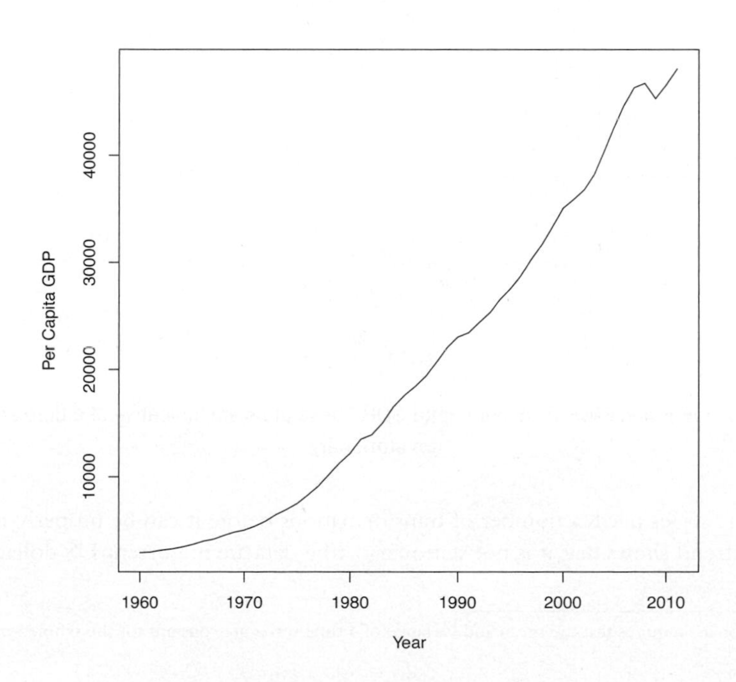

Figure 24.2 Time series plot of US per capita GDP.

Another way to assess a time series is to view its autocovariance function (ACF) and partial autocovariance function (PACF). In R this is done with the appropriately named **acf** and **pacf** functions.

The ACF shows the correlation of a time series with lags of itself. That is, how much the time series is correlated with itself at one lag, at two lags, at three lags and so on.

The PACF is a little more complicated. The autocorrelation at lag one can have lingering effects on the autocorrelation at lag two and onward. The partial autocorrelation is the amount of correlation between a time series and lags of itself that is not explained by a previous lag. So, the partial autocorrelation at lag two is the correlation between the time series and its second lag that is not explained by the first lag.

The ACF and PACF for the US per capita GDP data are shown in Figure 24.3. Vertical lines that extend beyond the horizontal line indicate autocorrelations and partial autocorrelations that are significant at those lags.

```
> acf(us)
> pacf(us)
```

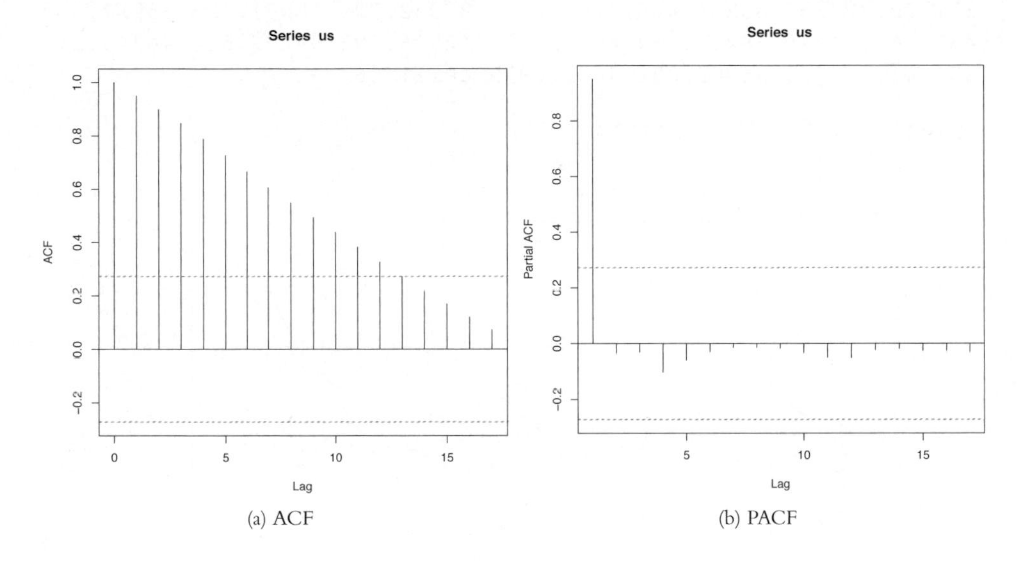

(a) ACF

(b) PACF

Figure 24.3 ACF and PACF of US per capita GDP. These plots are indicative of a time series that is not stationary.

This time series needs a number of transformations before it can be properly modeled. Its upward trend shows that it is not stationary[1] (the data are in current US dollars, so

1. Being stationary requires that the mean and variance of a time series are constant for the whole series.

inflation is not the cause). That can be fixed by diffing the series or applying some other transformation. Diffing is the process of subtracting one observation from another and can be done on any number of observations. For instance, we start with a series $x = [1\ 4\ 8\ 2\ 6\ 6\ 5\ 3]$. Diffing it yields $x^{(1)} = [3\ 4\ -6\ 4\ 0\ -1\ -2]$, which is the difference between successive elements. Diffing twice iteratively diffs the diffs, so $x^{(2)} = [1\ -10\ 10\ -4\ -1\ -1]$. Observe that for each level of diffing the there is one less element in the series. Doing this in R involves the **diff** function. The `differences` argument controls how many diffs are iteratively calculated. The `lag` determines which elements get subtracted from each other. A lag of 1 subtracts successive elements, while a lag of 2 subtracts elements that are two indices away from each other.

```
> x <- c(1 , 4 , 8 , 2 , 6 , 6 , 5, 3)
> # one diff
> diff(x, differences=1)

[1]  3  4 -6  4  0 -1 -2

> # two iterative diffs
> diff(x, differences=2)

[1]   1 -10  10  -4  -1  -1

> # equivalent to one diff
> diff(x, lag=1)

[1]  3  4 -6  4  0 -1 -2

> # diff elements that are two indices apart
> diff(x, lag=2)

[1]  7 -2 -2  4 -1 -3
```

Figuring out the correct number of diffs can be a tiresome process. Fortunately, the **forecast** package has a number of functions to make working with time series data easier, including determining the optimal number of diffs. The result is shown in Figure 24.4.

```
> library(forecast)
> ndiffs(x=us)

[1] 2

> plot(diff(us, 2))
```

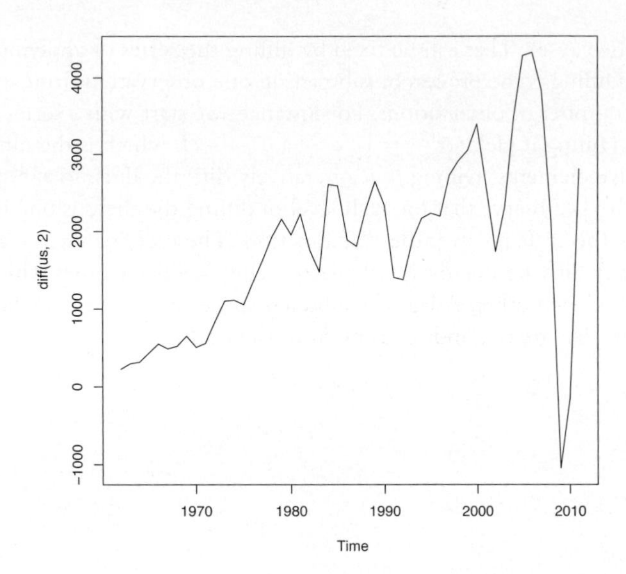

Figure 24.4 Plot of the US per capita GDP diffed twice.

While R offers individual **ar** and **ma** functions, a better option is the **arima** function, which can fit both AR and MA models and the combined ARMA model. It is even more robust in that it can diff the series and fit seasonal effects. Traditionally, the right order of each component of the model is determined by analyzing the ACF and PACF. This can be highly subjective, so fortunately **forecast** contains **auto.arima**, which will figure out the best specification.

```
> usBest <- auto.arima(x=us)
> usBest

Series:
ARIMA(2,2,1)

Coefficients:
         ar1      ar2      ma1
      0.4181  -0.2567  -0.8102
s.e.  0.1632   0.1486   0.1111

sigma^2 estimated as 286942:  log likelihood=-384.05
AIC=776.1   AICc=776.99   BIC=783.75
```

The function determined that an ARMA(2,1) (an AR(2) component and an MA(1) component) with two diffs is the optimal model based on minimum AICC (that is, AIC that is "corrected" to give a greater penalty to model complexity). The two diffs actually make this an ARIMA model rather than an ARMA model where the I stands for integrated. If this model is a good fit, then the residuals should resemble white noise.

Figure 24.5 shows the ACF and PACF of the residuals for the ideal model. They resemble the pattern for white noise, confirming our model selection.

```
> acf(usBest$residuals)
> pacf(usBest$residuals)
```

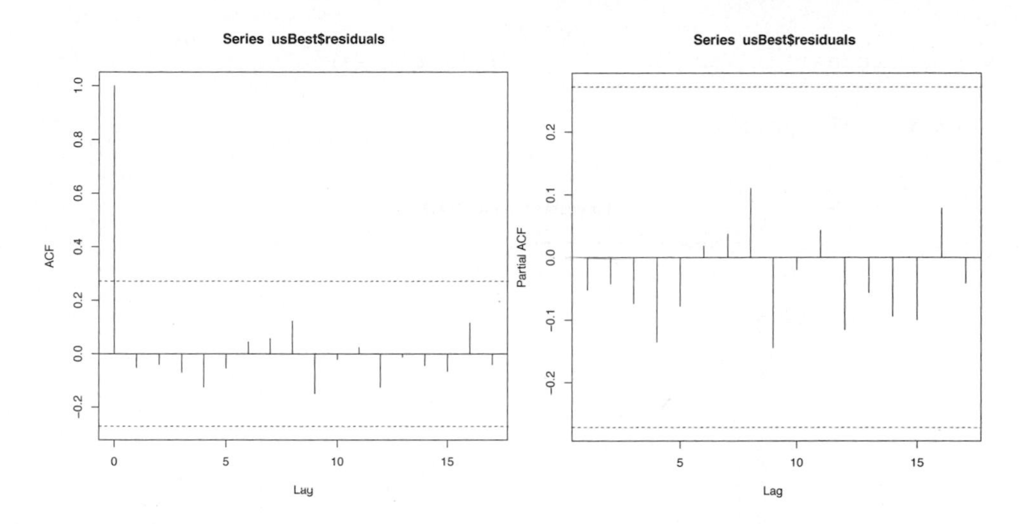

Figure 24.5 ACF and PACF plots for the residuals of ideal model chosen by `auto.arima`.

The coefficients for an ARIMA model are the AR and MA components.

```
> coef(usBest)

       ar1        ar2        ma1
 0.4181109 -0.2567494 -0.8102419
```

Making predictions based on an ARIMA model is much the same as with any other model type, using the **predict** function.

```
> # predict 5 years into the future and include the standard error
> predict(usBest, n.ahead=5, se.fit=TRUE)

$pred
Time Series:
Start = 2012
End = 2016
Frequency = 1
[1] 49292.41 50289.69 51292.41 52344.45 53415.70

$se
Time Series:
Start = 2012
```

```
End = 2016
Frequency = 1
[1]  535.6701 1014.2773 1397.6158 1731.1312 2063.2010
```

Visualizing this is easy enough but using the **forecast** function makes it even easier, as seen in Figure 24.6.

```
> # make a prediction for 5 years out
> theForecast <- forecast(object=usBest, h=5)
> # plot it
> plot(theForecast)
```

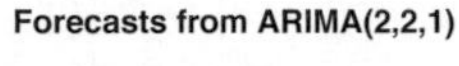

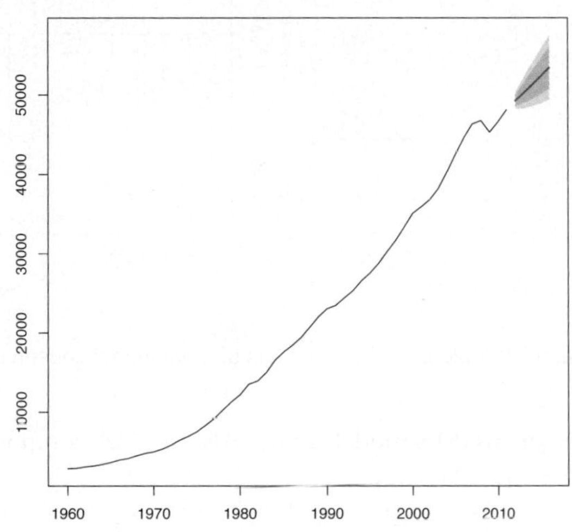

Figure 24.6 Five year prediction of US GDP. The think line is the point estimate and the shaded regions represent the confidence intervals.

24.2 VAR

When dealing with multiple time series where each depends on its own past, others' pasts and others' presents, things get more complicated. The first thing we will do is convert all of the GDP data into a multivariate time series. To do this we first cast the `data.frame` to wide format and then call **ts** to convert it. The result is shown in Figure 24.7.

```
> # load reshape2
> library(reshape2)
> # cast the data.frame to wide format
> gdpCast <- dcast(Year ~ Country,
+                  data=gdp[, c("Country", "Year", "PerCapGDP")],
+                  value.var="PerCapGDP")
> head(gdpCast)
```

```
  Year   Canada    China Germany   Israel     Japan Singapore
1 1960 2294.569 92.01123      NA 1365.683 478.9953  394.6489
2 1961 2231.294 75.87257      NA 1595.860 563.5868  437.9432
3 1962 2255.230 69.78987      NA 1132.383 633.6403  429.5377
4 1963 2354.839 73.68877      NA 1257.743 717.8669  472.1830
5 1964 2529.518 83.93044      NA 1375.943 835.6573  464.3773
6 1965 2739.586 97.47010      NA 1429.319 919.7767  516.2622
  United Kingdom United States
1       1380.306      2881.100
2       1452.545      2934.553
3       1513.651      3107.937
4       1592.614      3232.208
5       1729.400      3423.396
6       1850.955      3664.802

> # remove first 10 rows since Germany did not have
>
> # convert to time series
> gdpTS <- ts(data=gdpCast[, -1], start=min(gdpCast$Year),
+             end=max(gdpCast$Year))
>
> # build a plot and legend using base graphics
> plot(gdpTS, plot.type="single", col=1:8)
> legend("topleft", legend=colnames(gdpTS), ncol=2, lty=1,
+        col=1:8, cex=.9)
```

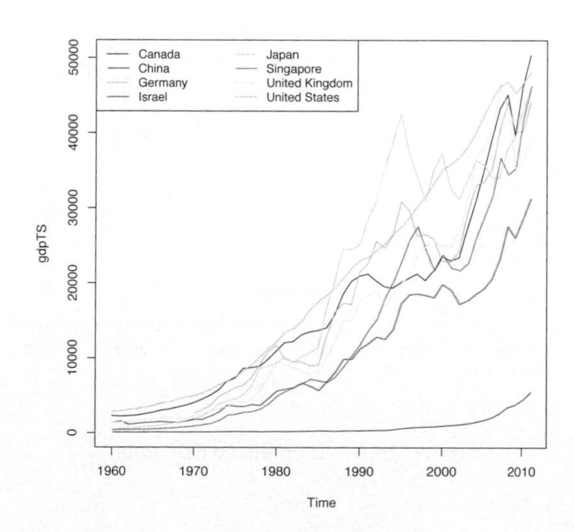

Figure 24.7 Time series plot of GDP data for all countries in the data. This is the same information as in Figure 24.1a, but this was built using base graphics.

Before proceeding we have to deal with the NAs for Germany. For some reason the World Bank does not have data on Germany's GDP before 1970. There are other resources, such as the St. Louis Federal Reserve Economic Data (FRED), but their data do not agree well with the World Bank data, so we remove Germany from our data.

```
> gdpTS <- gdpTS[, which(colnames(gdpTS) != "Germany")]
```

The most common way of fitting a model to multiple time series is to use a vector autoregressive (VAR) model. The equation for a VAR is

$$\mathbf{X}_t = \Phi_1 \mathbf{X}_{t-1} + \cdots + \Phi_p \mathbf{X}_{t-p} + \mathbf{Z}_t \tag{24.3}$$

where

$$\{\mathbf{Z}_t\} \sim \mathrm{WN}(\mathbf{0}, \mathbf{\Sigma}) \tag{24.4}$$

is white noise.

While **ar** can compute a VAR, it it often has problems with singular matrices when the AR order is high, so it is better to use **VAR** from the **vars** package. To check whether the data should be diffed, we use the **ndiffs** function on gdpTS and then apply that number of diffs. The diffed data is shown in Figure 24.8, which exhibits greater stationarity than Figure 24.7.

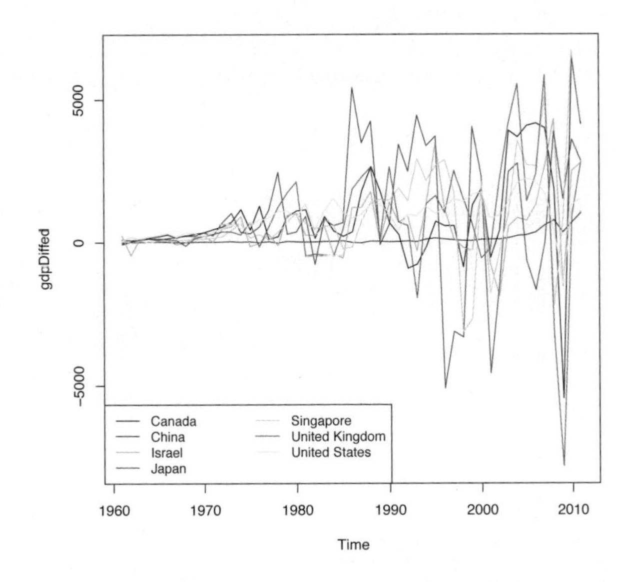

Figure 24.8 Differenced GDP data.

```
> numDiffs <- ndiffs(gdpTS)
> numDiffs

[1] 1
```

```
> gdpDiffed <- diff(gdpTS, differences=numDiffs)
> plot(gdpDiffed, plot.type="single", col=1:7)
> legend("bottomleft", legend=colnames(gdpDiffed), ncol=2, lty=1,
+        col=1:7, cex=.9)
```

Now that the data is prepared, we can fit a VAR using **VAR**. This essentially fits a separate regression using **lm** of each time series on the lags of itself and the other series. This is evidenced in the coefficient plot for the Canada and Japan models, shown in Figure 24.9.

```
> library(vars)
> # fit the model
> gdpVar <- VAR(gdpDiffed, lag.max=12)
> # chosen order
> gdpVar$p

AIC(n)
     6

> # names of each of the models
> names(gdpVar$varresult)

[1] "Canada"          "China"           "Israel"
[4] "Japan"           "Singapore"       "United.Kingdom"
[7] "United.States"

> # each model is actually an lm object
> class(gdpVar$varresult$Canada)

[1] "lm"

> class(gdpVar$varresult$Japan)

[1] "lm"

> # each model has its own coefficients
> head(coef(gdpVar$varresult$Canada))

       Canada.l1         China.l1        Israel.l1
     -1.07854513      -7.28241774       1.06538174
        Japan.l1     Singapore.l1 United.Kingdom.l1
     -0.45533608      -0.03827402       0.60149182

> head(coef(gdpVar$varresult$Japan))

       Canada.l1         China.l1        Israel.l1
       1.8045012      -19.7904918       -0.1507690
        Japan.l1     Singapore.l1 United.Kingdom.l1
       1.3344763        1.5738029        0.5707742
```

```
> library(coefplot)
> coefplot(gdpVar$varresult$Canada)
> coefplot(gdpVar$varresult$Japan)
```

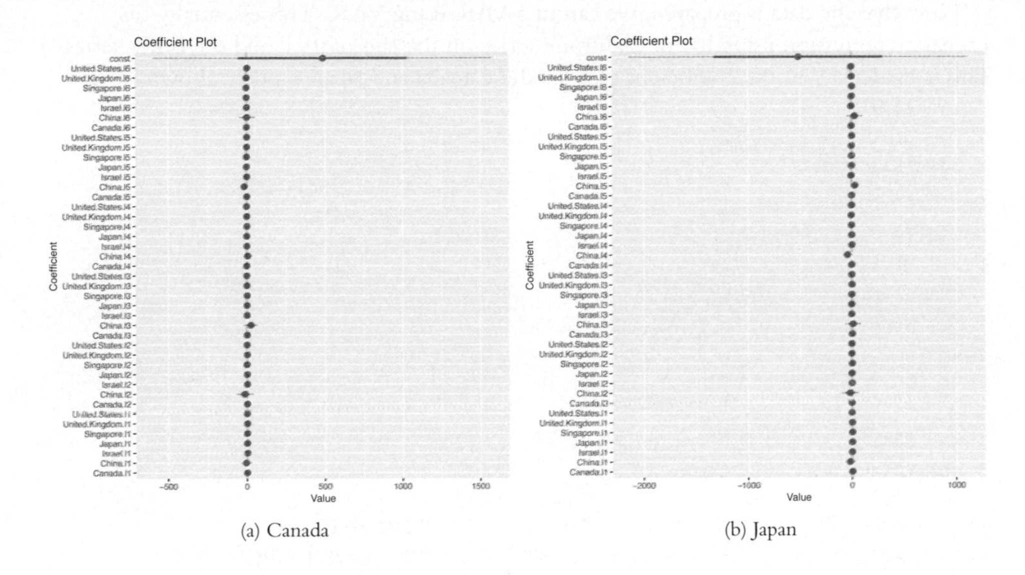

(a) Canada (b) Japan

Figure 24.9 Coefficient plots for VAR model of GDP data for Canada and Japan.

Predictions for this model are done just like with any other model, using the **predict** function.

```
> predict(gdpVar, n.ahead=5)

$Canada
          fcst      lower      upper        CI
[1,] -12459.46  -13284.63  -11634.30  825.1656
[2,]  15067.05   14106.02   16028.08  961.0344
[3,]  20632.99   19176.30   22089.69 1456.6943
[4,] -103830.42 -105902.11 -101758.73 2071.6904
[5,]  124483.19  119267.39  129699.00 5215.8046

$China
          fcst      lower      upper        CI
[1,]  -470.5917  -523.6101  -417.5733  53.01843
[2,]   899.5380   826.2362   972.8399  73.30188
[3,]  1730.8087  1596.4256  1865.1918 134.38308
[4,] -3361.7713 -3530.6042 -3192.9384 168.83288
[5,]  2742.1265  2518.9867  2965.2662 223.13974
```

```
$Israel
          fcst       lower      upper       CI
[1,]  -6686.711   -7817.289  -5556.133  1130.578
[2,] -39569.216  -40879.912 -38258.520  1310.696
[3,]  62192.139   60146.978  64237.300  2045.161
[4,] -96325.105 -101259.427 -91390.783  4934.322
[5,] -12922.005  -24003.839  -1840.171 11081.834

$Japan
            fcst       lower       upper       CI
[1,]  -14590.8574  -15826.761  -13354.954 1235.903
[2,]  -52051.5807  -53900.387  -50202.775 1848.806
[3,]    -248.4379   -3247.875    2750.999 2999.437
[4,]  -51465.6686  -55434.880  -47496.457 3969.212
[5,] -111005.8032 -118885.682 -103125.924 7879.879

$Singapore
          fcst      lower     upper        CI
[1,] -35923.80  -36071.93 -35775.67  148.1312
[2,]  54502.69   53055.85  55949.53 1446.8376
[3,] -43551.08  -47987.48 -39114.68 4436.3991
[4,] -99075.95 -107789.86 -90362.04 8713.9078
[5,] 145133.22  135155.64 155110.81 9977.5872

$United.Kingdom
          fcst     lower     upper       CI
[1,] -19224.96 -20259.35 -18190.56 1034.396
[2,]  31194.77  30136.87  32252.67 1057.903
[3,]  27813.08  24593.47  31032.68 3219.604
[4,] -66506.90 -70690.12 -62323.67 4183.226
[5,]  93857.98  88550.03  99165.94 5307.958

$United.States
           fcst     lower      upper        CI
[1,]  -657.2679 -1033.322  -281.2137  376.0542
[2,] 11088.0517 10614.924 11561.1792  473.1275
[3,]  2340.6277  1426.120  3255.1350  914.5074
[4,] -5790.0143 -7013.843 -4566.1855 1223.8288
[5,] 24306.5309 23013.525 25599.5373 1293.0064
```

24.3 GARCH

A problem with ARMA models is that they do not handle extreme events or high volatility well. To overcome this, a good tool to use is generalized autoregressive conditional heteroskedasticity or the GARCH family of models, which in addition to modelling the mean of the process also model the variance.

The model for the variance in a GARCH(m, s) is

$$\epsilon_t = \sigma_t e_t \tag{24.5}$$

where

$$\sigma_t^2 = \alpha_0 + \alpha_1\epsilon_{t-1}^2 + \cdots + \alpha_m\epsilon_{t-m}^2 + \beta_1\sigma_{t-1}^2 + \cdots + \beta_s\sigma_{t-s}^2 \tag{24.6}$$

and

$$e \sim \mathrm{GWN}(0, 1) \tag{24.7}$$

is generalized white noise.

For this example we download AT&T ticker data using the **quantmod** package.

```
> library(quantmod)
> load("data/att.rdata")
```

```
> library(quantmod)
> att <- getSymbols("T", auto.assign=FALSE)
```

This loads the data into an `xts` object from the **xts** package, which is a more robust time series object that, among many other improvements, can handle irregularly spaced events. These objects even have improved plotting over `ts`, as seen in Figure 24.10.

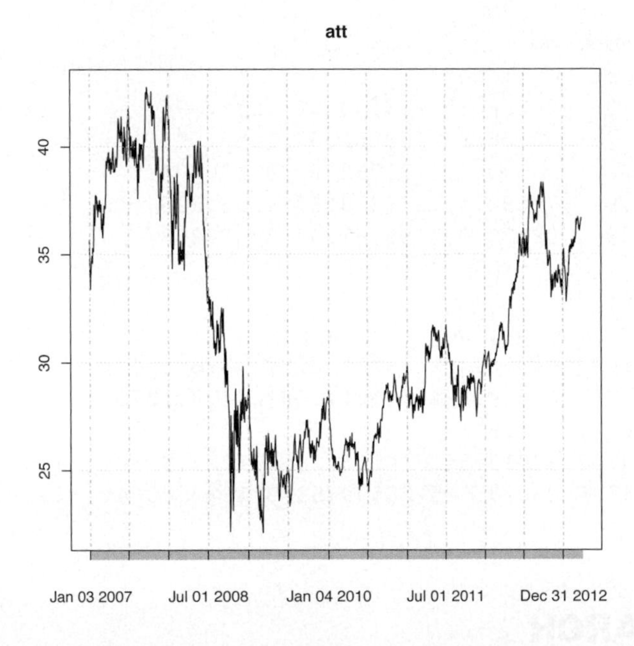

Figure 24.10 Time series plot of AT&T ticker data.

```
> library(xts)
> # show data
> head(att)

           T.Open T.High T.Low T.Close T.Volume T.Adjusted
2007-01-03  35.67  35.78 34.78   34.95 33694300      25.06
2007-01-04  34.95  35.24 34.07   34.50 44285400      24.74
2007-01-05  34.40  34.54 33.95   33.96 36561800      24.35
2007-01-08  33.40  34.01 33.21   33.81 40237400      24.50
2007-01-09  33.85  34.41 33.66   33.94 40082600      24.59
2007-01-10  34.20  35.00 31.94   34.03 29964300      24.66

> plot(att)
```

For those used to financial terminal charts, the **chartSeries** function should be comforting. It created the chart shown in Figure 24.11.

```
> chartSeries(att)
> addBBands()
> addMACD(32, 50, 12)
```

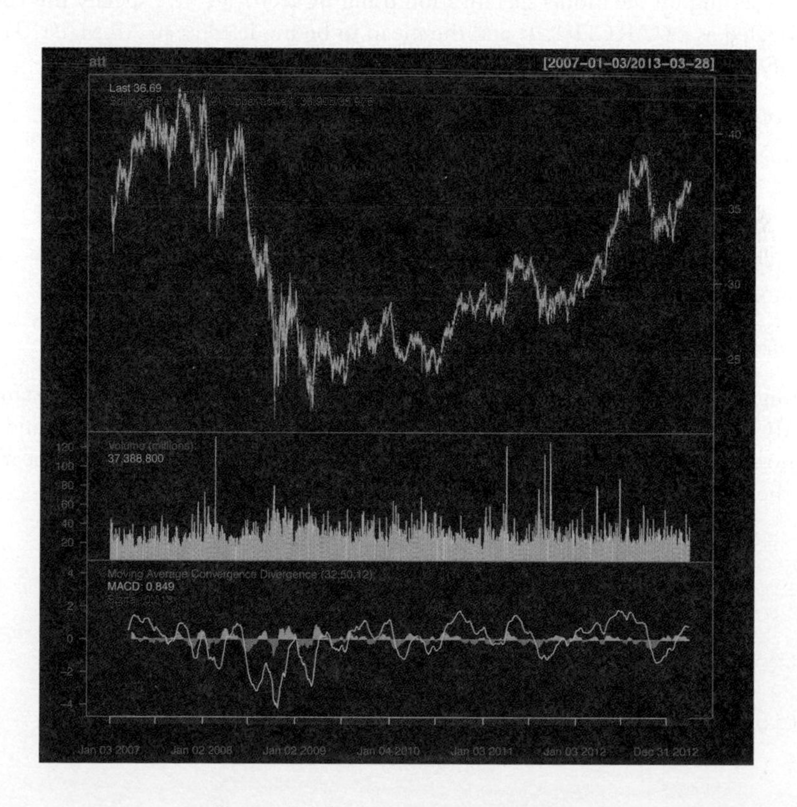

Figure 24.11 Series chart for AT&T.

We are only interested in the closing price, so we create a variable holding just that.

```
> attClose <- att$T.Close
> class(attClose)

[1] "xts" "zoo"

> head(attClose)

           T.Close
2007-01-03   34.95
2007-01-04   34.50
2007-01-05   33.96
2007-01-08   33.81
2007-01-09   33.94
2007-01-10   34.03
```

The package most widely considered to be the best for fitting GARCH models is **rugarch**. There are other packages for fitting GARCH models, such as **tseries**, **fGarch** and **bayesGARCH**, but we will focus on **rugarch**.

Generally, a GARCH(1,1) will be sufficient so we will fit that model to the data. The first step is setting up the model specification using **ugarchspec**. We specify the volatility to be modeled as a GARCH(1, 1) and the mean to be modeled as an ARMA(1, 1). We also specify that the innovation distribution should be the t distribution.

```
> library(rugarch)
> attSpec <- ugarchspec(variance.model=list(model="sGARCH",
+                                           garchOrder=c(1, 1)),
+                       mean.model=list(armaOrder=c(1, 1)),
+                       distribution.model="std")
```

The next step is to fit the model using **ugarchfit**.

```
> attGarch <- ugarchfit(spec=attSpec, data=attClose)
```

Printing the model spits out a lot of information, including the coefficients, standard errors, AIC and BIC. Most of this, such as the statistics on residuals, tests, AIC and BIC are diagnostic measures on the quality of the fit. The optimal parameters, seen near the top, are the crux of the model.

```
> attGarch

*---------------------------------*
*          GARCH Model Fit        *
*---------------------------------*

Conditional Variance Dynamics
```

```
---------------------------------------
GARCH Model : sGARCH(1,1)
Mean Model : ARFIMA(1,0,1)
Distribution : std

Optimal Parameters
------------------------------------
        Estimate  Std. Error   t value Pr(>|t|)
mu     34.966061    0.381089  91.75300 0.000000
ar1     0.996957    0.001288 774.08104 0.000000
ma1    -0.010240    0.026747  -0.38283 0.701846
omega   0.001334    0.000703   1.89752 0.057760
alpha1  0.069911    0.015443   4.52716 0.000006
beta1   0.925054    0.015970  57.92518 0.000000
shape   7.586620    1.405315   5.39852 0.000000

Robust Standard Errors:
        Estimate  Std. Error   t value Pr(>|t|)
mu     34.966061    0.043419 805.30860 0.000000
ar1     0.996957    0.001203 828.40704 0.000000
ma1    -0.010240    0.028700  -0.35678 0.721255
omega   0.001334    0.000829   1.60983 0.107435
alpha1  0.069911    0.019342   3.61450 0.000301
beta1   0.925054    0.020446  45.24344 0.000000
shape   7.586620    1.329563   5.70610 0.000000

LogLikelihood : -776.0465

Information Criteria
------------------------------------

Akaike       0.99751
Bayes        1.02140
Shibata      0.99747
Hannan-Quinn 1.00639

Weighted Ljung-Box Test on Standardized Residuals
------------------------------------
                        statistic p-value
Lag[1]                     0.5461  0.4599
Lag[2*(p+q)+(p+q)-1][5]    2.6519  0.6922
Lag[4*(p+q)+(p+q)-1][9]    4.5680  0.5549
d.o.f=2
H0 : No serial correlation
```

```
Weighted Ljung-Box Test on Standardized Squared Residuals
------------------------------------
                        statistic p-value
Lag[1]                   0.004473  0.9467
Lag[2*(p+q)+(p+q)-1][5]  3.119353  0.3857
Lag[4*(p+q)+(p+q)-1][9]  4.604070  0.4898
d.o.f=2

Weighted ARCH LM Tests
------------------------------------
            Statistic Shape Scale P-Value
ARCH Lag[3]     1.751 0.500 2.000  0.1857
ARCH Lag[5]     2.344 1.440 1.667  0.4001
ARCH Lag[7]     2.967 2.315 1.543  0.5198

Nyblom stability test
------------------------------------
Joint Statistic:  1.5862
Individual Statistics:
mu     0.27197
ar1    0.09594
ma1    0.25152
omega  0.13852
alpha1 0.62839
beta1  0.53037
shape  0.46974

Asymptotic Critical Values (10% 5% 1%)
Joint Statistic:         1.69 1.9 2.35
Individual Statistic:    0.35 0.47 0.75

Sign Bias Test
------------------------------------
                   t-value   prob sig
Sign Bias           0.8341 0.4043
Negative Sign Bias  0.8170 0.4141
Positive Sign Bias  0.4020 0.6877
Joint Effect        3.0122 0.3897


Adjusted Pearson Goodness-of-Fit Test:
------------------------------------
  group statistic p-value(g-1)
1    20     15.99       0.6581
2    30     23.71       0.7432
```

```
3    40     31.78       0.7873
4    50     46.62       0.5700


Elapsed time : 0.3694999
```

Figure 24.12 shows a time series plot and the ACF of the residuals from the model.

```
> # attGarch is an S4 object so its slots are accessed by @
> # the slot fit is a list, its elements are ac-
cessed by the dollar sign
> plot(attGarch@fit$residuals, type="l")
> plot(attGarch, which=10)
```

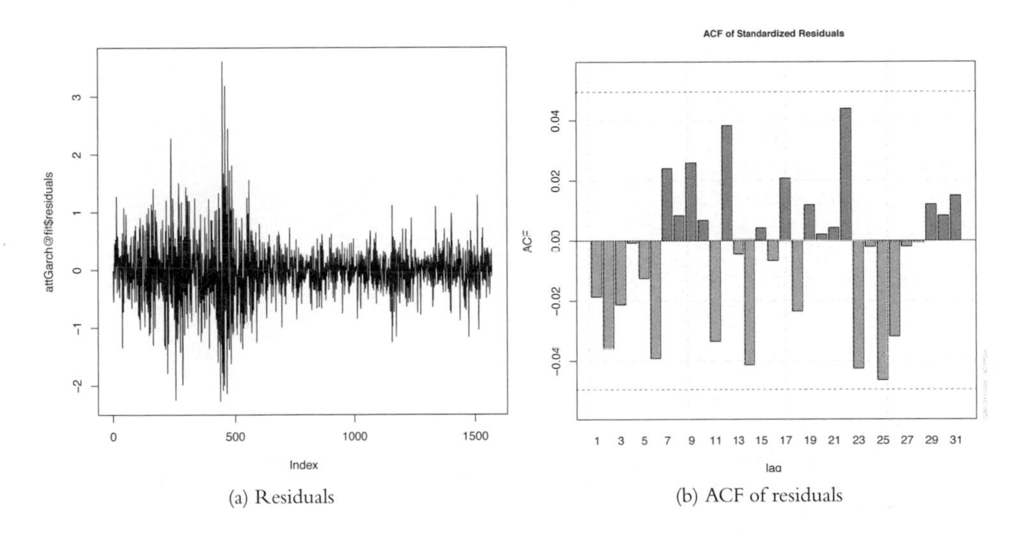

(a) Residuals

(b) ACF of residuals

Figure 24.12 Residual plots from GARCH model on AT&T data.

To judge the quality of this model, we build a few models with different mean specifications—all GARCH(1, 1)—and compare their AICs.

```
> # ARMA(1,1)
> attSpec1 <- ugarchspec(variance.model=list(model="sGARCH",
+                                            garchOrder=c(1, 1)),
+                        mean.model=list(armaOrder=c(1, 1)),
+                        distribution.model="std")
> # ARMA(0,0)
> attSpec2 <- ugarchspec(variance.model=list(model="sGARCH",
+                                            garchOrder=c(1, 1)),
+                        mean.model=list(armaOrder=c(0, 0)),
+                        distribution.model="std")
```

```
> # ARMA(0,2)
> attSpec3 <- ugarchspec(variance.model=list(model="sGARCH",
+                                            garchOrder=c(1, 1)),
+                        mean.model=list(armaOrder=c(0, 2)),
+                        distribution.model="std")
> # ARMA(1,2)
> attSpec4 <- ugarchspec(variance.model=list(model="sGARCH",
+                                            garchOrder=c(1, 1)),
+                        mean.model=list(armaOrder=c(1, 2)),
+                        distribution.model="std")
>
> attGarch1 <- ugarchfit(spec=attSpec1, data=attClose)
> attGarch2 <- ugarchfit(spec=attSpec2, data=attClose)
> attGarch3 <- ugarchfit(spec=attSpec3, data=attClose)
> attGarch4 <- ugarchfit(spec=attSpec4, data=attClose)
>
> infocriteria(attGarch1)


Akaike       0.9975114
Bayes        1.0214043
Shibata      0.9974719
Hannan-Quinn 1.0063921

> infocriteria(attGarch2)


Akaike       5.111944
Bayes        5.129011
Shibata      5.111924
Hannan-Quinn 5.118288

> infocriteria(attGarch3)


Akaike       3.413075
Bayes        3.436968
Shibata      3.413035
Hannan-Quinn 3.421956

> infocriteria(attGarch4)


Akaike       0.9971012
Bayes        1.0244073
Shibata      0.9970496
Hannan-Quinn 1.0072505
```

This shows that the first and fourth models were the best, according to AIC and BIC and the other criteria.

Predicting with objects from **rugarch** is done through the **ugarchboot** function, which can then be plotted as seen in Figure 24.13.

```
> attPred <- ugarchboot(attGarch, n.ahead=50,
+                       method=c("Partial", "Full")[1])
> plot(attPred, which=2)
```

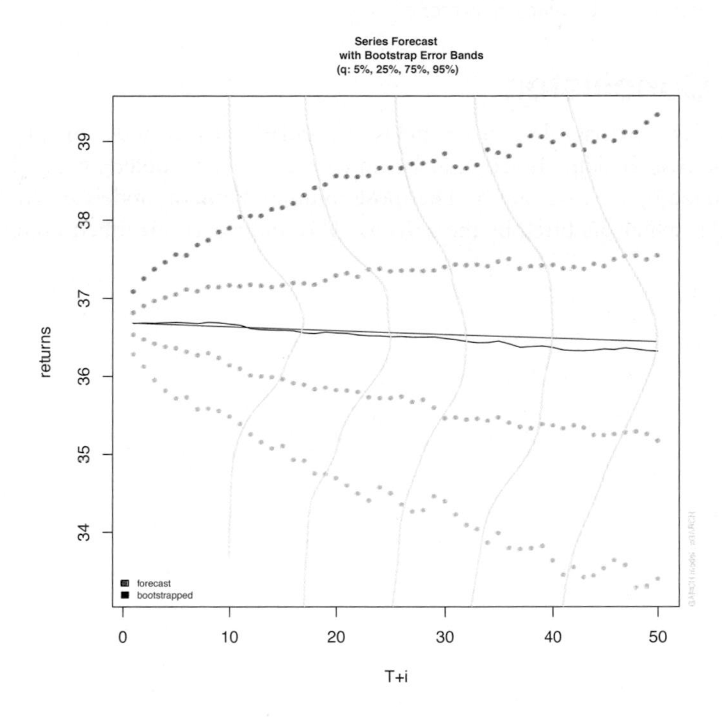

Figure 24.13 Predictions for GARCH model on AT&T data.

Because this is stock data, it is worth computing the model on the log returns instead of the actual closing prices.

```
> # diff the logs, drop the first one which is now NA
> attLog <- diff(log(attClose))[-1]
> # build the specification
> attLogSpec <- ugarchspec(variance.model=list(model="sGARCH",
+                                              garchOrder=c(1, 1)),
+                          mean.model=list(armaOrder=c(1, 1)),
+                          distribution.model="std")
> # fit the model
> attLogGarch <- ugarchfit(spec=attLogSpec, data=attLog)
> infocriteria(attLogGarch)
```

```
Akaike       -5.869386
Bayes        -5.845481
Shibata      -5.869426
Hannan-Quinn -5.860500
```

This led to a significant drop in AIC.

It is important to remember that the purpose of GARCH models is not to fit the signal better but to capture the volatility better.

24.4 Conclusion

Time series play a crucial role in many fields, particularly finance and some physical sciences. The basic building block in R for time series is the `ts` object, which has been greatly extended by the `xts` object. The most common types of models are ARMA, VAR and GARCH, which are fitted by the **arima**, **VAR** and **ugarchfit** functions, respectively.

25
Clustering

Clustering, which plays a big role in modern machine learning, is the partitioning of data into groups. This can be done in a number of ways, the two most popular being K-means and hierarchical clustering. In terms of a `data.frame`, a clustering algorithm finds out which rows are similar to each other. Rows that are grouped together are supposed to have high similarity to each other and low similarity with rows outside the grouping.

25.1 K-means

One of the more popular algorithms for clustering is K-means. It divides the observations into discrete groups based on some distance metric. For this example, we use the wine dataset from the University of California–Irvine Machine Learning Repository, available at `http://archive.ics.uci.edu/ml/datasets/Wine`.

```
> wineUrl <- 'http://archive.ics.uci.edu/ml/
                  machine-learning-databases/wine/wine.data'
> wine <- read.table(wineUrl, header=FALSE, sep=',',
+                    stringsAsFactors=FALSE,
+                    col.names=c('Cultivar', 'Alcohol', 'Malic.acid',
+                                'Ash', 'Alcalinity.of.ash',
+                                'Magnesium', 'Total.phenols',
+                                'Flavanoids', 'Nonflavanoid.phenols',
+                                'Proanthocyanin', 'Color.intensity',
+                                'Hue', 'OD280.OD315.of.diluted.wines',
+                                'Proline'
+ ))
```

```
> head(wine)

  Cultivar Alcohol Malic.acid  Ash Alcalinity.of.ash Magnesium
1        1   14.23       1.71 2.43              15.6       127
2        1   13.20       1.78 2.14              11.2       100
3        1   13.16       2.36 2.67              18.6       101
4        1   14.37       1.95 2.50              16.8       113
5        1   13.24       2.59 2.87              21.0       118
6        1   14.20       1.76 2.45              15.2       112
```

```
  Total.phenols Flavanoids Nonflavanoid.phenols Proanthocyanins
1          2.80       3.06                 0.28            2.29
2          2.65       2.76                 0.26            1.28
3          2.80       3.24                 0.30            2.81
4          3.85       3.49                 0.24            2.18
5          2.80       2.69                 0.39            1.82
6          3.27       3.39                 0.34            1.97
  Color.intensity  Hue OD280.OD315.of.diluted.wines Proline
1            5.64 1.04                         3.92    1065
2            4.38 1.05                         3.40    1050
3            5.68 1.03                         3.17    1185
4            7.80 0.86                         3.45    1480
5            4.32 1.04                         2.93     735
6            6.75 1.05                         2.85    1450
```

Because the first column is the cultivar, and that might be too correlated with group membership, we exclude that from the analysis.

```
> wineTrain <- wine[, which(names(wine) != "Cultivar")]
```

For K-means we need to specify the number of clusters, and then the algorithm assigns observations into that many clusters. There are heuristic rules for determining the number of clusters, which we will get to later. For now we will choose three. In R, K-means is done with the aptly named **kmeans** function. Its first two arguments are the data to be clustered, which must be all `numeric` (K-means does not work with categorical data), and the number of centers (clusters). Because there is a random component to the clustering, we set the seed to generate reproducible results.

```
> set.seed(278613)
> wineK3 <- kmeans(x=wineTrain, centers=3)
```

Printing the K-means objects displays the size of the clusters, the cluster mean for each column, the cluster membership for each row and similarity measures.

```
> wineK3

K-means clustering with 3 clusters of sizes 62, 47, 69

Cluster means:
   Alcohol Malic.acid      Ash Alcalinity.of.ash Magnesium
1 12.92984   2.504032 2.408065          19.89032 103.59677
2 13.80447   1.883404 2.426170          17.02340 105.51064
3 12.51667   2.494203 2.288551          20.82319  92.34783
  Total.phenols Flavanoids Nonflavanoid.phenols Proanthocyanins
1      2.111129   1.584032            0.3883871        1.503387
2      2.867234   3.014255            0.2853191        1.910426
3      2.070725   1.758406            0.3901449        1.451884
  Color.intensity       Hue OD280.OD315.of.diluted.wines   Proline
1        5.650323 0.8839677                     2.365484  728.3387
2        5.702553 1.0782979                     3.114043 1195.1489
3        4.086957 0.9411594                     2.490725  458.2319
```

```
Clustering vector:
  [1] 2 2 2 2 1 2 2 2 2 2 2 2 2 2 2 2 2 2 2 1 1 1 2 2 1 1 2 2 1 2 2 2
 [33] 2 2 2 1 1 2 2 1 1 2 2 1 1 2 2 2 2 2 2 2 2 2 2 2 2 2 2 3 1 3 1 3
 [65] 3 1 3 3 1 1 1 3 3 2 1 3 3 3 1 3 3 1 1 3 3 3 3 3 1 1 3 3 3 3 3 1
 [97] 1 3 1 3 1 3 3 3 1 3 3 3 3 1 3 3 1 3 3 3 3 3 3 3 1 3 3 3 3 3 3 3
[129] 3 3 1 3 3 1 1 1 1 3 3 3 1 1 3 3 1 1 3 1 1 3 3 3 3 1 1 1 3 1 1 1
[161] 3 1 3 1 1 3 1 1 1 1 3 3 1 1 1 1 1 3

Within cluster sum of squares by cluster:
[1]  566572.5 1360950.5  443166.7
 (between_SS / total_SS =  86.5 %)

Available components:

[1] "cluster"      "centers"      "totss"        "withinss"
[5] "tot.withinss" "betweenss"    "size"         "iter"
[9] "ifault"
```

Plotting the result of K-means clustering can be difficult because of the high dimensional nature of the data. To overcome this, the **plot.kmeans** function in **useful** performs multidimensional scaling to project the data into two dimensions and then color codes the points according to cluster membership. This is shown in Figure 25.1.

```
> library(useful)
> plot(wineK3, data=wineTrain)
```

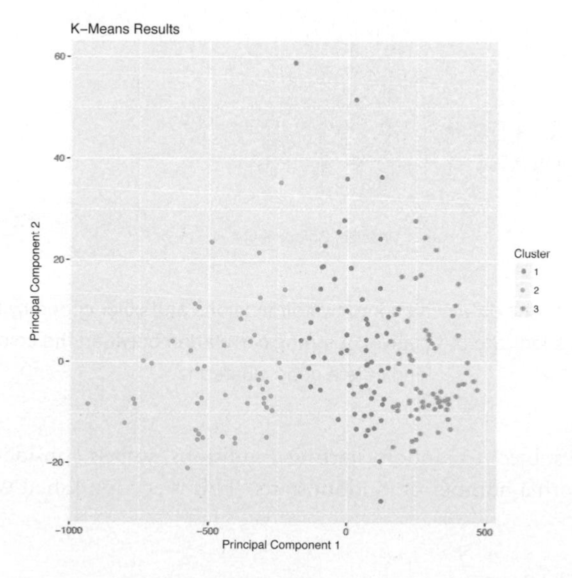

Figure 25.1 Plot of wine data scaled into two dimensions and color coded by results of K-means clustering.

If we pass the original wine data and specify that `Cultivar` is the true membership column, the shape of the points will be coded by `Cultivar`, so we can see how that compares to the colors in Figure 25.2. A strong correlation between the color and shape would indicate a good clustering.

```
> plot(wineK3, data=wine, class="Cultivar")
```

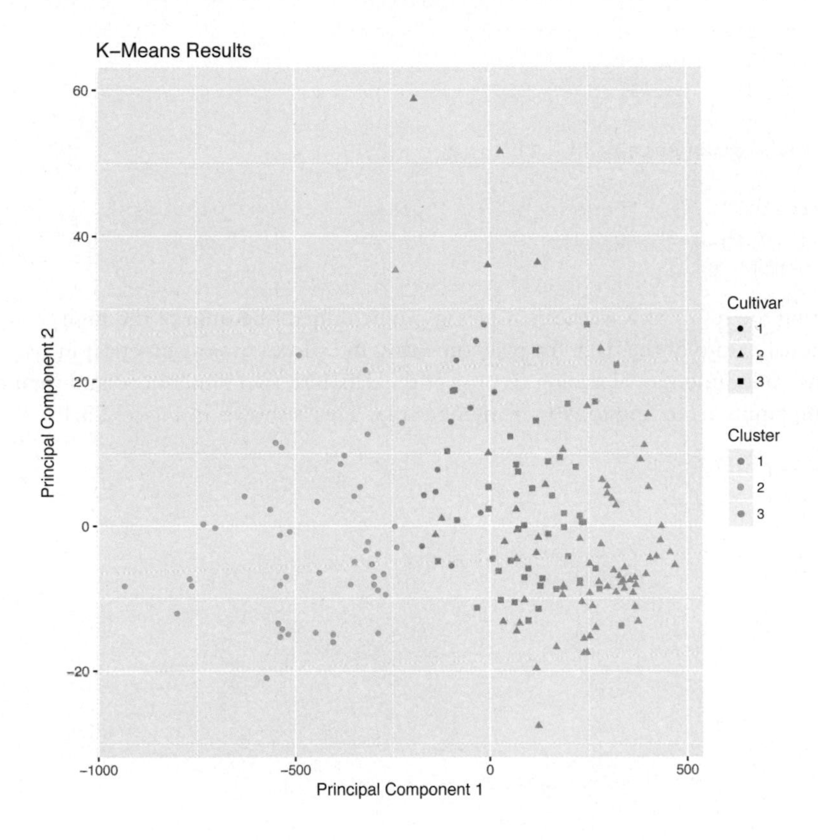

Figure 25.2 Plot of wine data scaled into two dimensions and color coded by results of K-means clustering. The shapes indicate the cultivar. A strong correlation between the color and shape would indicate a good clustering.

K-means can be subject to random starting conditions, so it is considered good practice to run it with a number of random starts. This is accomplished with the `nstart` argument.

```
> set.seed(278613)
> wineK3N25 <- kmeans(wineTrain, centers=3, nstart=25)
> # see the cluster sizes with 1 start
> wineK3$size
```

```
[1] 62 47 69

> # see the cluster sizes with 25 starts
> wineK3N25$size

[1] 62 47 69
```

For our data the results did not change. For other datasets the number of starts can have a significant impact.

Choosing the right number of clusters is important in getting a good partitioning of the data. According to David Madigan, the former chair of Department of Statistics and current Dean of Faculty of Arts and Sciences and Professor of Statistics at Columbia University, a good metric for determining the optimal number of clusters is Hartigan's rule (J. A. Hartigan is one of the authors of the most popular K-means algorithm). It essentially compares the ratio of the within-cluster sum of squares for a clustering with k clusters and one with $k + 1$ clusters, accounting for the number of rows and clusters. If that number is greater than 10, then it is worth using $k + 1$ clusters. Fitting this repeatedly can be a chore and computationally inefficient if not done right. The **useful** package has the **FitKMeans** function for doing just that. The results are plotted in Figure 25.3.

```
> wineBest <- FitKMeans(wineTrain, max.clusters=20, nstart=25,
+                       seed=278613)
> wineBest

   Clusters   Hartigan AddCluster
1         2 505.429310       TRUE
2         3 160.411331       TRUE
3         4 135.707228       TRUE
4         5  78.445289       TRUE
5         6  71.489710       TRUE
6         7  97.582072       TRUE
7         8  46.772501       TRUE
8         9  33.198650       TRUE
9        10  33.277952       TRUE
10       11  33.465424       TRUE
11       12  17.940296       TRUE
12       13  33.268151       TRUE
13       14   6.434996      FALSE
14       15   7.833562      FALSE
15       16  46.783444       TRUE
16       17  12.229408       TRUE
17       18  10.261821       TRUE
18       19 -13.576343      FALSE
19       20  56.373939       TRUE

> PlotHartigan(wineBest)
```

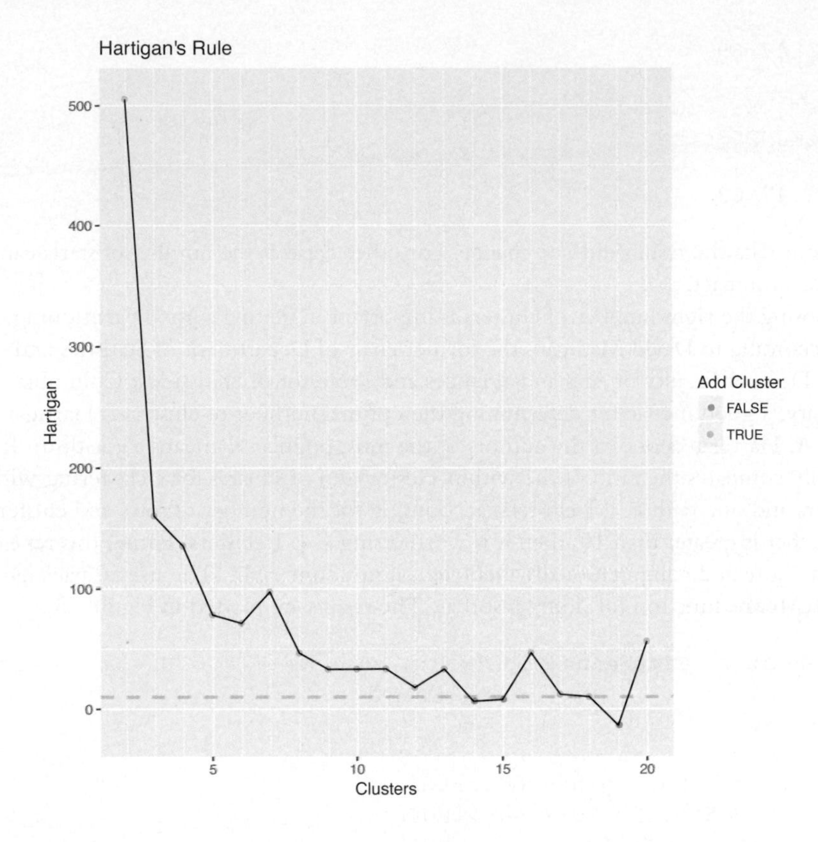

Figure 25.3 Plot of Hartigan's rule for a series of different cluster sizes.

According to this metric we should use 13 clusters. Again, this is just a rule of thumb and should not be strictly adhered to. Because we know there are three cultivars it would seem natural to choose three clusters because there are three cultivars. Then again, the results of the clustering with three clusters did only a fairly good job of aligning the clusters with the cultivars, so it might not be that good of a fit. Figure 25.4 shows the cluster assignment going down the left side and the cultivar across the top. Cultivar 1 is mostly alone in its own cluster, and cultivar 2 is just a little worse, while cultivar 3 is not clustered well at all. If this were truly a good fit, the diagonals would be the largest segments.

```
> table(wine$Cultivar, wineK3N25$cluster)


     1  2  3
  1 13 46  0
  2 20  1 50
  3 29  0 19
```

```
> plot(table(wine$Cultivar, wineK3N25$cluster),
+      main="Confusion Matrix for Wine Clustering",
+      xlab="Cultivar", ylab="Cluster")
```

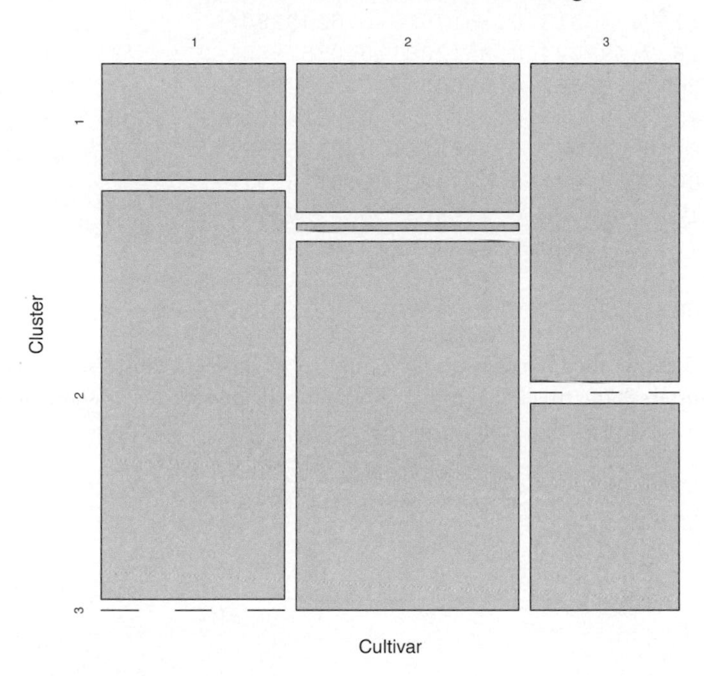

Figure 25.4 Confusion matrix for clustering of wine data by cultivars.

An alternative to Hartigan's rule is the Gap statistic, which compares the within-cluster dissimilarity for a clustering of the data with that of a bootstrapped sample of data. It is measuring the gap between reality and expectation. This can be calculated (for numeric data only) using **clusGap** in **cluster**. It takes a bit of time to run because it is doing a lot of simulations.

```
> library(cluster)
> theGap <- clusGap(wineTrain, FUNcluster=pam, K.max=20)
> gapDF <- as.data.frame(theGap$Tab)
> gapDF

       logW   E.logW       gap     SE.sim
1  9.655294 9.947093 0.2917988 0.03367473
2  8.987942 9.258169 0.2702262 0.03498740
3  8.617563 8.862178 0.2446152 0.03117947
4  8.370194 8.594228 0.2240346 0.03193258
```

```
5  8.193144 8.388382 0.1952376 0.03243527
6  7.979259 8.232036 0.2527773 0.03456908
7  7.819287 8.098214 0.2789276 0.03089973
8  7.685612 7.987350 0.3017378 0.02825189
9  7.591487 7.894791 0.3033035 0.02505585
10 7.496676 7.818529 0.3218525 0.02707628
11 7.398811 7.750513 0.3517019 0.02492806
12 7.340516 7.691724 0.3512081 0.02529801
13 7.269456 7.638362 0.3689066 0.02329920
14 7.224292 7.591250 0.3669578 0.02248816
15 7.157981 7.545987 0.3880061 0.02352986
16 7.104300 7.506623 0.4023225 0.02451914
17 7.054116 7.469984 0.4158683 0.02541277
18 7.006179 7.433963 0.4277835 0.02542758
19 6.971455 7.401962 0.4305071 0.02616872
20 6.932463 7.369970 0.4375070 0.02761156
```

Figure 25.5 shows the Gap statistic for a number of different clusters. The optimal number of clusters is the smallest number producing a gap within one standard error of the number of clusters that minimizes the gap.

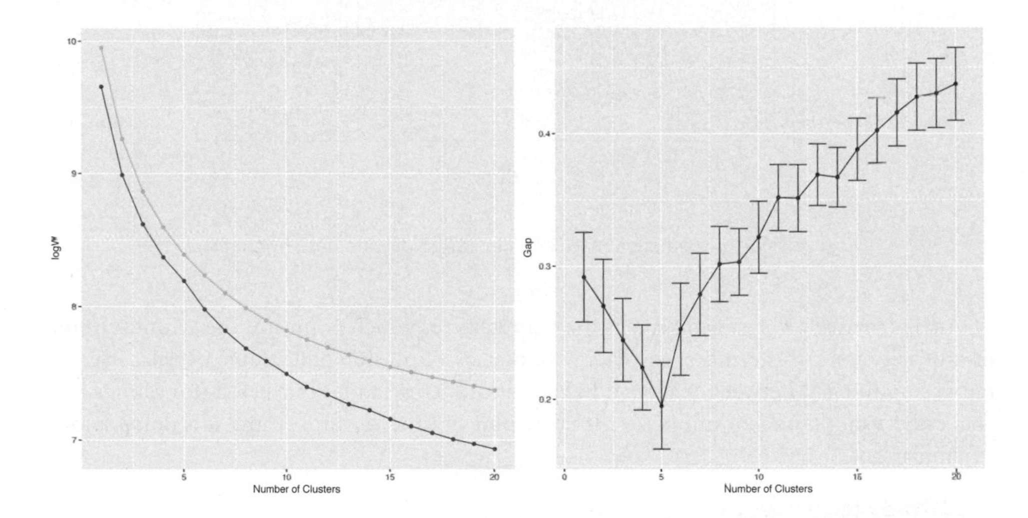

Figure 25.5 Gap curves for wine data. The blue curve is the observed within-cluster dissimilarity, and the green curve is the expected within-cluster dissimilarity. The red curve represents the Gap statistic (expected-observed) and the error bars are the standard error of the gap.

```
> # logW curves
> ggplot(gapDF, aes(x=1:nrow(gapDF))) +
+       geom_line(aes(y=logW), color="blue") +
+       geom_point(aes(y=logW), color="blue") +
```

```
+     geom_line(aes(y=E.logW), color="green") +
+     geom_point(aes(y=E.logW), color="green") +
+     labs(x="Number of Clusters")
>
> # gap curve
> ggplot(gapDF, aes(x=1:nrow(gapDF))) +
+     geom_line(aes(y=gap), color="red") +
+     geom_point(aes(y=gap), color="red") +
+     geom_errorbar(aes(ymin=gap-SE.sim, ymax=gap+SE.sim), color="red") +
+     labs(x="Number of Clusters", y="Gap")
```

For this data the minimum gap of 0.1952376 is for the clustering with five clusters. In this case there are no clusterings with fewer clusters that are within one standard error of the minimum. So, according to the Gap statistic, five clusters is optimal for this dataset.

25.2 PAM

Two problems with K-means clustering are that it does not work with categorical data and it is susceptible to outliers. An alternative is K-medoids. Instead of the center of a cluster being the mean of the cluster, the center is one of the actual observations in the cluster. This is akin to the median, which is likewise robust against outliers.

The most common K-medoids algorithm is Partitioning Around Medoids (PAM). The **cluster** package contains the **pam** function for performing Partitioning Around Medoids. For this example, we look at some data from the World Bank, including both numerical measures such as GDP and categorical information such as region and income level.

Now we use the country codes to download a number of indicators from the World Bank using **WDI**.

```
> indicators <- c("BX.KLT.DINV.WD.GD.ZS", "NY.GDP.DEFL.KD.ZG",
+                 "NY.GDP.MKTP.CD", "NY.GDP.MKTP.KD.ZG",
+                 "NY.GDP.PCAP.CD", "NY.GDP.PCAP.KD.ZG",
+                 "TG.VAL.TOTL.GD.ZS")
> library(WDI)
>
> # pull info on these indicators for all countries in our list
> # not all countries have information for every indicator
> # some countries do not have any data
> wbInfo <- WDI(country="all", indicator=indicators, start=2011,
+               end=2011, extra=TRUE)
> # get rid of aggregated info
> wbInfo <- wbInfo[wbInfo$region != "Aggregates", ]
> # get rid of countries where all the indicators are NA
> wbInfo <- wbInfo[which(rowSums(!is.na(wbInfo[, indicators])) > 0), ]
> # get rid of any rows where the iso is missing
> wbInfo <- wbInfo[!is.na(wbInfo$iso2c), ]
```

The data have a few missing values, but fortunately **pam** handles missing values well. Before we run the clustering algorithm we clean up the data some more, using the country names as the row names of the `data.frame` and ensuring the categorical variables are `factors` with the proper `levels`.

```
> # set rownames so we know the country without using that for clustering
> rownames(wbInfo) <- wbInfo$iso2c
> # refactorize region, income and lending
> # this accounts for any changes in the levels
> wbInfo$region <- factor(wbInfo$region)
> wbInfo$income <- factor(wbInfo$income)
> wbInfo$lending <- factor(wbInfo$lending)
```

Now we fit the clustering using **pam** from the **cluster** package. Figure 25.6 shows a silhouette plot of the results. As with K-means, the number of clusters need to be specified when using PAM. We could use methods like the Gap statistic, though we will choose 12 clusters, as this is slightly less than the square root of the number of rows of data, which is a simple heuristic for the number of clusters. Each line represents an observation, and each grouping of lines is a cluster. Observations that fit the cluster well have large positive lines, and observations that do not fit well have small or negative lines. A bigger average width for a cluster means a better clustering.

```
> # find which columns to keep
> # not those in this vector
> keep.cols <- which(!names(wbInfo) %in% c("iso2c", "country", "year",
+                                          "capital", "iso3c"))
> # fit the clustering
> wbPam <- pam(x=wbInfo[, keep.cols], k=12,
+              keep.diss=TRUE, keep.data=TRUE)
>
> # show the medoid observations
> wbPam$medoids

   BX.KLT.DINV.WD.GD.ZS NY.GDP.DEFL.KD.ZG NY.GDP.MKTP.CD
PT          5.507851973         0.6601427   2.373736e+11
HT          2.463873387         6.7745103   7.346157e+09
BY          7.259657119        58.3675854   5.513208e+10
BE         19.857364384         2.0299163   5.136611e+11
MX          1.765034004         5.5580395   1.153343e+12
GB          1.157530889         2.6028860   2.445408e+12
IN          1.741905033         7.9938177   1.847977e+12
CN          3.008038634         7.7539567   7.318499e+12
DE          1.084936891         0.8084950   3.600833e+12
NL          1.660830419         1.2428287   8.360736e+11
JP          0.001347863        -2.1202280   5.867154e+12
US          1.717849686         2.2283033   1.499130e+13
   NY.GDP.MKTP.KD.ZG NY.GDP.PCAP.CD NY.GDP.PCAP.KD.ZG
PT        -1.6688187     22315.8420       -1.66562016
HT         5.5903433       725.6333        4.22882080
BY         5.3000000      5819.9177        5.48896865
BE         1.7839242     46662.5283        0.74634396
MX         3.9106137     10047.1252        2.67022734
GB         0.7583280     39038.4583        0.09938161
IN         6.8559233      1488.5129        5.40325582
CN         9.3000000      5444.7853        8.78729922
```

```
DE          3.0288866  44059.8259          3.09309213
NL          0.9925175  50076.2824          0.50493944
JP         -0.7000000  45902.6716         -0.98497734
US          1.7000000  48111.9669          0.96816270
   TG.VAL.TOTL.GD.ZS region  longitude latitude income lending
PT          58.63188      2  -9.135520  38.7072      2       4
HT          49.82197      3 -72.328800  18.5392      3       3
BY         156.27254      2  27.576600  53.9678      6       2
BE         182.42266      2   4.367610  50.8371      2       4
MX          61.62462      3 -99.127600  19.4270      6       2
GB          45.37562      2  -0.126236  51.5002      2       4
IN          40.45037      6  77.225000  28.6353      4       1
CN          49.76509      1 116.286000  40.0495      6       2
DE          75.75581      2  13.411500  52.5235      2       4
NL         150.41895      2   4.890950  52.3738      2       4
JP          28.58185      1 139.770000  35.6700      2       4
US          24.98827      5 -77.032000  38.8895      2       4
```

```
> # make a silhouette plot
> plot(wbPam, which.plots=2, main="")
```

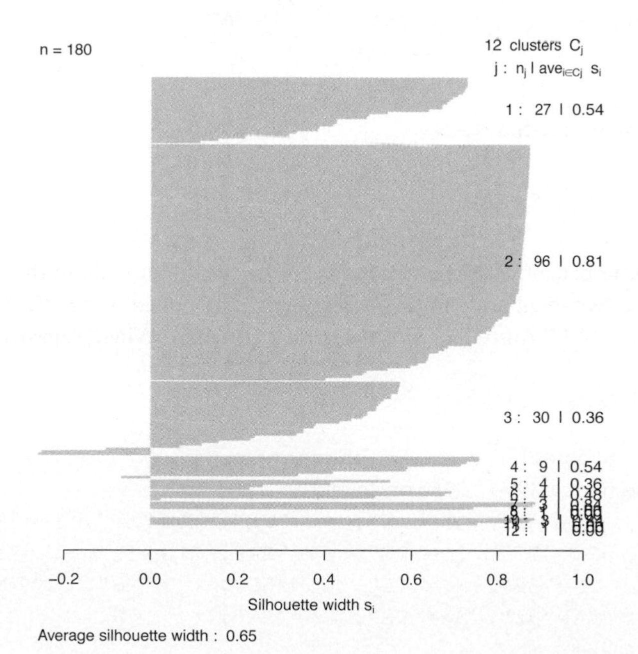

Figure 25.6 Silhouette plot for country clustering. Each line represents an observation, and each grouping of lines is a cluster. Observations that fit the cluster well have large positive lines and observations that do not fit well have small or negative lines. A bigger average width for a cluster means a better clustering.

Because we are dealing with country level information, it would be informative to view the clustering on a world map. As we are working with World Bank data, we will use the World Bank shapefile of the world. It can be downloaded in a browser as we would any other file or by using R. While this may be slower than using a browser, it can be nice if we have to programmatically download many files.

```
> download.file(url="http://jaredlander.com/data/worldmap.zip",
+               destfile="data/worldmap.zip", method="curl")
```

The file needs to be unzipped, which can be done through the operating system or in R.

```
> unzip(zipfile="data/worldmap.zip", exdir="data")
```

Of the four files, we only need to worry about the one ending in `.shp` because R will handle the rest. We read it in using **readShapeSpatial** from **maptools**.

```
           name              CntryName FipsCntry
0 Fips Cntry:                  Aruba        AA
1 Fips Cntry:      Antigua & Barbuda        AC
2 Fips Cntry: United Arab Emirates          AE
3 Fips Cntry:            Afghanistan        AF
4 Fips Cntry:                Algeria        AG
5 Fips Cntry:             Azerbaijan        AJ

> library(maptools)
> world <- readShapeSpatial(
+     "data/world_country_admin_boundary_shapefile_with_fips_codes.shp"
+     )
> head(world@data)
```

There are some blatant discrepancies between the two-digit code in the World Bank shapefile and the two-digit code in the World Bank data pulled using **WDI**. Notably, Austria should be "AT," Australia "AU," Myanmar (Burma) "MM," Vietnam "VN" and so on.

```
> library(dplyr)
> world@data$FipsCntry <- as.character(
+     recode(world@data$FipsCntry,
+            AU="AT", AS="AU", VM="VN", BM="MM", SP="ES",
+            PO="PT", IC="IL", SF="ZA", TU="TR", IZ="IQ",
+            UK="GB", EI="IE", SU="SD", MA="MG", MO="MA",
+            JA="JP", SW="SE", SN="SG")
+ )
```

In order to use **ggplot2** we need to convert this shapefile object into a `data.frame`, which requires a few steps. First we create a new column, called `id`, from the row names of the data. Then we use the **tidy** function from the **broom** package, written by David Robinson, to convert it into a `data.frame`. The **broom** package is a great general purpose tool for converting R objects, such as `lm` models and `kmeans` clusterings, into nice, rectangular `data.frames`.

```
> # make an id column using the rownames
> world@data$id <- rownames(world@data)
> # convert into a data.frame
> library(broom)
> world.df <- tidy(world, region="id")
> head(world.df)

       long      lat order  hole piece group id
1 -69.88223 12.41111     1 FALSE     1   0.1  0
2 -69.94695 12.43667     2 FALSE     1   0.1  0
3 -70.05904 12.54021     3 FALSE     1   0.1  0
4 -70.05966 12.62778     4 FALSE     1   0.1  0
5 -70.03320 12.61833     5 FALSE     1   0.1  0
6 -69.93224 12.52806     6 FALSE     1   0.1  0
```

Before we can join this to the clustering, we need to join `FipsCntry` back into `world.df`.

```
> world.df <- left_join(world.df,
+                       world@data[, c("id", "CntryName", "FipsCntry")],
+                       by="id")
> head(world.df)

       long      lat order  hole piece group id CntryName FipsCntry
1 -69.88223 12.41111     1 FALSE     1   0.1  0     Aruba        AA
2 -69.94695 12.43667     2 FALSE     1   0.1  0     Aruba        AA
3 -70.05904 12.54021     3 FALSE     1   0.1  0     Aruba        AA
4 -70.05966 12.62778     4 FALSE     1   0.1  0     Aruba        AA
5 -70.03320 12.61833     5 FALSE     1   0.1  0     Aruba        AA
6 -69.93224 12.52806     6 FALSE     1   0.1  0     Aruba        AA
```

Now we can take the steps of joining in data from the clustering and the original World Bank data.

```
> clusterMembership <- data.frame(FipsCntry=names(wbPam$clustering),
+                                 Cluster=wbPam$clustering,
+                                 stringsAsFactors=FALSE)
> head(clusterMembership)

   FipsCntry Cluster
AE        AE       1
AF        AF       2
AG        AG       2
AL        AL       2
AM        AM       2
AO        AO       3
```

```
> world.df <- left_join(world.df, clusterMembership, by="FipsCntry")
> world.df$Cluster <- as.character(world.df$Cluster)
> world.df$Cluster <- factor(world.df$Cluster, levels=1:12)
```

Building the plot itself requires a number of **ggplot2** commands to format it correctly. Figure 25.7 shows the map, color coded by cluster membership; the gray countries either do not have World Bank information or were not properly matched up between the two datasets.

```
> ggplot() +
+     geom_polygon(data=world.df, aes(x=long, y=lat, group=group,
+                                     fill=Cluster, color=Cluster)) +
+     labs(x=NULL, y=NULL) + coord_equal() +
+     theme(panel.grid.major=element_blank(),
+           panel.grid.minor=element_blank(),
+           axis.text.x=element_blank(), axis.text.y=element_blank(),
+           axis.ticks=element_blank(), panel.background=element_blank())
```

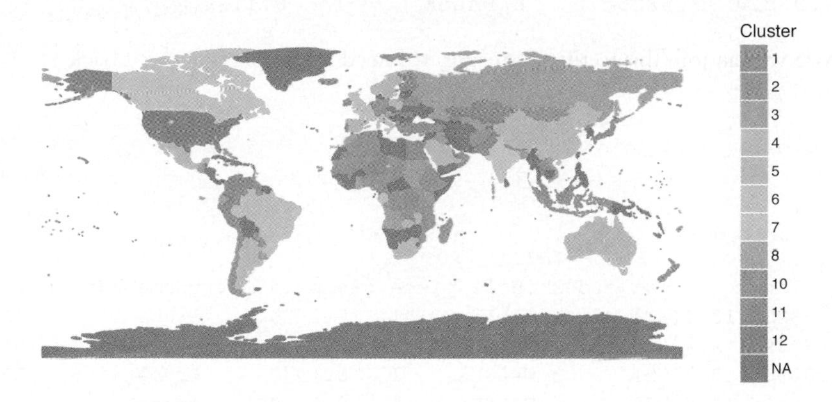

Figure 25.7 Map of PAM clustering of World Bank data. Gray countries either do not have World Bank information or were not properly matched up between the two datasets.

Much like with K-means, the number of clusters in a K-medoids clustering must be specified. Something similar to Hartigan's Rule can be built using the dissimilarity information returned by **pam**.

```
> wbPam$clusinfo

     size     max_diss      av_diss     diameter   separation
[1,]   27 122871463849  46185193372 200539326122 1.967640e+10
[2,]   96  22901202940   7270137217  31951289020 3.373324e+09
[3,]   30  84897264072  21252371506 106408660458 3.373324e+09
[4,]    9 145646809734  59174398936 251071168505 4.799168e+10
[5,]    4 323538875043 146668424920 360634547126 2.591686e+11
```

```
 [6,] 4 327624060484 152576296819 579061061914 3.362014e+11
 [7,] 3 111926243631  40573057031 121719171093 2.591686e+11
 [8,] 1            0            0            0 1.451345e+12
 [9,] 1            0            0            0 8.278012e+11
[10,] 3  61090193130  23949621648  71848864944 1.156755e+11
[11,] 1            0            0            0 1.451345e+12
[12,] 1            0            0            0 7.672801e+12
```

25.3 Hierarchical Clustering

Hierarchical clustering builds clusters within clusters, and does not require a pre-specified number of clusters like K-means and K-medoids do. A hierarchical clustering can be thought of as a tree and displayed as a dendrogram; at the top there is just one cluster consisting of all the observations, and at the bottom each observation is an entire cluster. In between are varying levels of clustering.

Using the wine data, we can build the clustering with **hclust**. The result is visualized as a dendrogram in Figure 25.8. While the text is hard to see, it labels the observations at the end nodes.

```
> wineH <- hclust(d=dist(wineTrain))
> plot(wineH)
```

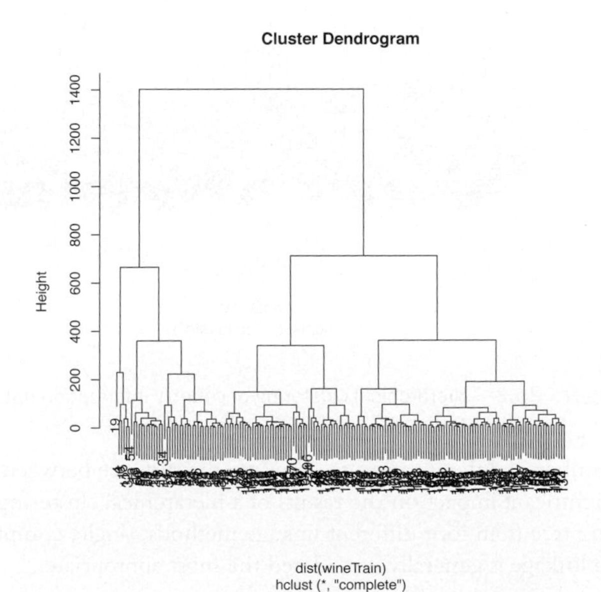

Figure 25.8 Hierarchical clustering of wine data.

Hierarchical clustering also works on categorical data like the country information data. However, its dissimilarity `matrix` must be calculated differently. The dendrogram is shown in Figure 25.9.

```
> # calculate distance
> keep.cols <- which(!names(wbInfo) %in% c("iso2c", "country", "year",
+                                          "capital", "iso3c"))
> wbDaisy <- daisy(x=wbInfo[, keep.cols])
>
> wbH <- hclust(wbDaisy)
> plot(wbH)
```

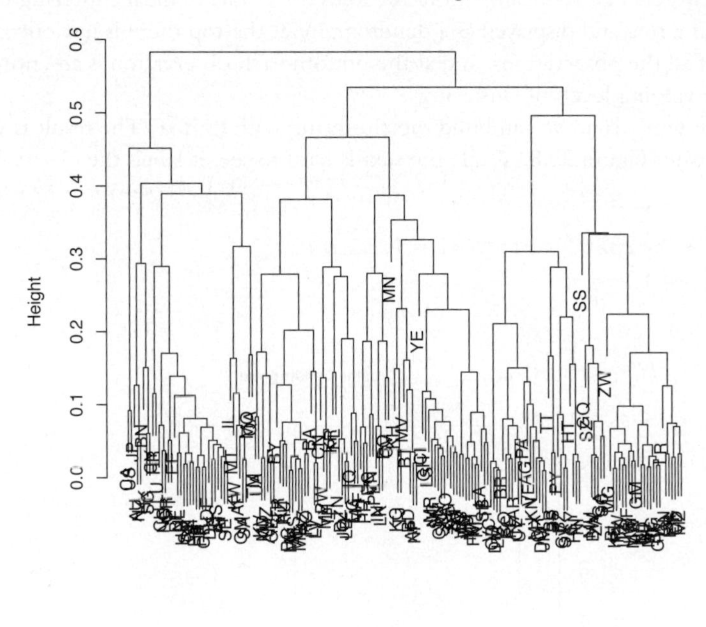

Figure 25.9 Hierarchical clustering of country information data.

There are a number of different ways to compute the distance between clusters and they can have a significant impact on the results of a hierarchical clustering. Figure 25.10 shows the resulting tree from four different linkage methods: single, complete, average and centroid. Average linkage is generally considered the most appropriate.

```
> wineH1 <- hclust(dist(wineTrain), method="single")
> wineH2 <- hclust(dist(wineTrain), method="complete")
> wineH3 <- hclust(dist(wineTrain), method="average")
> wineH4 <- hclust(dist(wineTrain), method="centroid")
```

```
>
> plot(wineH1, labels=FALSE, main="Single")
> plot(wineH2, labels=FALSE, main="Complete")
> plot(wineH3, labels=FALSE, main="Average")
> plot(wineH4, labels=FALSE, main="Centroid")
```

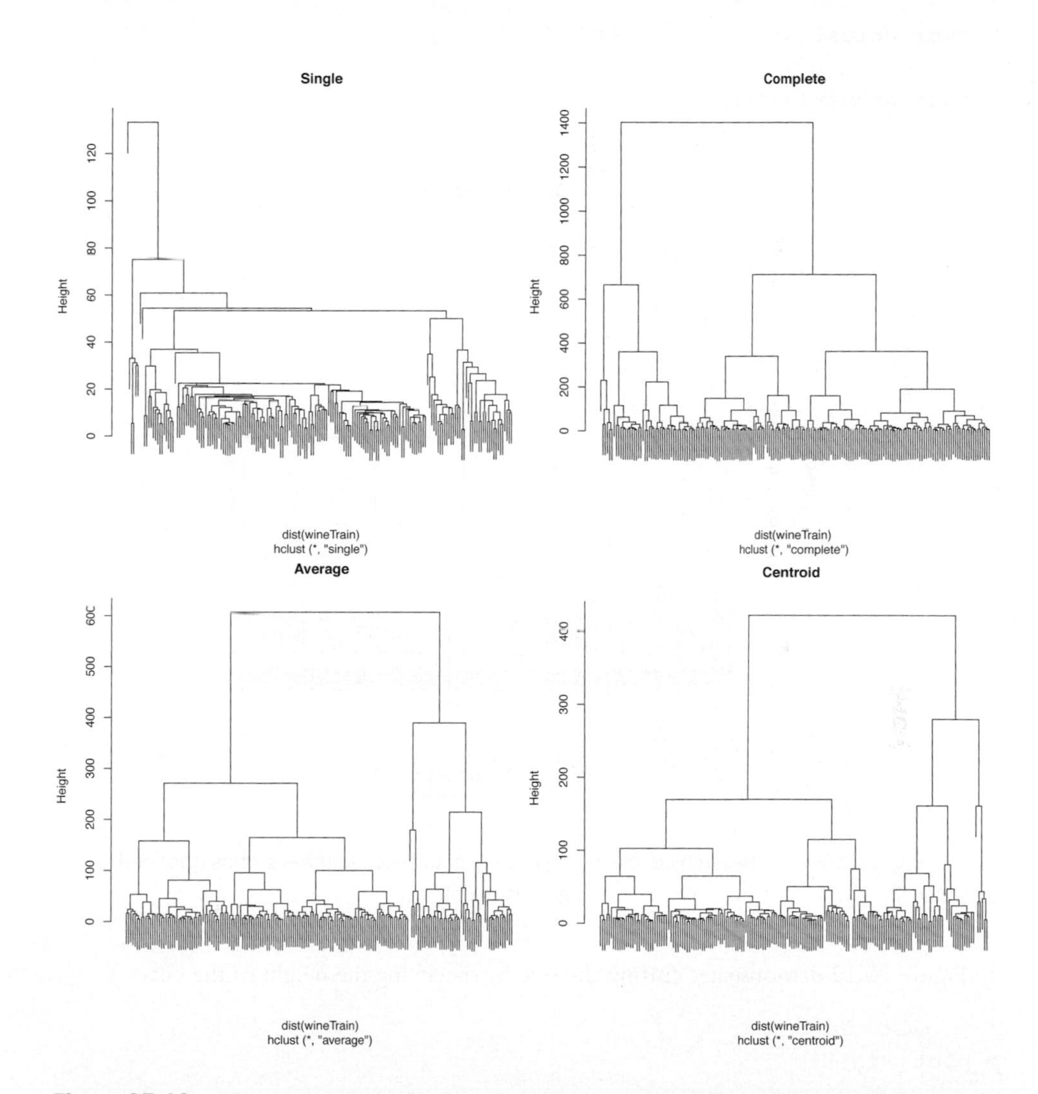

Figure 25.10 Wine hierarchical clusters with different linkage methods. Clockwise from top left: single, complete, centroid, average.

Cutting the resulting tree produced by hierarchical clustering splits the observations into defined groups. There are two ways to cut it: either specifying the number of clusters, which determines where the cuts take place, or specifying where to make the cut, which determines the number of clusters. Figure 25.11 demonstrates cutting the tree by specifying the number of clusters.

```
> # plot the tree
> plot(wineH)
> # split into 3 clusters
> rect.hclust(wineH, k=3, border="red")
> # split into 13 clusters
> rect.hclust(wineH, k=13, border="blue")
```

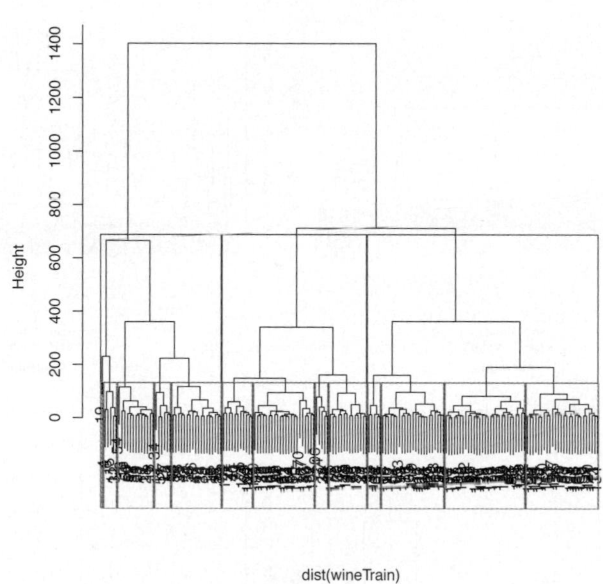

Figure 25.11 Hierarchical clustering of wine data split into three groups (red) and 13 groups (blue).

Figure 25.12 demonstrates cutting the tree by specifying the height of the cuts.

```
> # plot the tree
> plot(wineH)
> # split into 3 clusters
> rect.hclust(wineH, h=200, border="red")
> # split into 13 clusters
> rect.hclust(wineH, h=800, border="blue")
```

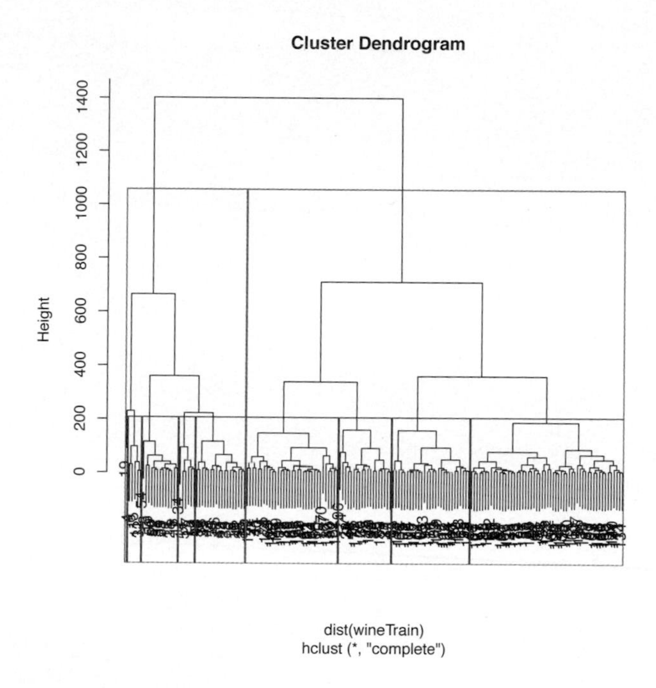

Figure 25.12 Hierarchical clustering of wine data split by the height of cuts.

25.4 Conclusion

Clustering is a popular technique for segmenting data. The primary options for clustering in R are **kmeans** for K-means, **pam** in **cluster** for K-medoids and **hclust** for hierarchical clustering. Speed can sometimes be a problem with clustering, especially hierarchical clustering, so it is worth considering replacement packages like **fastcluster**, which has a drop-in replacement function, **hclust**, which operates just like the standard **hclust**, only faster.

26 Model Fitting with Caret

Fitting models often involves brute-force iteration over many different values for numerous parameters and deciding upon the setting that creates the "best" model. While this could be coded from scratch, **caret** performs parameter tuning automatically. Additionally, it provides a standardized interface for all the models it supports, which makes the modelling process that much easier when trying multiple models.

26.1 Caret Basics

The first question a lot of people have about **caret** is in regard to its spelling since it is spelled like neither the vegetable nor the unit of weight. That is because **caret** is short for **C**lassification **A**nd **RE**gression **T**raining. It is written, with great effort, primarily by Max Kuhn.

While **caret** can do many things—including preprocessing the data, splitting the data and visualization—we focus on its capability to choose parameter values based on some measure of model performance. As discussed in Section 21.3, cross-validation is a popular method of determing the quality of a model. Here we use **caret** to fit many iterations of a model, each with different parameter settings, and assess the quality of the models with cross-validation. The parameter values that result in the best cross-validation score is regarded as the best model.

There are a large number of models that can be tuned using **caret**. More accurately, any model function in R can be tuned with **caret**, and hundreds of those have already been built into **caret**. All of these functions can be accessed using the `formula` interface. This is helpful when using a function like **glmnet** that requires `matrices` as inputs.

26.2 Caret Options

Model training is performed with the **train** function in **caret**. Data is provided either as a predictor variable and a response variable or as a `formula` and `data.frame`. Max Kuhn has spent a great deal of time researching the best way to provide data to various models.[1] Some models do better when categorical data are treated as a `matrix` of indicator

1. His video at the 2016 New York R Conference can be seen at `https://youtu.be/ul2zLF61CyY`.

variables, while others do better when the categorical data are represented as a factor vector. When fitting models with **train**, using the formula method will turn categorical data into a matrix of indicator variables, and not using the formula interface will use factor variables.

After the data, the method argument specifies, as a character vector, the type of model or models to be trained. An extensive list of available models is found at https://topepo.github.io/caret/available-models.html. The metric argument specifies the statistic that is used to determine the optimal model, such as "RMSE" for regression and "Accuracy" for classification.

Computational controls are set with the **trainControl** function which is passed to the trControl argument of **train**. In addition to the **train** arguments, model function specific arguments can be given to **train** and are passed on to the modelling function. However, tunable parameters for models cannot be passed directly to **train**. They must be included in a data.frame that is passed to the tuneGrid argument.

26.2.1 caret Training Controls

The **trainControl** function sets options controling the computation of model fitting and assessment. While there are reasonable default settings, it is best to explicitely set them. There are many arguments, though the ones discussed here are usually sufficient for control over the modelling process.

Model quality is assessed via repeated model fitting (achieved through resampling of some form) and comparing using some type of model quality metric. The method argument takes a character specifying the type of resampling with the most common being "boot" for the bootstrap and "repeatedcv" for repeated cross-validation. When using the bootstrap the number argument specifies the number of iterations. When using cross-validation the number argument indicates the number of folds, and the repeats argument specifies the number of times to run the k-fold cross-validation.

Performance is judged by a function supplied to the summaryFunction argument, such as **twoClassSummary** assessing binary classification with Area Under the Curve (AUC) or **postResample** for assessing regression with root mean squared error (RMSE). Running **train** in parallel is quite simple. If the allowParallel argument is set to TRUE and a parallel backend is loaded, then **train** will atuomatically work in parallel. An example set of controls is shown in the following code:

```
> library(caret)
> ctrl <- trainControl(method = "repeatedcv",
+                      repeats=3,
+                      number=5,
+                      summaryFunction=defaultSummary,
+                      allowParallel=TRUE)
```

26.2.2 Caret Search Grid

The biggest benefit of using **caret** is to select optimal model parameters. In the case of **xgboost** this could be the maximum tree depth and amount of shrinkage. For **glmnet** this could be the size of the penalty term or the the mixture between ridge and lasso. The **train** function iterates over a set of possible parameters, stored in a `data.frame`, which are supplied to the `tuneGrid` argument, fits a model to each set of parameters and then assesses the quality of the model. This is called a grid search.

Each column of the `data.frame` represents a tuning parameter and each row is a set of parameters. As an example, generalized additive models (GAMs), as fit by the **mgcv** package, have two tuning parameters: `select` for adding an extra penalty to each term and `method` for setting the parameter estimation method. An example tuning grid for the **gam** function follows:

```
> gamGrid <- data.frame(select=c(TRUE, TRUE, FALSE, FALSE),
+                       method=c('GCV.Cp', 'REML', 'GCV.Cp', 'REML'),
+                       stringsAsFactors=FALSE)
> gamGrid

  select method
1   TRUE GCV.Cp
2   TRUE   REML
3  FALSE GCV.Cp
4  FALSE   REML
```

This grid will cause **train** to fit four models. The first has `select=TRUE` and `method='GCV.Cp'`. The second model has `select=TRUE` and `method='REML'` and so on with the other combinations.

26.3 Tuning a Boosted Tree

The first model we tune with **caret** is a boosted tree as covered in Section 23.5. For the examples in this chapter we return to the American Community Survey (ACS) data.

```
> acs <- tibble::as_tibble(
+     read.table(
+         "http://jaredlander.com/data/acs_ny.csv",
+         sep=",", header=TRUE, stringsAsFactors=FALSE
+     )
+ )
```

We create a new variable called `Income` that is a `factor` with `levels` “Below” and “Above” $150,000. Since **train** is going to load **plyr**, it is important we load it before **dplyr**, so we do that here even though we will not use the package.

```
> library(plyr)
> library(dplyr)
> acs <- acs %>%
+     mutate(Income=factor(FamilyIncome >= 150000,
+                          levels=c(FALSE, TRUE),
+                          labels=c('Below', 'Above')))
```

Under direct usage **xgboost** requires a `matrix` of predictors and a response `vector`, but **caret** enables the `formula` interface—even for models that ordinarily do not—so we take advantage of that capability.

```
> acsFormula <- Income ~ NumChildren +
+     NumRooms + NumVehicles + NumWorkers + OwnRent +
+     ElectricBill + FoodStamp + HeatingFuel
```

To evaluate the best fitting set of parameters we use five-fold repeated cross-validation with two repeats. We set `summaryFunction` to **twoClassSummary** and `classProb` to `TRUE` so that AUC will be used to evaluate the models. Even though it is easy for **caret** to fit the models in parallel, **xgboost** has its own parallelism, so we set `allowParallel` to `FALSE`.

```
> ctrl <- trainControl(method = "repeatedcv",
+                      repeats=2,
+                      number=5,
+                      summaryFunction=twoClassSummary,
+                      classProbs=TRUE,
+                      allowParallel=FALSE)
```

As of early 2017 there are seven tuning parameters for **xgboost**. The `nrounds` argument determines the number of boosting iterations and `max_depth` sets the maximum complexity of the trees. The learning rate is determined by `eta`, which controls the amount of shrinkage. Leaf splitting is determined by `gamma` and `min_child_weight`. The sampling rate for columns and rows are set by `colsample_bytree` and `subsample`, respectively. We use **expand.grid** to get every combination of the values set in the following code:

```
> boostGrid <- expand.grid(nrounds=100, #the maximum number of iterations
+                          max_depth=c(2, 6, 10),
+                          eta=c(0.01, 0.1), # shrinkage
+                          gamma=c(0),
+                          colsample_bytree=1,
+                          min_child_weight=1,
+                          subsample=0.7)
```

With the computation controls and tuning grid set we can train the model. We provide the **train** function the `formula` and data and specify "xgbTree" as the `method`. We also supply the controls and tuning grid. The `nthread` argument is passed through to **xgboost** to dictate the number of processor threads to run in parallel. Even with parallelization it can take a bit of time to fit all of these models, so we must have some patience. There are a number of stochastic pieces to this process, so we set the random seed for reproducibility.

```
> set.seed(73615)
> boostTuned <- train(acsFormula, data=acs,
+                     method="xgbTree",
+                     metric="ROC",
+                     trControl=ctrl,
+                     tuneGrid=boostGrid, nthread=4)
```

The returned object has many slots, including the search grid appended with the resulting quality metrics.

```
> boostTuned$results %>% arrange(ROC)

   eta max_depth gamma colsample_bytree min_child_weight subsample
1 0.01         2     0                1                1       0.7
2 0.10        10     0                1                1       0.7
3 0.01        10     0                1                1       0.7
4 0.01         6     0                1                1       0.7
5 0.10         6     0                1                1       0.7
6 0.10         2     0                1                1       0.7
  nrounds       ROC      Sens      Spec       ROCSD      SensSD
1     100 0.7261711 1.0000000 0.0000000 0.010465376 0.000000000
2     100 0.7377721 0.9522002 0.1818182 0.009782476 0.003538260
3     100 0.7486185 0.9679318 0.1521358 0.009455179 0.004366311
4     100 0.7504831 0.9807206 0.1059146 0.009736577 0.004450671
5     100 0.7560484 0.9666667 0.1599124 0.009505135 0.004260313
6     100 0.7602718 0.9766227 0.1292442 0.008331900 0.002959298
      SpecSD
1 0.00000000
2 0.01345420
3 0.01342891
4 0.01177458
5 0.01555843
6 0.01080588
```

This is easier to see in graphical form, so we plot it. Figure 26.1 shows that a `max_depth` of 2 and an `eta` of 0.1 results in the best receiver operating characteristic (ROC).

```
> plot(boostTuned)
```

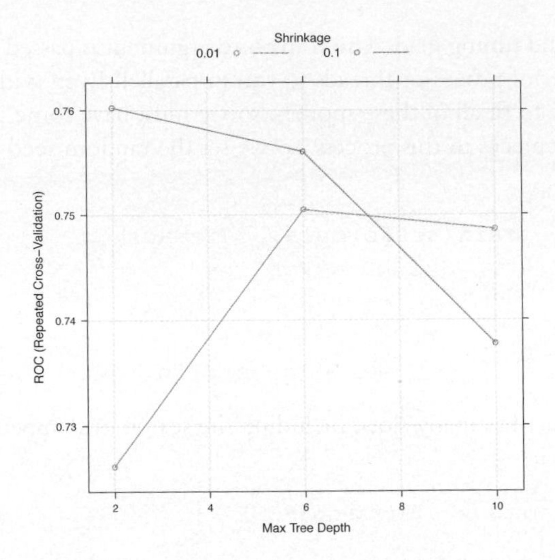

Figure 26.1 Plot showing that for the American Community Survey data, `max_depth=2` and `eta=0.1` leads to the best ROC.

For most models, **caret** provides the best model in the `finalModel` slot. It is not always best practice to access this model directly, though in this case we are able to plot the model as shown in Figure 26.2.

```
> xgb.plot.multi.trees(boostTuned$finalModel,
+                      feature_names=boostTuned$coefnames)
```

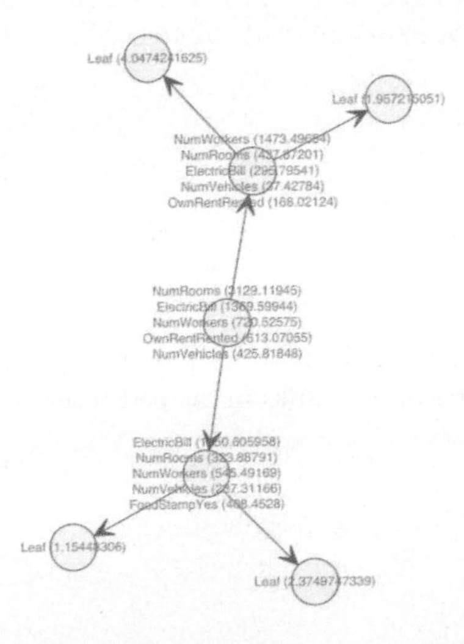

Figure 26.2 Visualization of boosted tree decided by `caret` to be best, given the parameter grid.

Most interaction with the model is best done with the interface provided by **caret**. For instance, prediction is handled by **caret**-specific **predict** functions. This case illustrates why it is good to use the **predict** function from **caret** rather than from **xgboost** because **caret** will handle the categorical variables as is appropriate, whereas **xgboost** expects us to process the data first. To see this we read another set of ACS data and make predictions using **caret**.

```
> acsNew <- read.table('http://www.jaredlander.com/data/acsNew.csv',
+                      header=TRUE, sep=',', stringsAsFactors=FALSE)
```

With two-class classification **caret** can predict the dominant class or show the probability of each class. In this case they are all predicted to be "Below."

```
> predict(boostTuned, newdata=acsNew, type='raw') %>% head

[1] Above Below Below Below Below Below
Levels: Below Above

> predict(boostTuned, newdata=acsNew, type='prob') %>% head

      Below      Above
1 0.4974915 0.50250852
2 0.5924829 0.40751714
3 0.5721835 0.42781645
4 0.9204149 0.07958508
5 0.8550579 0.14494210
6 0.7603117 0.23968834
```

26.4 Conclusion

While R has always been great for statistical modelling, **caret** created a uniform interface to hundreds of models, enabling a consistent experience regardless of the underlying function. Of even bigger benefit, **caret** also provides cross-validation and parameter tuning to determine the "best" model. Further, **caret** provides mechanisms for creatnig test and train datasets and numerous model validation metrics. All of this makes **caret** a great choice for fitting models.

26.4 Conclusion

27

Reproducibility and Reports with knitr

Successfully delivering the results of an analysis can be just as important as the analysis itself, so it is important to communicate them in an effective way. This can be a written report, a Web site of results, a slideshow or a dashboard. In this chapter we focus on reports, which are made remarkably easy using **knitr**, a package written by Yihui Xie. Chapter 28 covers writing Web pages and slideshows with RMarkdown and Chapter 29 goes over building Shiny dashboards. **knitr** was initially created as a replacement for **Sweave** for the creation of PDF documents using LaTeX interweaved with R code and the generated results. It has since added the capability to work with Markdown for generating a broad range of documents.

The combination of **knitr** and RStudio is so powerful that it was possible to write this entire book inside the RStudio IDE using **knitr** to insert and run R code and graphics.

27.1 Installing a LaTeX Program

LaTeX (pronounced "lay-tech") is a markup language based on the TeX typesetting system created by Donald Knuth. It is regularly used for writing scientific papers and books, including this one. Like any other program, LaTeX must be installed before it can be used.

Each of the operating systems uses a different LaTeX distribution. Table 27.1 lists OS-specific distributions and download locations.

Table 27.1 LaTeX distributions and their locations

OS	Distribution	URL
Windows	MiKTeX	`http://miktex.org/`
Mac	MacTeX	`http://www.tug.org/mactex/`
Linux	TeX Live	`http://www.tug.org/texlive/`

27.2 LaTeX Primer

This is not intended to be anywhere near a comprehensive lesson in LaTeX, but it should be enough to get started with making documents. LaTeX documents should be saved with a .tex extension to identify them as such. While RStudio is intended for working with R, it is a suitable text editor for LaTeX and is the environment we will be using.

The very first line in a LaTeX file declares the type of document, the most common being "article" and "book." This is done with \documentclass{...}, replacing ...with the desired document class. Other popular document classes are "report," "beamer," "memoir" and "letter."

Immediately following the declaration of the documentclass is the preamble. This is where commands that affect the document go, such as what packages to load (LaTeX packages) using \usepackage{...} and making an index with \makeindex.

In order to include images, it is advisable to use the graphicx package. This enables us to specify the type of image file that will be used by entering \DeclareGraphicsExtensions{.png, .jpg}, which means LaTeX will first search for files ending in .png and then search for files ending in .jpg. This will be explained more when dealing with images later.

This is also where the title, author and date are declared with \title, \author and \date, respectively. New shortcuts can be created here such as \newcommand{\dataframe}{\texttt{data.frame}}, so that every time \dataframe{} is typed it will be printed as data.frame, which appears in a typewriter font because of the \texttt{...}.

The actual document starts with \begin{document} and ends with \end{document}. That is where all the content goes. So far our LaTeX document looks like the following example.

```
\documentclass{article}
% this is a comment
% all content following a % on a line will be commented out as if
it never existed to latex

\usepackage{graphicx} % use graphics
\DeclareGraphicsExtensions{.png,.jpg} % search for png then jpg

% define shortcut for dataframe
\newcommand{\dataframe}{\texttt{data.frame}}

\title{A Simple Article}
\author{Jared P. Lander\\ Lander Analytics}
% the \\ puts what follows on the next line
\date{December 22nd, 2016}

\begin{document}
\maketitle
```

(Continues)

(Continued)

```
Some Content

\end{document}
```

Content can be split into sections using `\section{Section Name}`. All text following this command will be part of that section until another `\section{...}` is reached. Sections (and subsections and chapters) are automatically numbered by LaTeX. If given a label using `\label{...}` they can be referred to using `\ref{...}`. The table of contents is automatically numbered and is created using `\tableofcontents`. We can now further build out our document with some sections and a table of contents. Normally, LaTeX must be run twice for cross references and the table of contents but RStudio, and most other LaTeX editors, will do that automatically.

```
\documentclass{article}
% this is a comment
% all content following a % on a line will be commented out as if
it never existed to latex

\usepackage{graphicx} % use graphics
\DeclareGraphicsExtensions{.png,.jpg} % search for png then jpg

% define shortcut for dataframe
\newcommand{\dataframe}{\texttt{data.frame}}

\title{A Simple Article}
\author{Jared P. Lander\\ Lander Analytics}
% the \\ puts what follows on the next line
\date{December 22nd, 2016}

\begin{document}
\maketitle % create the title page
\tableofcontents % build table of contents

\section{Getting Started}
\label{sec:GettingStarted}
This is the first section of our article. The only thing it will
talk about is building \dataframe{}s and not much else.

A new paragraph is started simply by leaving a blank line. That
is all that is required. Indenting will happen automatically.
```

(Continues)

(Continued)

```
\section{More Information}
\label{sec:MoreInfo}
Here is another section. In Section \ref{sec:GettingStarted} we
learned some basics and now we will see just a little more. Suppose
this section is getting too long so it should be broken up into
subsections.

\subsection{First Subsection}
\label{FirstSub}
Content for a subsection.

\subsection{Second Subsection}
\label{SecondSub}
More content that is nested in Section \ref{sec:MoreInfo}

\section{Last Section}
\label{sec:LastBit}
This section was just created to show how to stop a preceding
subsection, section or chapter. Note that chapters are only
available in books, not articles.

\makeindex % create the index

\end{document}
```

While there is certainly a lot more to be learned about LaTeX, this should provide enough of a start for using it with **knitr**. A great reference is the "Not So Short Introduction to LaTeX," which can be found at `http://tobi.oetiker.ch/lshort/lshort.pdf`.

27.3 Using `knitr` with LaTeX

Writing a LaTeX document with R code is fairly straightforward. Regular text is written using normal LaTeX conventions, and the R code is delineated by special commands. All R code is preceded by `<<label-name,option1='value1',option2='value2'>>=` and is followed by `@`. While editing, RStudio nicely colors the background of the editor according to what is being written, LaTeX or R code. This is seen in Figure 27.1, and is called a "chunk."

These documents are saved as `.Rnw` files. During the knitting process an `.Rnw` file is converted to a .tex file, which is then compiled to a PDF. If using the consolethis occurs when the **knit** function is called, pass the `.Rnw` file as the first argument. In RStudio this is done by clicking the Compile PDF button in the toolbar or pressing `Ctrl+Shift+K` on the keyboard.

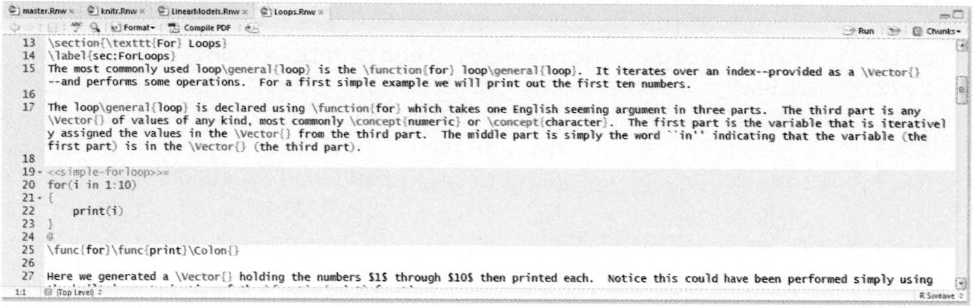

```
13 \section{\texttt{For} Loops}
14 \label{sec:ForLoops}
15 The most commonly used loop\general{loop} is the \function{for} loop\general{loop}.  It iterates over an index--provided as a \Vector{}
   --and performs some operations.  For a first simple example we will print out the first ten numbers.
16
17 The loop\general{loop} is declared using \function{for} which takes one English seeming argument in three parts.  The third part is any
   \Vector{} of values of any kind, most commonly \concept{numeric} or \concept{character}.  The first part is the variable that is iterativel
   y assigned the values in the \Vector{} from the third part.  The middle part is simply the word ``in'' indicating that the variable (the
   first part) is in the \Vector{} (the third part).
18
19 <<simple-forloop>>=
20 for(i in 1:10)
21 {
22     print(i)
23 }
24 @
25 \func{for}\func{print}\Colon{}
26
27 Here we generated a \Vector{} holding the numbers $1$ through $10$ then printed each.  Notice this could have been performed simply using
```

Figure 27.1 Screenshot of LaTeX and R code in RStudio text editor. Notice that the code section is gray.

Chunks are the workforce of **knitr** and are essential to understand. A typical use is to show both the code and results. It is possible to do one or the other, or neither as well, but for now we will focus on getting code printed and evaluated. Suppose we want to illustrate loading **ggplot2**, viewing the head of the `diamonds` data, and then fit a regression. The first step is to build a chunk.

```
<<load-and-model-diamonds>>=

# load ggplot
library(ggplot2)

# load and view the diamonds data
data(diamonds)
head(diamonds)

# fit the model
mod1 <- lm(price ~ carat + cut, data=diamonds)
# view a summary
summary(mod1)

@
```

This will then print both the code and the result in the final document as shown next.

```
> # load ggplot
> library(ggplot2)
>
> # load and view the diamonds data
> data(diamonds)
> head(diamonds)

# A tibble: 6 × 10
```

```
  carat       cut color clarity depth table price     x     y     z
  <dbl>     <ord> <ord>   <ord> <dbl> <dbl> <int> <dbl> <dbl> <dbl>
1  0.23     Ideal     E     SI2  61.5    55   326  3.95  3.98  2.43
2  0.21   Premium     E     SI1  59.8    61   326  3.89  3.84  2.31
3  0.23      Good     E     VS1  56.9    65   327  4.05  4.07  2.31
4  0.29   Premium     I     VS2  62.4    58   334  4.20  4.23  2.63
5  0.31      Good     J     SI2  63.3    58   335  4.34  4.35  2.75
6  0.24 Very Good     J    VVS2  62.8    57   336  3.94  3.96  2.48

> # fit the model
> mod1 <- lm(price ~ carat + cut, data=diamonds)
> # view a summary
> summary(mod1)


Call:
lm(formula = price ~ carat + cut, data = diamonds)

Residuals:
     Min       1Q   Median       3Q      Max
-17540.7   -791.6    -37.6    522.1  12721.4

Coefficients:
            Estimate Std. Error  t value Pr(>|t|)
(Intercept) -2701.38      15.43 -175.061  < 2e-16 ***
carat        7871.08      13.98  563.040  < 2e-16 ***
cut.L        1239.80      26.10   47.502  < 2e-16 ***
cut.Q        -528.60      23.13  -22.851  < 2e-16 ***
cut.C         367.91      20.21   18.201  < 2e-16 ***
cut^4          74.59      16.24    4.593 4.37e-06 ***
---
Signif. codes:  0 '***' 0.001 '**' 0.01 '*' 0.05 '.' 0.1 ' ' 1

Residual standard error: 1511 on 53934 degrees of freedom
Multiple R-squared:  0.8565,Adjusted R-squared:  0.8565
F-statistic: 6.437e+04 on 5 and 53934 DF,  p-value: < 2.2e-16
```

So far, the only thing supplied to the chunk was the label, in this case "diamonds-model." It is best to avoid periods and spaces in chunk labels. Options can be passed to the chunk to control display and evaluation and are entered after the label, separated by commas. Some common **knitr** chunk options are listed in Table 27.2. These options can be strings, numbers, TRUE/FALSE or any R object that evaluates to one of these.

Displaying images is made incredibly easy with **knitr**. Simply running a command that generates a plot inserts the image immediately following that line of code, with further code and results printed after that.

Table 27.2 **Common knitr chunk options**

Option	Effect
eval	Results printed when TRUE.
echo	Code printed when TRUE.
include	When FALSE, code is evaluated but neither the code nor results are printed.
cache	If the code has not changed, the results will be available but not evaluated again in order to save compilation time.
fig.cap	Caption text for images. Images will automatically be put into a special figure environment and be given a label based on the chunk label.
fig.scap	The short version of the image caption to be used in the list of captions.
out.width	Width of displayed image.
fig.show	Controls when images are shown. 'as.is' prints them when they appear in code and 'hold' prints them all at the end.
dev	Type of image to be printed, such as .png, .jpg, etc.
engine	knitr can handle code in other languages like Python, Bash, Perl, C++ and SAS.
prompt	Specifies the prompt character put before lines of code. If FALSE, there will be no prompt.
comment	For easier reproducibility, result lines can be commented out.

The following chunk will print the following:

1. The expression `1 + 1`
2. The result of that expression, 2
3. The code `plot(1:10)`
4. The image resulting from that code
5. The expression `2 + 2`
6. The result of that expression, 4

```
<<inline-plot-knitr>>=

1 + 1
plot(1:10)
2 + 2

@
```

```
> 1 + 1

[1] 2

> plot(1:10)
```

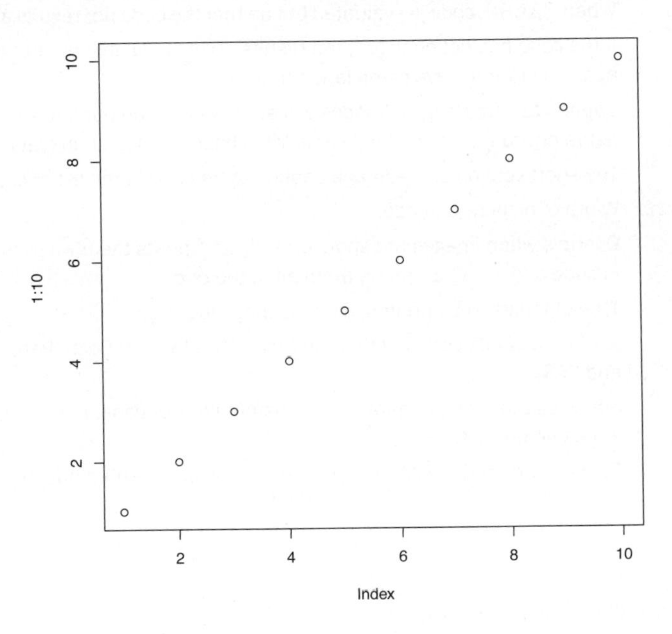


```
> 2 + 2

[1] 4
```

Adding the `fig.cap` option will put the image in a figure environment, which gets placed in a convenient spot with a caption. Running the same chunk with `fig.cap` set to `"Simple plot of the numbers 1 through 10."` will display the following:

1. The expression `1 + 1`
2. The result of that expression, 2
3. The code `plot(1:10)`
4. The expression `2 + 2`
5. The result of that expression, 4

The image, along with the caption, will be placed where there is room, which very well could be in between lines of code. Setting `out.width` to `'.75\\linewidth'` (including the quote marks) will make the image's width 75% of the width of the line. While `\linewidth` is a LaTeX command, because it is in an R string the backslash (\) needs to be escaped with another backslash. The resulting plot is shown in Figure 27.2.

```
<<figure-plot,fig.cap="Simple plot of the numbers 1 through 10.",
  fig.scap="Simple plot of the numbers 1 through 10",
  out.width='.75\\linewidth'>>=

1 + 1
plot(1:10)
2 + 2

@
```

```
> 1 + 1

[1] 2

> plot(1:10)
```

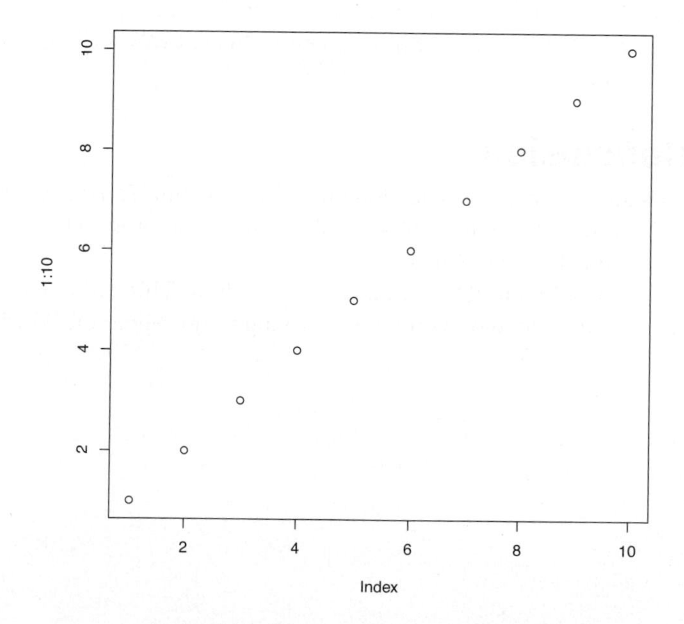

Figure 27.2 Simple plot of the numbers 1 through 10.

```
> 2 + 2

[1] 4
```

This just scratches the surface of what is possible with LaTeX and **knitr**. More information can be found on Yihui's site at `http://yihui.name/knitr/`. When using **knitr** it is considered good form to use a formal citation of the form `Yihui Xie (2013). knitr: A general-purpose package for dynamic report generation in R. R package version 1.2.` Proper citations can be found, for some packages, using the **citation** function.

```
> citation(package="knitr")


To cite the 'knitr' package in publications use:

  Yihui Xie (2016). knitr: A General-Purpose Package for
  Dynamic Report Generation in R. R package version 1.15.1.

  Yihui Xie (2015) Dynamic Documents with R and knitr. 2nd
  edition. Chapman and Hall/CRC. ISBN 978-1498716963

  Yihui Xie (2014) knitr: A Comprehensive Tool for
  Reproducible Research in R. In Victoria Stodden, Friedrich
  Leisch and Roger D. Peng, editors, Implementing Reproducible
  Computational Research. Chapman and Hall/CRC.
  ISBN 978-1466561595
```

27.4 Conclusion

Writing reproducible, and maintainable, documents from within R has never been easier, thanks to Yihui's **knitr** package. It enables seamless integration of R code, with results including images and either LaTeX text.

On top of that, the RStudio IDE is a fantastic text editor. This entire book was written using **knitr** from within RStudio, without ever having to use Microsoft Word or a LaTeX editor.

28

Rich Documents with RMarkdown

RMarkdown has quickly become the preferred means of communicating results from R, eclipsing LaTeX due to RMarkdown's simplicity to write and that it compiles to many different formats, including HTML, PDF, Microsoft Word, slideshows and Shiny apps. The RMarkdown format is extendable and templated, enabling customized documents. The workflow for writing RMarkdown is similar to that for LaTeX as detailed in Chapter 27: Normal text (flavored with Markdown) is written and R code is put in chunks. The style of the chunk delineation is different, but the idea is the same. The files are typically saved as `.Rmd` files.

28.1 Document Compilation

RMarkdown depends on **knitr** for evaluating R code and pandoc for converting between formats, both of which are installed with RMarkdown. Upon compilation **knitr** is used to process the code chunks into plain text in a temporary Markdown document. Pandoc is then run on this intermediate document to produce the output document.

Compiling the document in the console is done with the **render** function from the **rmarkdown** package and in RStudio with the Knit button or `Ctrl+Shift+K`.

Each output type corresponds to an R function. The most notable document types in the **rmarkdown** package are generated by **html_document**, **pdf_document**, **word_document** and **ioslides_presentation**. Other packages, such as **rticles**, **tufte** and **resumer** provide additional functions for other document types.

28.2 Document Header

The first part of an RMarkdown document is the `yaml`[1] header providing details about the document. The `yaml` header is delineated by three dashes before and after. Each

1. This was originally an acronym for "Yet another markup language" but has since become a recursive acronym for "YAML Ain't Markup Language."

line is a key-value pair specifying arguments for the document, such as `title`, `author`, `date` and `output` type. An example `yaml` header follows.

```
---
title: "Play Time"
author: "Jared P. Lander"
date: "December 22, 2016"
output: html_document
---
```

Different `yaml` tags are possible depending on the output type. One frustration of `yaml` in general is knowing what tags should be used, requiring users to scour documentation of varying quality. Thanks to the design of the **rmarkdown** package, `yaml` tags are the same as the function arguments that have to be documented in order to be accepted by CRAN. The tags for the document type, such as `html_document`, are the function names themselves. Functions that are included in the **rmarkdown** package can just be referenced by name. For functions in other packages, such as **jss_article** in **rticles**, the function name must be preceded by the package name as follows.

```
---
title: "Play Time"
author: "Jared P. Lander"
date: "December 22, 2016"
output: rticles::jss_article
---
```

Document specific arguments are included as sub-tags to the output tag. It is important to properly indent `yaml` sections, with at least two spaces (tabs will not work) indicating an indent. The following block specifies the generarion of an `HTML` document with numbered sections and a table of contents.

```
---
title: "Play Time"
author: "Jared P. Lander"
date: "December 22, 2016"
output:
  html_document:
    number_sections: yes
    toc: yes
---
```

28.3 Markdown Primer

Markdown syntax is designed for simplicity. It does not offer the control and flexibility of LaTeX or `HTML` but is much faster to write. In addition to being faster to write, Markdown is faster to learn, which is always nice. The following guidelines should be enough to get started with the basics of Markdown.

Line breaks are created by leaving a blank line between blocks of text or by ending a line with two or more spaces. Italics can be generated by putting an underscore (_) on both sides of a word, and bold is generated by putting two underscores on each side. Three underscores will make the text both italic and bold. Block quotes are indicated using right-angle brackets (>) at the start of each line.

Unordered lists are created by putting each element on its own line starting with a dash (-) or an asterisk (*). Ordered lists start each line with a number (any number or letter) and a period. Lists can be nested by indenting certain items in the list.

Headings (called headers in Markdown but not to be confused with HTML headers) are made by starting a line with a pound symbol (#). The number of pounds indicate the heading level, ranging from one to six. These are equivalent to the heading tags in `HTML`. When rendering to PDF the heading level dictates the section type such as a level-one heading for a section and a level-two heading for a subsection.

Links are created by putting the text to be displayed in square brackets ([ ]) and the linked URL in parentheses. Inserting images is also done with square brackets and parentheses and preceded by an exclamation mark (!). A sample Markdown document is shown next.

Equations are started and ended with two dollar signs ($). These should either be on their own lines or with at least a space between them and the equation. Equations are written using standard LaTeX math syntax. Inline equations are immediately preceded and followed with a single dollar sign with no spaces.

```
# Title - Also a Heading 1

_this will be italicized_

__this will be bolded__

## Heading 2

Build an unordered list

- Item 1
- Item 2
- Item 3

Build an ordered list mixed with an unordered list
```

(Continues)

(Continued)

```
1. An item
1. Another item
    - Sublist item
    - Another sublist item
    - one more item
1. Another ordered item

The following is a link

[My Website](http://www.jaredlander.com)

## Another Heading 2

This inserts an image

![Alt text goes in here](location-of-image.png)

#### Heading 4

A regular equation

$$
    \boldsymbol{\hat{\beta}} = (X^TX)^{-1}X^TY
$$

An inline equation: $\bar{x}=\frac{1}{n}\sum_{i=1}^n$ with no spaces

### Heading 3
> This is the start of a block quote
>
> This is the following line in the block quote
```

RStudio provides a handy quick reference guide to Markdown, accessed through the `Help` menu.

28.4 Markdown Code Chunks

Markdown documents that have chunks of R code are called RMarkdown and are saved with a `.Rmd` extension. Code chunks in RMarkdown behave similarly to chunks in **knitr** documents but are demarcated differently and have some added flexibility. The opening of a chunk is denoted with three back ticks (`), an opening curly brace ({), the letter `r`, a chunk label followed by comma-separated options, then a closing curly brace (}). The chunk is closed with three back ticks. All code and comments inside the chunk are treated as R code.

````
```{r simple-math-ex}

# this is a comment
1 + 1

```
````

This renders as the following code and result.

```
# this is a comment
1 + 1

## [1] 2
```

All of the standard chunk options can be used with the most common listed in Table 27.2.

Images are automatically generated when plotting code is executed in a code chunk. A caption is added if the `fig.cap` argument is supplied. A nice, default feature is that documents are self-contained so that even when the final output is `HTML` the images will be embedded (base 64 encoded) into the file so that only one file is needed, as opposed to separate files for each image.

The following chunk first displays the expression `1 + 1`, prints the result, `2`, displays the code `plot(1:10)` followed by the actual graph and finally displays `2 + 2` and the result, `4`.

````
```{r code-and-plot}

1 + 1
plot(1:10)
2 + 2

```
````

```
1 + 1

## [1] 2

plot(1:10)
```

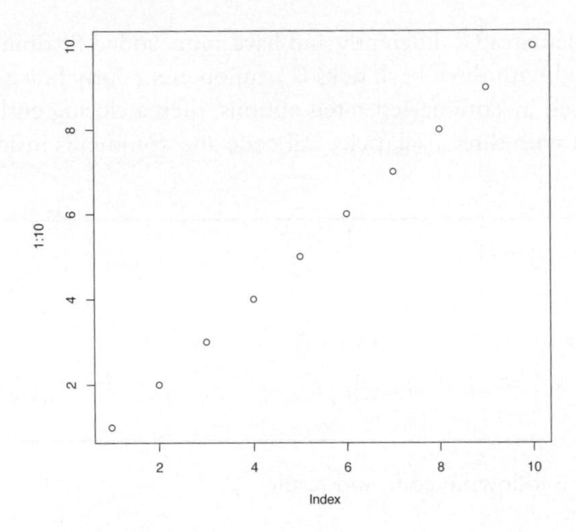


```
2 + 2

## [1] 4
```

When working with LaTeX documents **knitr** puts a prompt (>) in front of code, a plus sign (+) in front of continued lines and prints results directly. For RMarkdown documents, by default, there is no prompt in front of code and results are commented. This is because the output is usually intended for digital consumption, and this enables easy copy-and-pasting of code into the R console. This book includes the prompt and does not comment results so that the experience is similar to working with R, even at the expense of easy copy-and-pasting.

28.5 `htmlwidgets`

The **htmlwidgets** package enables the simple creation of R packages that provide R bindings for arbitrary JavaScript libraries. This provides R users access to a wide array of useful JavaScript libraries for visualizing data, all within R and without having to learn JavaScript. Popular packages include **DT** for tabular displays of data, **leaflet** for generating rich maps, **threejs** for 3D scatterplots and globes, **d3heatmap** for interactive heatmaps and **dygraphs** for time series charting.

These **htmlwidgets**-based packages generate `HTML` and JavaScript output, meaning they only have full functionality with `HTML`-based output types. Including the tag `always_allow_html: yes` in the `yaml` header will enable output types such as PDF to be rendered as an image. This depends on the **webshot** package and will only work after installing PhantomJS using `webshot::install_phantomjs`.

The functions in **htmlwidgets** packages can greatly enhance RMarkdown documents with their full power of JavaScript, particularly their interactivity. They can also be used in the console and in Shiny apps. If the functions are executed in the console within RStudio the resulting widget is displayed in the Viewer pane. If they are executed from the console in the terminal then a browser is launched displaying the widget. The widgets thrive in Shiny, where their interactivity is coupled with Shiny's reactivity for a rich user experience.

28.5.1 datatables

While visualizing data with graphs is usually the preferred method, sometimes data must be presented in tabular form. For static tables **kable** from the **knitr** package prints nice looking tables that are adapted to the type of output document. The following code generates Table 28.1 in PDF form. The design is similar, but aesthetically a little different, when the final document is `HTML`.

```
> knitr::kable(head(iris), caption='Tabular data printed using kable.')
```

Table 28.1 Tabular data printed using kable

Sepal.Length	Sepal.Width	Petal.Length	Petal.Width	Species
5.1	3.5	1.4	0.2	setosa
4.9	3.0	1.4	0.2	setosa
4.7	3.2	1.3	0.2	setosa
4.6	3.1	1.5	0.2	setosa
5.0	3.6	1.4	0.2	setosa
5.4	3.9	1.7	0.4	setosa

The **DT** package provides an interactive tabular experience through the `DataTables` JavaScript library. Since **DT** is based on **htmlwidgets**, its full interactivity is only experienced in `HTML`-based output, but screenshots of the resulting DataTable are automatically captured in PDF-based outputs. The following code generates Figure 28.1, showing a DataTable of the first 100 rows of the `diamonds` data.

```
> library(DT)
> data(diamonds, package='ggplot2')
> datatable(head(diamonds, 100))
```

Show 10 entries Search:

	carat	cut	color	clarity	depth	table	price	x	y	z
1	0.23	Ideal	E	SI2	61.5	55	326	3.95	3.98	2.43
2	0.21	Premium	E	SI1	59.8	61	326	3.89	3.84	2.31
3	0.23	Good	E	VS1	56.9	65	327	4.05	4.07	2.31
4	0.29	Premium	I	VS2	62.4	58	334	4.2	4.23	2.63
5	0.31	Good	J	SI2	63.3	58	335	4.34	4.35	2.75
6	0.24	Very Good	J	VVS2	62.8	57	336	3.94	3.96	2.48
7	0.24	Very Good	I	VVS1	62.3	57	336	3.95	3.98	2.47
8	0.26	Very Good	H	SI1	61.9	55	337	4.07	4.11	2.53
9	0.22	Fair	E	VS2	65.1	61	337	3.87	3.78	2.49
10	0.23	Very Good	H	VS1	59.4	61	338	4	4.05	2.39

Showing 1 to 10 of 100 entries Previous 1 2 3 4 5 … 10 Next

Figure 28.1 A JavaScript DataTable generated by the DT package.

The DataTable library has many extensions, plugins and options, most of which are implemented by the **DT** package. To make our table look nicer we turn off `rownames`; make each column searchable with the `filter` argument; enable the `Scroller` extension for better vertical scrolling; allow horizontal scrolling with `scrollX`; and set the displayed `dom` elements to be the table itself (`t`), table information (`i`) and the `Scroller` capability (`S`). Some of these are listed as arguments to the **datatable** function, and others are specified in a `list` provided to the `options` argument. This is seen in the following code which results in Figure 28.2. Deciphering what argument goes in which part of the function unfortunately requires scouring the **DT** documentation and vignettes and the `DataTables` documentation.

```
> datatable(head(diamonds, 100),
+           rownames=FALSE, extensions='Scroller', filter='top',
+           options = list(
+               dom = "tiS", scrollX=TRUE,
+               scrollY = 400,
+               scrollCollapse = TRUE
+           ))
```

carat	cut	color	clarity	depth	table	price	x	y	z
All	All	All	All	All	All	All	All	All	All
0.23	Ideal	E	SI2	61.5	55	326	3.95	3.98	2.43
0.21	Premium	E	SI1	59.8	61	326	3.89	3.84	2.31
0.23	Good	E	VS1	56.9	65	327	4.05	4.07	2.31
0.29	Premium	I	VS2	62.4	58	334	4.2	4.23	2.63
0.31	Good	J	SI2	63.3	58	335	4.34	4.35	2.75
0.24	Very Good	J	VVS2	62.8	57	336	3.94	3.96	2.48
0.24	Very Good	I	VVS1	62.3	57	336	3.95	3.98	2.47
0.26	Very Good	H	SI1	61.9	55	337	4.07	4.11	2.53
0.22	Fair	E	VS2	65.1	61	337	3.87	3.78	2.49
0.23	Very Good	H	VS1	59.4	61	338	4	4.05	2.39
0.3	Good	J	SI1	64	55	339	4.25	4.28	2.73

Showing 1 to 12 of 100 entries

Figure 28.2 A DataTable with numerous options set.

A `datatables` object can be passed, via a pipe, to formatting functions to customize the output. The following code builds a `datatables` object, formats the `price` column as currency rounded to the nearest whole number and color codes the rows depending on the value of the `cut` column. The result is seen in Figure 28.3.

```
> datatable(head(diamonds, 100),
+           rownames=FALSE, extensions='Scroller', filter='top',
+           options = list(
+               dom = "tiS", scrollX=TRUE,
+               scrollY = 400,
+               scrollCollapse = TRUE
+           )) %>%
+     formatCurrency('price', digits=0) %>%
+     formatStyle(columns='cut', valueColumns='cut', target='row',
+                 backgroundColor=styleEqual(levels=c('Good', 'Ideal'),
+                                            values=c('red', 'green'))
+                 )
+     )
```

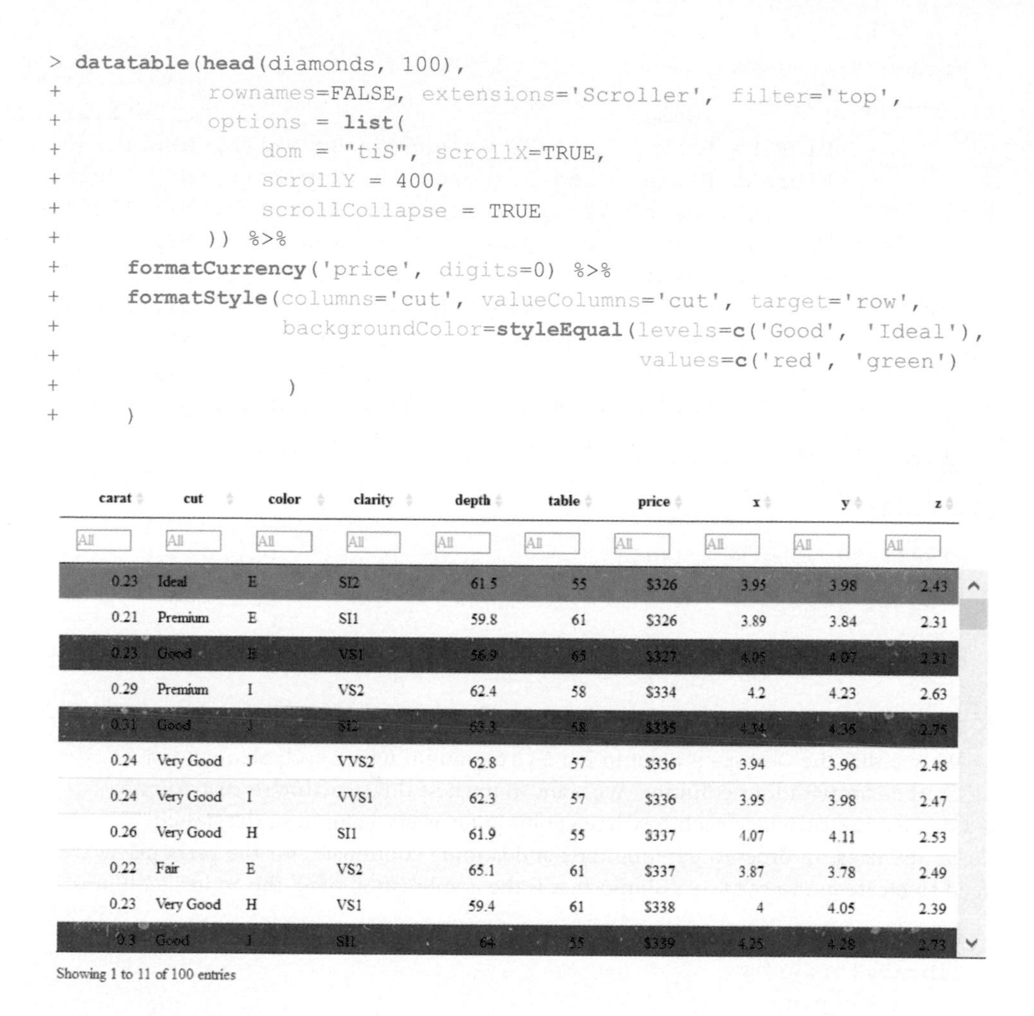

carat	cut	color	clarity	depth	table	price	x	y	z
All	All	All	All	All	All	All	All	All	All
0.23	Ideal	E	SI2	61.5	55	$326	3.95	3.98	2.43
0.21	Premium	E	SI1	59.8	61	$326	3.89	3.84	2.31
0.23	Good	E	VS1	56.9	65	$327	4.05	4.07	2.31
0.29	Premium	I	VS2	62.4	58	$334	4.2	4.23	2.63
0.31	Good	J	SI2	63.3	58	$335	4.34	4.35	2.75
0.24	Very Good	J	VVS2	62.8	57	$336	3.94	3.96	2.48
0.24	Very Good	I	VVS1	62.3	57	$336	3.95	3.98	2.47
0.26	Very Good	H	SI1	61.9	55	$337	4.07	4.11	2.53
0.22	Fair	E	VS2	65.1	61	$337	3.87	3.78	2.49
0.23	Very Good	H	VS1	59.4	61	$338	4	4.05	2.39
0.3	Good	J	SI1	64	55	$339	4.25	4.28	2.73

Showing 1 to 11 of 100 entries

Figure 28.3 A DataTable with numerous options and formatting.

28.5.2 leaflet

As shown in Figure 25.7, R can make detailed and attractive static maps. This capability has been extended to interactive maps thanks to the **leaflet** package. This package creates maps based on the OpenStreetMap (or other map provider) that are scrollable and zoomable. It can also use shapefiles, GeoJSON, TopoJSON and raster images to build up the map. To see this in action we plot our favorite pizza places on a map.

First we read the JSON file holding the list of favorite pizza places.

```
> library(jsonlite)
> pizza <- fromJSON('http://www.jaredlander.com/data/PizzaFavorites.json')
```

```
> pizza
                      Name                                      Details
1            Di Fara Pizza     1424 Avenue J, Brooklyn, NY, 11230
2            Fiore's Pizza   165 Bleecker St, New York, NY, 10012
3               Juliana's  19 Old Fulton St, Brooklyn, NY, 11201
4      Keste Pizza & Vino   271 Bleecker St, New York, NY, 10014
5   L & B Spumoni Gardens      2725 86th St, Brooklyn, NY, 11223
6 New York Pizza Suprema       413 8th Ave, New York, NY, 10001
7             Paulie Gee's 60 Greenpoint Ave, Brooklyn, NY, 11222
8                  Ribalta       48 E 12th St, New York, NY, 10003
9               Totonno's  1524 Neptune Ave, Brooklyn, NY, 11224

> class(pizza$Details)

[1] "list"

> class(pizza$Details[[1]])

[1] "data.frame"

> dim(pizza$Details[[1]])

[1] 1 4
```

We see that the `Details` column is a `list`-column where each element is a `data.frame` with four columns. We want to un-nest this structure so that `pizza` is a `data.frame` where each row has a column for every column in the nested `data.frames`. In order to get longitude and latitude coordinates for the pizza places we need to create a `character` column that is the combination of all the address columns.

```
> library(dplyr)
> library(tidyr)
> pizza <- pizza %>%
+     # unnest the data.frame
+     unnest() %>%
+     # Rename the Address column Street
+     rename(Street=Address) %>%
+     # create new column to hod entire address
+     unite(col=Address,
+           Street, City, State, Zip,
+           sep=', ', remove=FALSE)
> pizza

                      Name                                      Address
1            Di Fara Pizza     1424 Avenue J, Brooklyn, NY, 11230
2            Fiore's Pizza   165 Bleecker St, New York, NY, 10012
3               Juliana's  19 Old Fulton St, Brooklyn, NY, 11201
4      Keste Pizza & Vino   271 Bleecker St, New York, NY, 10014
5   L & B Spumoni Gardens      2725 86th St, Brooklyn, NY, 11223
```

```
6 New York Pizza Suprema       413 8th Ave, New York, NY, 10001
7           Paulie Gee's 60 Greenpoint Ave, Brooklyn, NY, 11222
8                Ribalta      48 E 12th St, New York, NY, 10003
9             Totonno's  1524 Neptune Ave, Brooklyn, NY, 11224
            Street     City State   Zip
1     1424 Avenue J Brooklyn    NY 11230
2   165 Bleecker St New York    NY 10012
3  19 Old Fulton St Brooklyn    NY 11201
4   271 Bleecker St New York    NY 10014
5      2725 86th St Brooklyn    NY 11223
6       413 8th Ave New York    NY 10001
7 60 Greenpoint Ave Brooklyn    NY 11222
8      48 E 12th St New York    NY 10003
9  1524 Neptune Ave Brooklyn    NY 11224
```

The **RDSTK** provides the **street2coordinates** function to geocode addresses. We build a helper function to geocode an address and extract just the `latitude` and `longitude` columns.

```
> getCoords <- function(address)
+ {
+     RDSTK::street2coordinates(address) %>%
+         dplyr::select_('latitude', 'longitude')
+ }
```

We then apply this function to each address and bind the results back to the `pizza` `data.frame`.

```
> library(dplyr)
> library(purrr)
> pizza <- bind_cols(pizza, pizza$Address %>% map_df(getCoords))
> pizza

                    Name                                 Address
1          Di Fara Pizza     1424 Avenue J, Brooklyn, NY, 11230
2          Fiore's Pizza   165 Bleecker St, New York, NY, 10012
3              Juliana's  19 Old Fulton St, Brooklyn, NY, 11201
4     Keste Pizza & Vino   271 Bleecker St, New York, NY, 10014
5  L & B Spumoni Gardens      2725 86th St, Brooklyn, NY, 11223
6 New York Pizza Suprema       413 8th Ave, New York, NY, 10001
7           Paulie Gee's 60 Greenpoint Ave, Brooklyn, NY, 11222
8                Ribalta      48 E 12th St, New York, NY, 10003
9              Totonno's  1524 Neptune Ave, Brooklyn, NY, 11224
            Street     City State   Zip latitude longitude
1     1424 Avenue J Brooklyn    NY 11230 40.62503 -73.96214
2   165 Bleecker St New York    NY 10012 40.72875 -74.00005
3  19 Old Fulton St Brooklyn    NY 11201 40.70282 -73.99418
```

```
4    271 Bleecker St New York     NY 10014 40.73147 -74.00314
5       2725 86th St Brooklyn     NY 11223 40.59431 -73.98152
6        413 8th Ave New York     NY 10001 40.75010 -73.99515
7 60 Greenpoint Ave Brooklyn      NY 11222 40.72993 -73.95823
8      48 E 12th St New York      NY 10003 40.73344 -73.99177
9  1524 Neptune Ave Brooklyn      NY 11224 40.57906 -73.98327
```

Now that we have data with coordinates we can build a map with markers showing our points of interest. The **leaflet** function initializes the map. Running just that renders a blank map. Passing that object, via pipe, into **addTiles** draws a map, based on OpenStreetMap tiles, at minimum zoom and centered on the Prime Meridian since we did not provide any data. Passing that to the **addMarkers** function adds markers at the specified longitude and latitude of our favorite pizza places. The columns holding the information are specified using the formula interface. Clicking on the markers reveals a popup displaying the name and street address of a pizza place. In an `HTML`-based document this map can be zoomed and dragged just like any other interactive map. In a PDF document it appears as an image as in Figure 28.4.

```
> library(leaflet)
> leaflet() %>%
+ addTiles() %>%
+ addMarkers(lng=~longitude, lat=~latitude,
+            popup=~sprintf('%s<br/>%s', Name, Street),
+            data=pizza
+     )
```

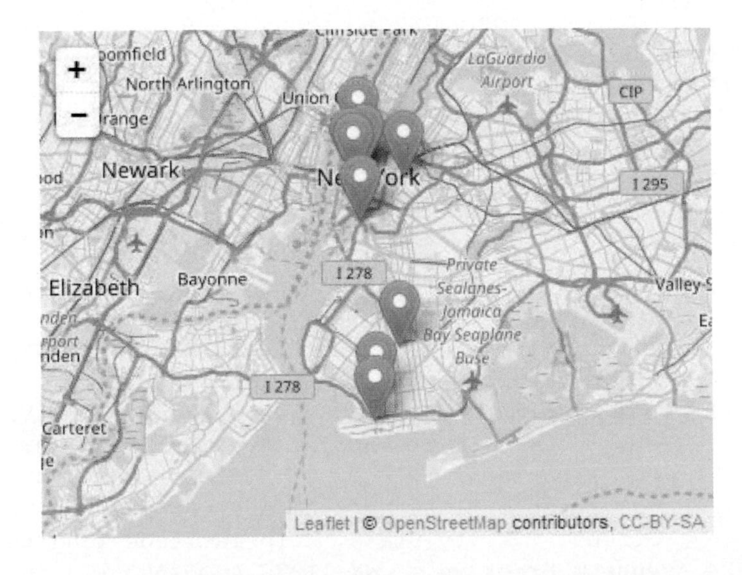

Figure 28.4 A `leaflet` map of pizza places in New York.

28.5.3 dygraphs

Plotting time series can be done with **ggplot2**, **quantmod** and many other packages, but **dygraphs** creates interactive plots. To illustrate, we look at the GDP data from the World Bank as seen in Section 24.1, this time with fewer countries and starting with 1970. We use the **WDI** package to access data through the World Bank's API.

```
> library(WDI)
> gdp <- WDI(country=c("US", "CA", "SG", "IL"),
+            indicator=c("NY.GDP.PCAP.CD"),
+            start=1970, end=2011)
> # give it good names
> names(gdp) <- c("iso2c", "Country", "PerCapGDP", "Year")
```

This gives us GDP data in the long format. We convert it to wide format using **spread** from the **tidyr** package.

```
> head(gdp, 15)

   iso2c Country PerCapGDP Year
1     CA  Canada  4047.268 1970
2     CA  Canada  4503.181 1971
3     CA  Canada  5048.482 1972
4     CA  Canada  5764.261 1973
5     CA  Canada  6915.889 1974
6     CA  Canada  7354.268 1975
7     CA  Canada  8624.614 1976
8     CA  Canada  8731.679 1977
9     CA  Canada  8931.293 1978
10    CA  Canada  9831.079 1979
11    CA  Canada 10933.732 1980
12    CA  Canada 12075.025 1981
13    CA  Canada 12217.373 1982
14    CA  Canada 13113.169 1983
15    CA  Canada 13506.372 1984

> gdpWide <- gdp %>%
+     dplyr::select(Country, Year, PerCapGDP) %>%
+     tidyr::spread(key=Country, value=PerCapGDP)
>
> head(gdpWide)

  Year   Canada   Israel Singapore United States
1 1970 4047.268 1806.423  925.0584      4997.757
2 1971 4503.181 1815.936 1070.7664      5360.178
3 1972 5048.482 2278.840 1263.8942      5836.224
4 1973 5764.261 2819.451 1684.3411      6461.736
5 1974 6915.889 3721.525 2339.3890      6948.198
6 1975 7354.268 3570.763 2488.3415      7516.680
```

With the time element in the first column and each time series represented as a single column, we use **dygraphs** to make an interactive JavaScript plot, shown in Figure 28.5.

```
> library(dygraphs)
> dygraph(gdpWide, main='Yearly Per Capita GDP',
+         xlab='Year', ylab='Per Capita GDP') %>%
+     dyOptions(drawPoints = TRUE, pointSize = 1) %>%
+     dyLegend(width=400)
```

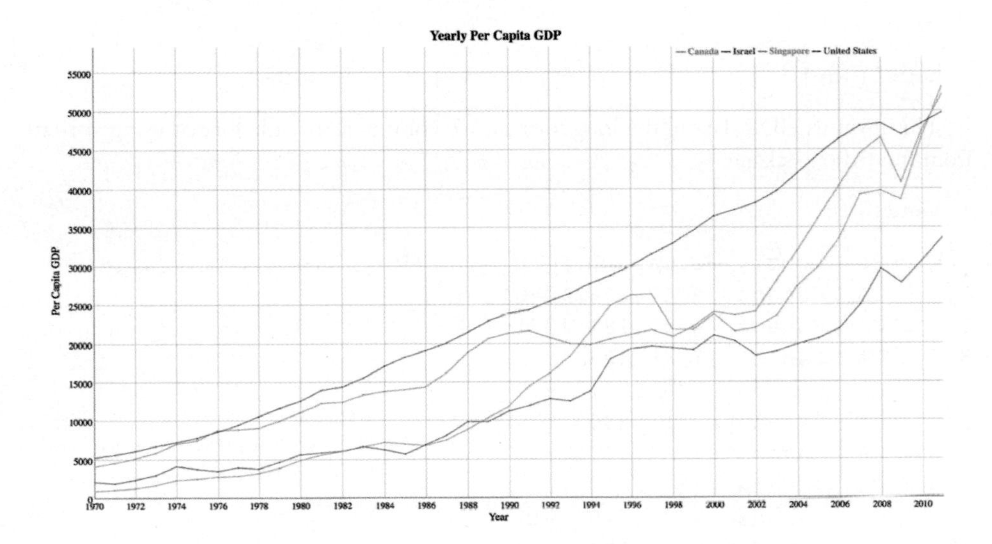

Figure 28.5 Interactive time series graph of per capita GDP.

Hovering over lines of the graph will highlight synchronized points on each line and display the values in the legend. Drawing a rectangle in the graph will zoom into the data. We add a range selection that can be dragged to show different part of the graph with **dyRangeSelector** as shown in Figure 28.6.

```
> dygraph(gdpWide, main='Yearly Per Capita GDP',
+         xlab='Year', ylab='Per Capita GDP') %>%
+     dyOptions(drawPoints = TRUE, pointSize = 1) %>%
+     dyLegend(width=400) %>%
+     dyRangeSelector(dateWindow=c("1990", "2000"))
```

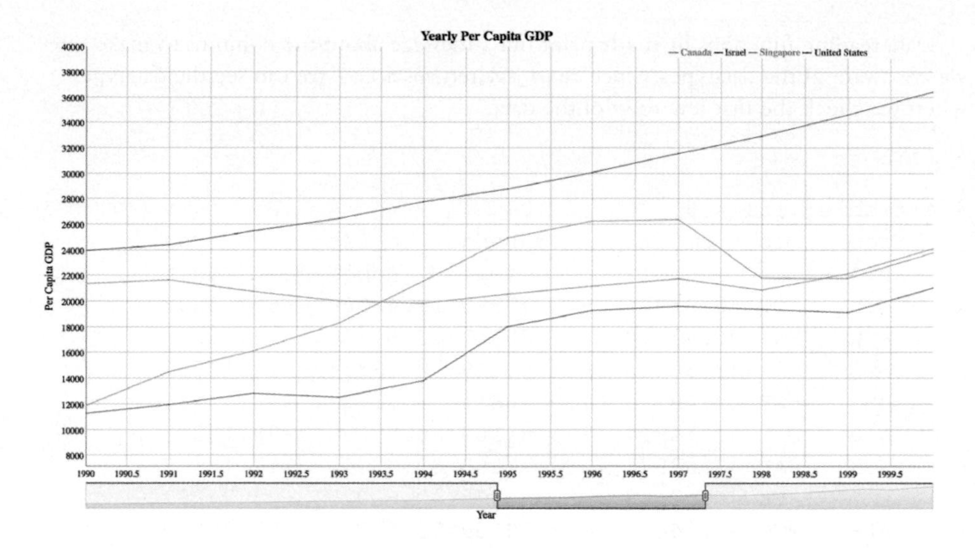

Figure 28.6 Interactive time series graph of per capita GDP with range selector.

28.5.4 threejs

The **threejs**, by Bryan Lewis, has functions for building 3D scatterplots and globes that can be spun around to view different angles. To see this we draw arcs between origin and destination cities of flights that were in the air in the afternoon of January 2, 2017. The dataset contains the airport codes and coordinates of the airports on both ends of the route.

```
> library(readr)
> flights <- read_tsv('http://www.jaredlander.com/data/Flights_Jan_2.tsv')

Parsed with column specification:
  cols(
    From = col_character(),
    To = col_character(),
    From_Lat = col_double(),
    From_Long = col_double(),
    To_Lat = col_double(),
    To_Long = col_double()
  )
```

Data reading functions in **readr** print out a message about the columns to make sure we are aware of the datatypes. Since **read_tsv** returns a `tbl` we can see the datatypes when we check the first few rows of the data.

```
> flights

# A tibble: 151 × 6
    From    To From_Lat  From_Long   To_Lat    To_Long
   <chr> <chr>    <dbl>      <dbl>    <dbl>      <dbl>
1    JFK   SDQ 40.63975  -73.77893 18.42966  -69.66893
2    RSW   EWR 26.53617  -81.75517 40.69250  -74.16867
3    BOS   SAN 42.36435  -71.00518 32.73356 -117.18967
4    RNO   LGB 39.49911 -119.76811 33.81772 -118.15161
5    ALB   FLL 42.74827  -73.80169 26.07258  -80.15275
6    JFK   SAN 40.63975  -73.77893 32.73356 -117.18967
7    FLL   JFK 26.07258  -80.15275 40.63975  -73.77893
8    ALB   MCO 42.74827  -73.80169 28.42939  -81.30899
9    LAX   JFK 33.94254 -118.40807 40.63975  -73.77893
10   SJU   BDL 18.43942  -66.00183 41.93889  -72.68322
# ... with 141 more rows
```

The dataset is already in proper form to draw arcs between destinations and origins. It is also prepared to plot points for the airports, but airports are in the dataset multiple times, so the plot will simply overlay the points. It will be more useful to have counts for the number of times an airport appears so that we can draw one point with a height determined by the number of flights originating from each airport.

```
> airports <- flights %>%
+     count(From_Lat, From_Long) %>%
+     arrange(desc(n))
> airports

Source: local data frame [49 x 3]
Groups: From_Lat [49]

   From_Lat  From_Long     n
      <dbl>      <dbl> <int>
1  40.63975  -73.77893    25
2  26.07258  -80.15275    16
3  42.36435  -71.00518    15
4  28.42939  -81.30899    11
5  18.43942  -66.00183     7
6  40.69250  -74.16867     5
7  26.53617  -81.75517     4
8  26.68316  -80.09559     4
9  33.94254 -118.40807     4
10 12.50139  -70.01522     3
# ... with 39 more rows
```

The first argument to **globejs** is the image to use as a surface map for the globe. The default image is nice, but NASA has a high-resolution "blue marble" image we use.

```
> earth <- "http://eoimages.gsfc.nasa.gov/images/imagerecords/
            73000/73909/world.topo.bathy.200412.3x5400x2700.jpg"
```

Now that the data are prepared and we have a nice image for the surface map, we can draw the globe. The first argument, `img`, is the image to use, which we saved to the `earth` object. The next two arguments, `lat` and `long`, are the coordinates of points to draw. The `value` argument controls how tall to draw the points. The `arcs` argument takes a four-column `data.frame` where the first two columns are the origin latitude and longitude and the second two columns are the destination latitude and longitude. The rest of the arguments customize the look and feel of the globe. The following code generates Figure 28.7.

```
> library(threejs)
> globejs(img=earth, lat=airports$From_Lat, long=airports$From_Long,
+         value=airports$n*5, color='red',
+         arcs=flights %>%
+             dplyr::select(From_Lat, From_Long, To_Lat, To_Long),
+         arcsHeight=.4, arcsLwd=4, arcsColor="#3e4ca2", arcsOpacity=.85,
+         atmosphere=TRUE, fov=30, rotationlat=.5, rotationlong=-.05)
```

Figure 28.7 Globe, drawn with `threejs`, showing flight paths.

28.5.5 d3heatmap

Heatmaps display the intensity of numeric data and are particularly helpful with correlation matrices. We revisit the `economics` data from Section 18.2 to build an interactive heatmap. We first build a correlation matrix of the `numeric` columns, then call **d3heatmap**, written by Tal Galili, which builds the heatmap and clusters the variables, displaying a dendrogram for the clustering. The result is seen in Figure 28.8. Hovering over individual cells shows more information about the data, and dragging a box zooms in on the plot.

```
> library(d3heatmap)
> data(economics, package='ggplot2')
> econCor <- economics %>% select_if(is.numeric) %>% cor
> d3heatmap(econCor, xaxis_font_size='12pt', yaxis_font_size='12pt',
+           width=600, height=600)
```

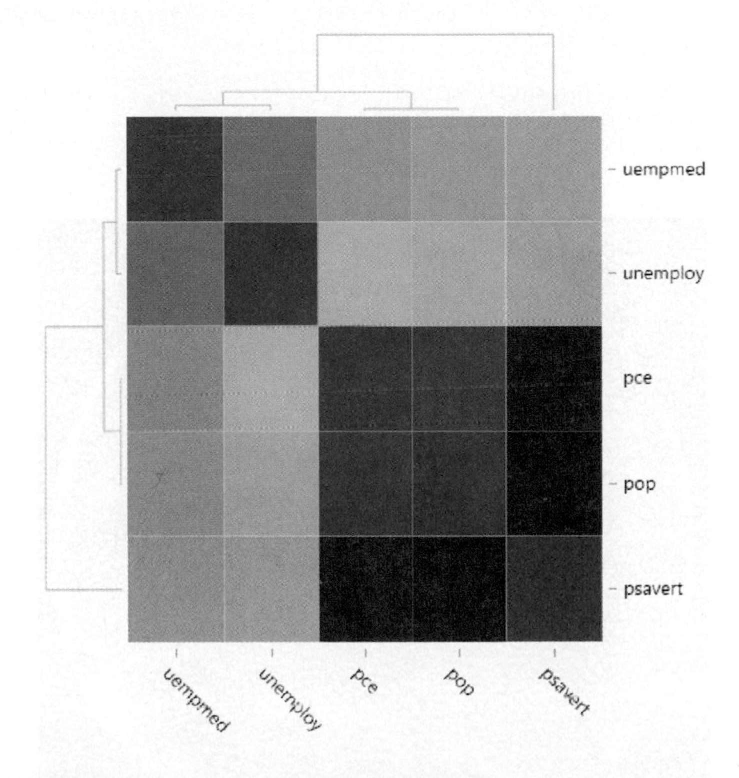

Figure 28.8 Correlation heatmap of the economics data built with `d3heatmap`.

28.6 RMarkdown Slideshows

Creating reproducible presentations without leaving the friendly confines of the R environment has long been possible using LaTeX's Beamer mode, which creates a PDF where each page is a slide. However, writing all that LaTeX code can be unnecessarily time consuming. A simpler option is to write an RMarkdown document and compile it into an HTML5 slideshow such as ioslides or revealjs. RMarkdown also has built-in support for generating a Beamer slideshow, thus avoiding LaTeX while achieving the same output.

Setting the yaml output tag to **ioslides_presentation** will make the document render as an ioslides presentation. The function for rendering revealjs slideshows is in the **revealjs** package, not the **rmarkdown** package, so the output tag is set to revealjs::revealjs_presentation. A similar pattern works for other output types.

Slides are indicated by the level-two header (##)[2]. Text written on the same line as the slide indicator is printed as the slide title. The class and ID of the slide can be provided, using standard CSS notation, inside curly braces after the (optional) title text. For example, {.vcenter .flexbox #SlideID} sets the class of a slide to both vcenter and flexbox and sets the ID to SlideID.

Aside from these caveats, and a few others, regular Markdown should be used. Code for an example slideshow follows.

````
---
title: "Slide Test"
author: "Jared P. Lander"
date: "December 22, 2016"
output: ioslides_presentation
---

## First Slide
A list of things to cover
- First Item
- Second Item
- Third Item

## Some R Code
The code below will generate some results and a plot.

```{r figure-plot,fig.cap="Simple plot of the numbers 1 through
10.", fig.scap="Simple plot of the numbers 1 through 10",
out.width='50%', fig.show='hold'}
1 + 1
plot(1:10)
2 + 2
```
````

(Continues)

2. Like many aspects of RMarkdown this is customizable.

(Continued)

```
## Another Slide
Some more information goes here


## Some Links
[My Website](http://www.jaredlander.com)


[R Bloggers](http://www.r-bloggers.com)
```

28.7 Conclusion

RMarkdown has revolutionized writing documents that interweave text with R code and results. It is faster and easier to write than LaTeX and enables a wide range of document types to be created with a simple change of one `yaml` tag. RMarkdown documents and presentations are a great way to share code and workflow, deliver scientific findings and present compelling results.

29

Interactive Dashboards with Shiny

Displaying data and analysis is an important part of the data science process. R has long had great visualization capabilities thanks to built in graphics and **ggplot2**. With Shiny from RStudio, we can now build dashboards, all with R code. There are many other dashboard tools that integrate with R such as SiSense, PowerBI, Qlik and Tableau, but they are limited by the types of data and objects that can be passed back and forth to R. Shiny is built with R natively, so the dashboard can be backed by any data munging, modelling, processing and visualization that can be done in R.

Shiny enables R programmers to develop Web-based dashboards without having to learn `HTML` and `JavaScript`, though knowing those tools helps. Using Shiny can be simple but the code takes some getting used to and can feel awkward at first. The most important thing to think of at first is inputs and outputs: Users provide inputs through a UI and the app produces an output that is sent back to the UI. To see this we first explore building Shiny documents using RMarkdown and then look at building it the more traditional way by separating out the UI components and the back end server components.

Note that due to the limitations of the paper, ePub and PDF formats of this book, the results presented in this chapter may appear different than those rendered on a computer, though the ideas will be the same.

29.1 Shiny in RMarkdown

The simplest way to build a Shiny app is by using an RMarkdown document, which is covered in Chapter 28. Code chunks are created the same way as with RMarkdown, but the results of the chunks are now interactive.

Just like with a regular RMarkdown document the top of the document contains a `yaml` header providing details about the document. The `yaml` header is immediately preceded and followed by three dashes. At a minimum the header should contain the title of the document and a tag indicating that the runtime is Shiny. Other recommended tags are for the author, output and date.

```
---
title: "Simple Shiny Document"
author: "Jared P. Lander"
date: "November 3, 2016"
output: html_document
runtime: shiny
---
```

That last tag, `runtime: shiny`, tells the RMarkdown processor that this document should be rendered as in interactive Shiny HTML document rather than an ordinary HTML document. If nothing else is added to the document and we build it, we simply see the title, author and date as in Figure 29.1.

Simple Shiny Document

Jared P. Lander

November 3, 2016

Figure 29.1 Shiny document resulting from only specifying header information.

Building the document can be done in the console with the **run** function from the **rmarkdown** package.

```
> rmarkdown::run('ShinyDocument.Rmd')
```

Inside RStudio the document can be built by clicking the `Run Document` button as seen in Figure 29.2 or by typing `ctrl+shift+K`. Using the RStudio button or shortcut keys runs the Shiny app in a separate process so the console can still be used.

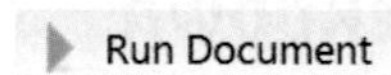

Figure 29.2 RStudio Run Shiny Document Button.

The first UI element we add is a dropdown selector. This is placed in a code chunk, just like any other RMarkdown code chunk. The code in these chunks can be either displayed or hidden with the `echo` chunk argument. The **selectInput** function builds an HTML select object. The first argument, `inputId`, specifies the HTML ID for the object and will be used by other functions in the Shiny app to access information in the select object. The `label` argument specifies what is displayed to the user. Possible selection options are listed, in a `list` (optionally named) or `vector`. The `list` names represent what the user sees and the values are the actual value that will be selected. The result of the following code chunk is seen in Figure 29.3.

````
```{r build-selector,echo=FALSE}

selectInput(inputId='ExampleDropDown',
            label='Please make a selection',
            choices=list('Value 1'=1,
                         'Value 2'=2,
                         'Value 3'=3))

```
````

Please make a selection

Value 1

Value 1
Value 2
Value 3

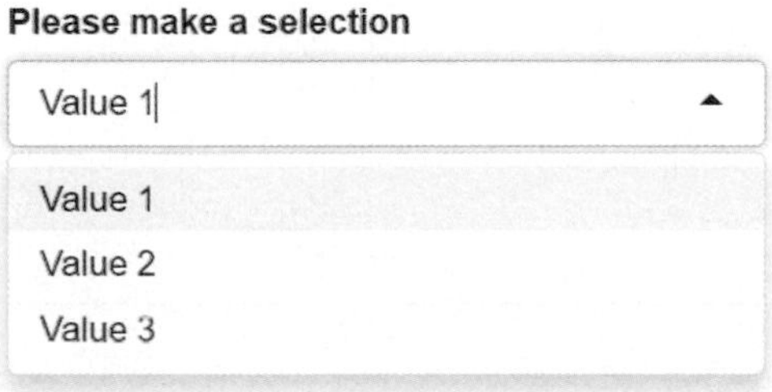

Figure 29.3 Shiny dropdown select.

Including **selectInput** in an RMarkdown document merely generates HTML code as can be seen by running the command in the console.

```
> selectInput(inputId='ExampleDropDown', label='Please make a selection',
+             choices=list('Value 1'=1,
+                          'Value 2'=2,
+                          'Value 3'=3))

<div class="form-group shiny-input-container">
    <label class="control-label" for="ExampleDropDown">
        Please make a selection
    </label>
    <div>
        <select id="ExampleDropDown">
            <option value="1" selected>Value 1</option>
            <option value="2">Value 2</option>
            <option value="3">Value 3</option>
        </select>
        <script type="application/json"
        data-for="ExampleDropDown"
        data-nonempty="">{}
        </script>
    </div>
</div>
```

Now that we have the capability to control an input we should do something with the value selected. As a simple example we simply print out the choice with **renderPrint**. The primary argument to **renderPrint** is the item to be printed. This could be a simple string of text, but that would defeat the purpose of using Shiny. It is really useful when used to print the result of an expression or an input. To print the input we identified as `ExampleDropDown`, we need to access it first. All the inputs are stored in a `list` called `input`.[1] The names of the individual inputs in this `list` are the `inputId`s supplied in the input functions, such as **selectInput** and are accessed with the `$` operator just like ordinary `list`s.

The following chunk creates both the dropdown select and prints the selected value as seen in Figure 29.4. Changing the selection alters what is printed on the fly.

```
```{r select-print-drop-down,echo=FALSE}

selectInput(inputId='ExampleDropDown', label='Please make a selection',
            choices=list('Value 1'=1,
                         'Value 2'=2,
                         'Value 3'=3))

renderPrint(input$ExampleDropDown)

```
```

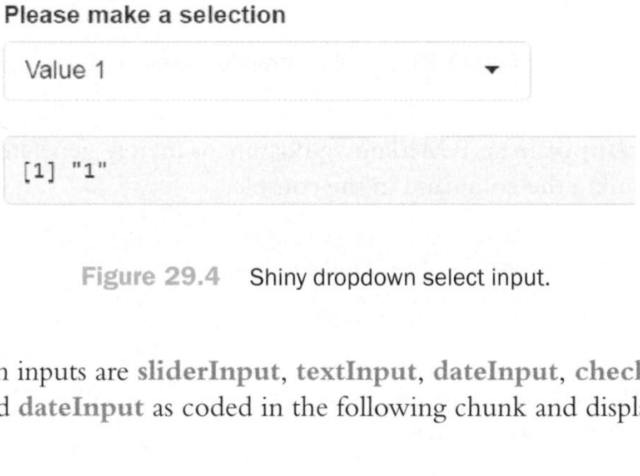

Please make a selection

Value 1

[1] "1"

Figure 29.4 Shiny dropdown select input.

Other common inputs are **sliderInput**, **textInput**, **dateInput**, **checkboxInput**, **radioButtons** and **dateInput** as coded in the following chunk and displayed in Figure 29.5.

```
```{r common-inputs,echo=FALSE}

sliderInput(inputId='SliderSample', label='This is a slider',
            min=0, max=10, value=5)
textInput(inputId='TextSample', label='Space to enter text')
```

(Continues)

1. When building a full Shiny app the name of this input `list` can be changed, though this is not standard practice.
```

(Continued)

```
checkboxInput(inputId='CheckSample', label='Single check box')
checkboxGroupInput(inputId='CheckGroupSample',
                   label='Multiple check boxes',
                   choices=list('A', 'B', 'C'))
radioButtons(inputId='RadioSample', label='Radio button',
             choices=list('A', 'B', 'C'))
dateInput(inputId='DateChoice', label='Date Selector')

```
```

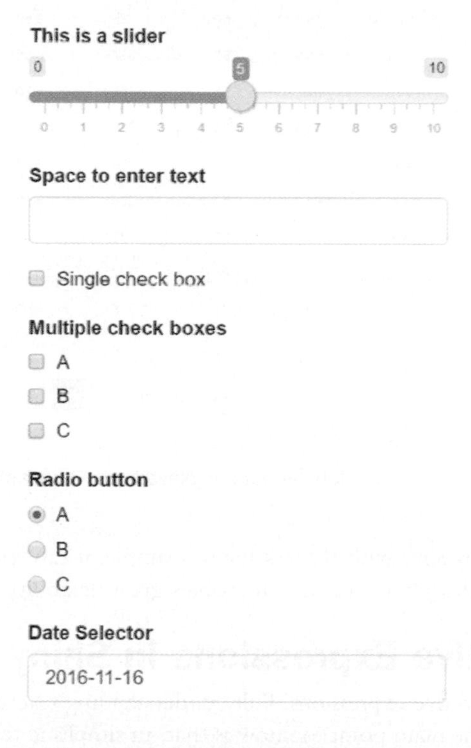

Figure 29.5 Common Shiny inputs.

All of the values set through these inputs can be used in other R code by accessing their corresponding elements in the `input list` and displayed with the appropriate render functions such as **renderPrint**, **renderText**, **renderDataTable** and **renderPlot**.

For instance, we can render data with **renderDataTable**, which uses the DataTables `JavaScript` library, via the **htmlwidgets** package, to display the data. The following code chunk results in Figure 29.6.
```

```
```{r shiny-datatable-diamonds,echo=FALSE}

data(diamonds, package='ggplot2')
renderDataTable(diamonds)

```
```

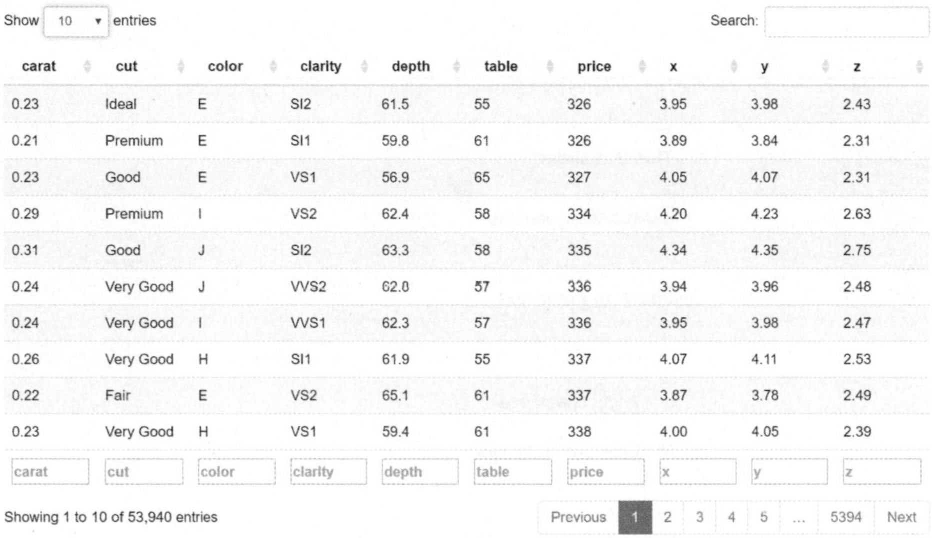

Show 10 entries Search:

carat	cut	color	clarity	depth	table	price	x	y	z
0.23	Ideal	E	SI2	61.5	55	326	3.95	3.98	2.43
0.21	Premium	E	SI1	59.8	61	326	3.89	3.84	2.31
0.23	Good	E	VS1	56.9	65	327	4.05	4.07	2.31
0.29	Premium	I	VS2	62.4	58	334	4.20	4.23	2.63
0.31	Good	J	SI2	63.3	58	335	4.34	4.35	2.75
0.24	Very Good	J	VVS2	62.8	57	336	3.94	3.96	2.48
0.24	Very Good	I	VVS1	62.3	57	336	3.95	3.98	2.47
0.26	Very Good	H	SI1	61.9	55	337	4.07	4.11	2.53
0.22	Fair	E	VS2	65.1	61	337	3.87	3.78	2.49
0.23	Very Good	H	VS1	59.4	61	338	4.00	4.05	2.39
carat	cut	color	clarity	depth	table	price	x	y	z

Showing 1 to 10 of 53,940 entries Previous 1 2 3 4 5 ... 5394 Next

Figure 29.6 Tabular data displayed with DataTables.

While building Shiny apps with RMarkdown is simple, it can result in complex layouts, especially when using flexdashboard, which enables great flexibility.

29.2 Reactive Expressions in Shiny

Shiny is powered by reactive expressions. Fully understanding reactive expressions can take quite some time, but the main point to know is that, in simplistic terms, they listen for and react to changes in variables. For our purposes we consider user inputs and programmatic outputs such as rendered text and plots.

The elements of the `input list` are by their very nature reactive. This is seen by calling **renderText** with the value set in a **textInput** as its first argument. As the input changes so does the printed output as seen in the following code chunk and Figure 29.7.

```
```{r text-input-output,echo=FALSE}

textInput(inputId='TextInput', label='Enter Text')
```

(Continues)
```

(Continued)

```
renderText(input$TextInput)
```
```

Enter Text

This text entry is displayed below

This text entry is displayed below

Figure 29.7 Shiny text input and rendered text output.

Using the element of the `input list` is the easiest way to make use of a reactive expression. Sometimes, however, we need to store an element of `input` in a variable and act on it later. Trying to code this as regular R code results in the error seen in Figure 29.8.

````
```{r render-date,echo=FALSE}

library(lubridate)

dateInput(inputId='DateChoice', label='Choose a date')

theDate <- input$DateChoice

renderText(sprintf('%s %s, %s',
                   month(theDate, label=TRUE, abbr=FALSE),
                   day(theDate),
                   year(theDate)))

```
````

**Error:** Operation not allowed without an active reactive context. (You tried to do something that can only be done from inside a reactive expression or observer.)

Figure 29.8 Error resulting from improper use of reactive expressions.

This error occurred because we saved a reactive expression, `input$DateChoice` to a static variable, `theDate`, and then tried using that static variable in a reactive context, **renderText**. To fix this, we pass `input$DateChoice` into **reactive** and save it to
````

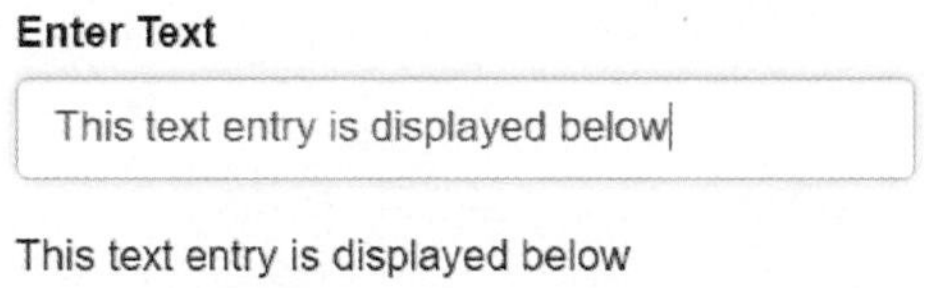

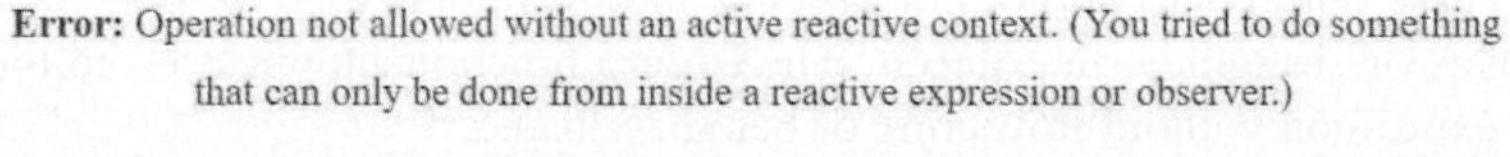

theDate. This makes theDate a reactive expression, allowing it to change as inputs change. To access the content in theDate we treat it like a function and call it with trailing parentheses as in the following chunk with the results in Figure 29.9. If the first argument to **reactive** is multiple lines of code, they should all be enclosed inside curly braces ({ and }).

```
```{r render-date-reactive,echo=FALSE}

library(lubridate)

dateInput(inputId='DateChoice', label='Choose a date')

theDate <- reactive(input$DateChoice)

renderText(sprintf('%s %s, %s',
                   month(theDate(), label=TRUE, abbr=FALSE),
                   day(theDate()),
                   year(theDate())))
```
```

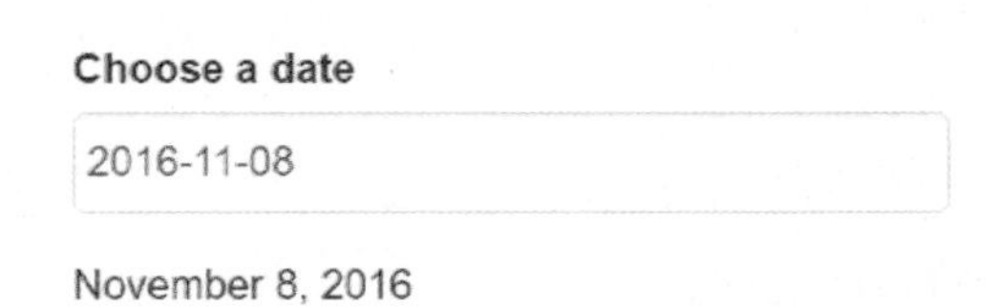

Figure 29.9 Using reactive expressions to store adaptable variables.

Reactive expressions form the backbone of Shiny and enable complex interactivity. Using them properly takes some time to get acquainted with as it is quite different from traditional R programming. In addition to **reactive**, **observe** and **isolate** also help with reactive programming. Objects created by **reactive** are used for their values, which update given any change in inputs. Objects created by **observe** only update when specifically invoked. These objects also do not hold a result and so are only used for side effects such as creating plots or changing other objects. The **isolate** function allows access to the value in a reactive expression without it reacting or being evaluated.

29.3 Server and UI

So far we have used a single RMarkdown document to build a simple Shiny app. However, the more robust method is to define discrete UI and server components, where the UI piece controls what the user sees in the browser and the server piece controls calculations and interactions. The most traditional way to build an app is to have a directory for the app and within that directory there is a `ui.r` file and a `server.r` file.

Before we create a UI file we write a server in the `server.r` file. At first it will do nothing, only existing so that we can run the app and see the bare bones UI. At a minimum, the server file needs to instate a server with the **shinyServer** function. The sole argument to **shinyServer** is a function with at at least two arguments—`input` and `output`—and an optional third argument—`session`.[2] The function can be defined inline, as is done here, or as its own piece of code. The `input` argument is the same `input` as in Section 29.1. Code in the **shinyServer** function will access input elements through this `list`. The `output` argument is a `list` that stores rendered R objects that can be accessed by the UI. For the most part, the `session` argument can be ignored and is most useful when working with modules. This blank **shinyServer** function has no effect other than enabling the app to be built and ran.

```
library(shiny)

shinyServer(function(input, output, session)
{

})
```

Then we look at a the UI setup in the `ui.r` file. There are numerous ways to lay out a Shiny app, and `shinydashboard` is easy to code yet results in an attractive app. The **shinydashboard** package provides the **dashboardPage** function which holds all the pieces together. The main components of a dashboard are the header, sidebar, and body as illustrated in Figure 29.10.

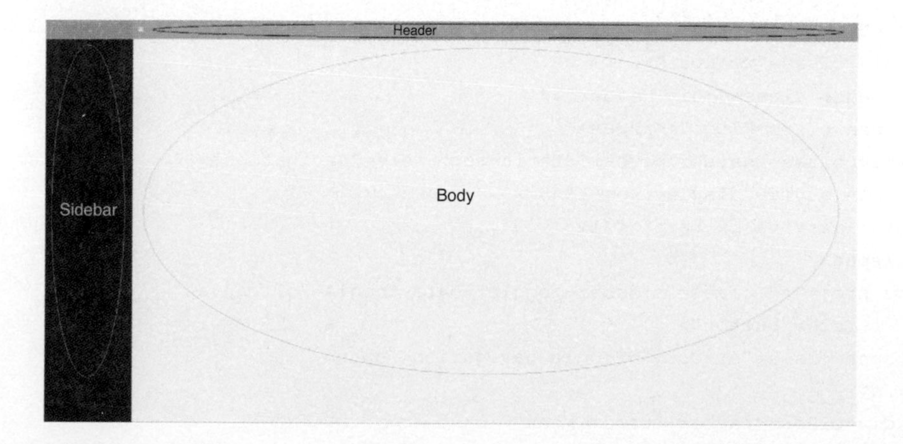

Figure 29.10 Locations of the header, sidebar and body in shinydashboard.

2. These arguments can have other names, but there is not much good reason for using different names.

The app in Figure 29.10 was generated by putting an empty header, sidebar and body as arguments to **dashboardPage**. The `title` argument specifies the title displayed in the browser taskbar.

```
library(shiny)
library(shinydashboard)

dashboardPage(
    header=dashboardHeader(),
    sidebar=dashboardSidebar(),
    body=dashboardBody(),
    title='Example Dashboard'
)
```

Like all UI objects, these functions only generate `HTML` code, which can be seen by running the code in the console.

```
> library(shiny)
> library(shinydashboard)
>
> dashboardPage(
+      header=dashboardHeader(),
+      sidebar=dashboardSidebar(),
+      body=dashboardBody(),
+      title='Example Dashboard'
+ )
```

```
<body class="skin-blue" style="min-height: 611px;">
<div class="wrapper">
<header class="main-header">
<span class="logo"></span>
<nav class="navbar navbar-static-top" role="navigation">
<span style="display:none;">
<i class="fa fa-bars"></i>
</span>
<a href="#" class="sidebar-toggle" data-toggle="offcanvas"
    role="button">
<span class="sr-only">Toggle navigation</span>
</a>
<div class="navbar-custom-menu">
<ul class="nav navbar-nav"></ul>
</div>
</nav>
</header>
```

(*Continues*)

(*Continued*)

```
<aside class="main-sidebar">
<section class="sidebar"></section>
</aside>
<div class="content-wrapper">
<section class="content"></section>
</div>
</div>
</body>
```

As we add more objects to the app, the UI code can get quite complicated, so rather than put all the code for the header, sidebar and body in one place, we save each to an object and then use that object inside the **dashboardPage** call.

We keep the header simple, providing just the name of the dashboard that is to be displayed. We store it in the `dashHeader` object, which we will supply to the `header` argument of **dashboardPage**.

```
> dashHeader <- dashboardHeader(title='Simple Dashboard')
```

The sidebar can be used for many things, and a common use is for navigation. We create clickable links using **menuItems** inside of a **sidebarMenu**. These **menuItems** point to **tabItems** in the body. We create a **sidebarMenu** with two **menuItems**, where one points to the home tab and the other points to a tab demonstrating some graphs. Each **menuItem** function takes at minimum two arguments, `text` and `tabName`, which specify the text to display and the linked tab, respectively. An optional argument, `icon`, specifies a pictograph to be displayed to the right of the text. The pictographs are generated using the **icon** function, which can draw from Font Awesome[3] and Glyphicons.[4]

```
> dashSidebar <- dashboardSidebar(
+     sidebarMenu(
+         menuItem('Home',
+                  tabName='HomeTab',
+                  icon=icon('dashboard')
+         ),
+         menuItem('Graphs',
+                  tabName='GraphsTab',
+                  icon=icon('bar-chart-o')
+         )
+     )
+ )
```

Putting this code together generates the dashboard seen in Figure 29.11.

3. http://fontawesome.io/icons/
4. http://getbootstrap.com/components/#glyphicons

```
library(shiny)
library(shinydashboard)

dashHeader <- dashboardHeader(title='Simple Dashboard')

dashSidebar <- dashboardSidebar(
    sidebarMenu(
        menuItem('Home',
                 tabName='HomeTab',
                 icon=icon('dashboard')
        ),
        menuItem('Graphs',
                 tabName='GraphsTab',
                 icon=icon('bar-chart-o')
        )
    )
)

dashboardPage(
    header=dashHeader,
    sidebar=dashSidebar,
    body=dashboardBody(),
    title='Example Dashboard'
)
```

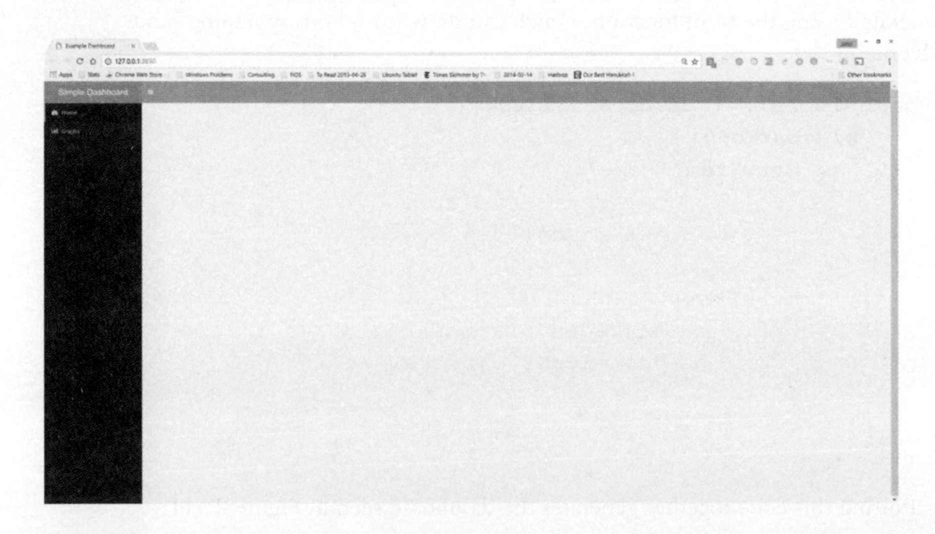

Figure 29.11 Shiny dashboard with a simple header and a sidebar with links to Home and Graphs tabs.

The links in the sidebar point to tabs in the body. These tabs are built using the **tabItem** function, which are inputs to the **tabItems** function. Each **tabItem** has a `tabName` that is pointed to by the corresponding **menuItem** function. After the `tabName`, other UI objects are specified. For example, we use **h1** to create first-level header text, **p** for a paragraph and **em** for emphasized text.

A great number of `HTML` tags are available in Shiny, through the **htmltools** package. They can be displayed with the following code.

```
> names(htmltools::tags)

  [1] "a"           "abbr"        "address"     "area"
  [5] "article"     "aside"       "audio"       "b"
  [9] "base"        "bdi"         "bdo"         "blockquote"
 [13] "body"        "br"          "button"      "canvas"
 [17] "caption"     "cite"        "code"        "col"
 [21] "colgroup"    "command"     "data"        "datalist"
 [25] "dd"          "del"         "details"     "dfn"
 [29] "div"         "dl"          "dt"          "em"
 [33] "embed"       "eventsource" "fieldset"    "figcaption"
 [37] "figure"      "footer"      "form"        "h1"
 [41] "h2"          "h3"          "h4"          "h5"
 [45] "h6"          "head"        "header"      "hgroup"
 [49] "hr"          "html"        "i"           "iframe"
 [53] "img"         "input"       "ins"         "kbd"
 [57] "keygen"      "label"       "legend"      "li"
 [61] "link"        "mark"        "map"         "menu"
 [65] "meta"        "meter"       "nav"         "noscript"
 [69] "object"      "ol"          "optgroup"    "option"
 [73] "output"      "p"           "param"       "pre"
 [77] "progress"    "q"           "ruby"        "rp"
 [81] "rt"          "s"           "samp"        "script"
 [85] "section"     "select"      "small"       "source"
 [89] "span"        "strong"      "style"       "sub"
 [93] "summary"     "sup"         "table"       "tbody"
 [97] "td"          "textarea"    "tfoot"       "th"
[101] "thead"       "time"        "title"       "tr"
[105] "track"       "u"           "ul"          "var"
[109] "video"       "wbr"
```

On the Graphs page we build a dropdown selector. We hard code it with the some names of columns from the `diamonds` data. This could be done programmatically, but for now hard coding will suffice. We also use **plotOutput** to indicate where a plot will be displayed. It will be blank until we build the plot on the server side.

```
> dashBody <- dashboardBody(
+     tabItems(
+         tabItem(tabName='HomeTab',
```

```
+                   h1('Landing Page!'),
+                   p('This is the landing page for the dashboard.'),
+                   em('This text is emphasized')
+               ),
+               tabItem(tabName='GraphsTab',
+                   h1('Graphs!'),

+                   selectInput(inputId='VarToPlot',
+                               label='Choose a Variable',
+                               choices=c('carat', 'depth',
+                                         'table', 'price'),
+                               selected='price'),
+                   plotOutput(outputId='HistPlot')
+               )
+       )
+ )
```

Notice that in UI functions individual items are separated by commas. This is because they are all arguments to functions, which may be arguments to other functions. This creates deeply nested code, which is why we break them into discrete portions and save them to objects that are then inserted into the appropriate functions.

All of this UI code forms the two pages of the dashboard as seen in Figures 29.12 and 29.13.

```
library(shiny)
library(shinydashboard)

dashHeader <- dashboardHeader(title='Simple Dashboard')

dashSidebar <- dashboardSidebar(
    sidebarMenu(
        menuItem('Home', tabName='HomeTab',
                 icon=icon('dashboard')
        ),
        menuItem('Graphs', tabName='GraphsTab',
                 icon=icon('bar-chart-o')
        )
    )
)

dashBody <- dashboardBody(
    tabItems(
        tabItem(tabName='HomeTab',
                h1('Landing Page!'),
                p('This is the landing page for the dashboard.'),
```

(*Continues*)

(Continued)

```
                em('This text is emphasized')
            ),
            tabItem(tabName='GraphsTab',
                h1('Graphs!'),
                selectInput(inputId='VarToPlot',
                            label='Choose a Variable',
                            choices=c('carat', 'depth',
                                      'table', 'price'),
                            selected='price'),
                plotOutput(outputId='HistPlot')
            )
        )
)

dashboardPage(
    header=dashHeader,
    sidebar=dashSidebar,
    body=dashBody,
    title='Example Dashboard'
)
```

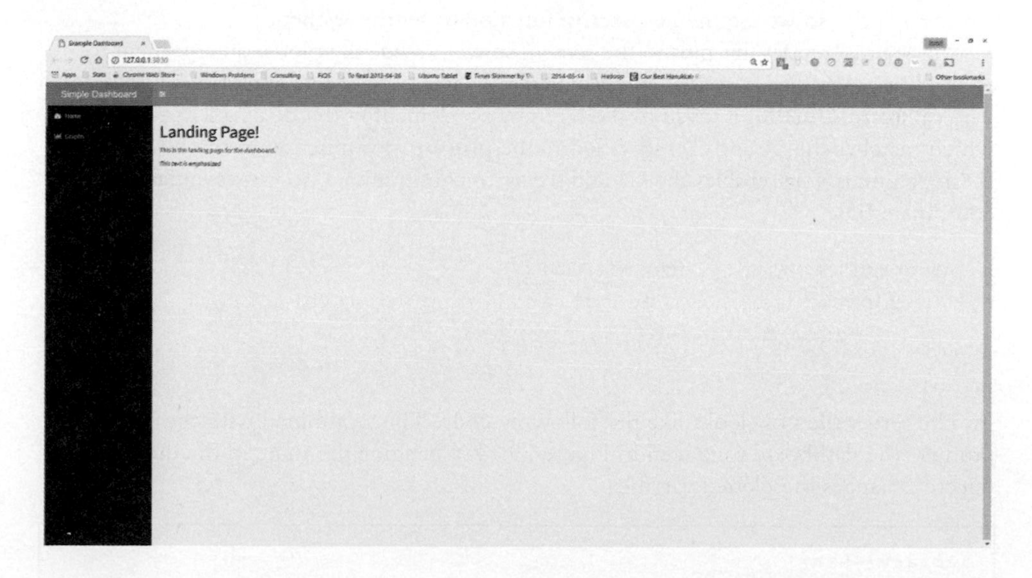

Figure 29.12 Shiny dashboard home page.

Figure 29.13 Shiny dashboard graphs page.

Now that we have a placeholder for a graph and a dropdown selector to choose a variable to plot, we need to create the plot on the server side. The plot itself is built using standard **ggplot2** code. The variable to plot is specified using `input$VarToPlot`, which is set by the dropdown selector with ID `"VarToPlot"` and contains the selected value as a `character`, so we use the **aes_string** function to set the aesthetic.

We want to render this plot to the screen, so we provide the plot as an argument to **renderPlot**. We wrap the code in curly braces ({ and }) to enable a multiline expression. The call to **renderPlot** is saved to the `Histplot` element of the `output list`, which matches the `outputID` specified in the **plotOutput** function in the UI. This ID matching is what enables the UI and server to communicate, so it is essential to coordinate IDs.

```
> output$HistPlot <- renderPlot({
+     ggplot(diamonds, aes_string(x=input$VarToPlot)) +
+         geom_histogram(bins=30)
+     })
```

The server file now looks like the following code. This, combined with the UI file, produce the dashboard page seen in Figure 29.14. Changing the value in the dropdown selector changes the plotted variable.

```
library(shiny)
library(ggplot2)
data(diamonds, package='ggplot2')
```

(Continues)

(Continued)

```
shinyServer(function(input, output, session)
{
    output$HistPlot <- renderPlot({
        ggplot(diamonds, aes_string(x=input$VarToPlot)) +
            geom_histogram(bins=30)
    })
})
```

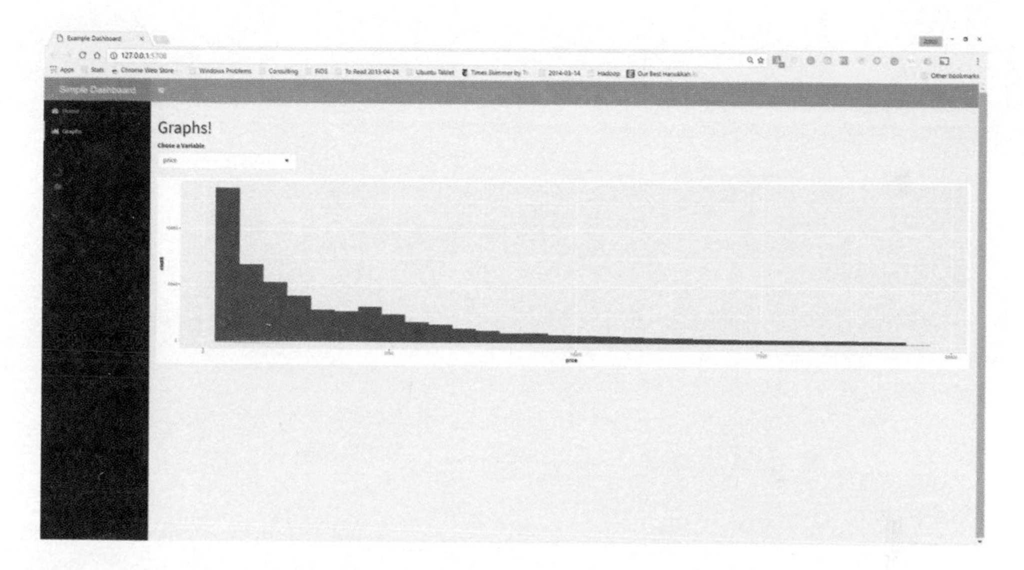

Figure 29.14 Shiny dashboard graphs page with histogram based on the dropdown selection.

A frequent point of confusion when building Shiny apps is where to separate objects with commas. Items input into the UI functions are function arguments, so they need to be separated by commas. Objects built in the server are normal expressions inside a function, so they are coded like regular R code and do not have comma separation.

29.4 Conclusion

Shiny is a powerful tool for building Web-based dashboards, all with R code. At first glance, the best part of Shiny is that everything can be done in R preventing the need to learn new tools, but that is only half the power of Shiny. Since everything is written in R, the dashboards can make use of the entire R ecosystem and compute statistics and models not possible in most dashboard tools. This capability brings machine learning, data science and even AI to accessible dashboards that everyone can understand. This is a powerful advance in data visualization and presentation.

We have just scratched the surface of Shiny apps, and there is a lot more to learn, including in-depth understanding of reactivity, modules, how to separate code into multiple files and much more. Shiny is constantly evolving and growing, but the most important thing to understand is the relationship between inputs and outputs and between the UI and server.

30

Building R Packages

As of early February 2017, there were over 10,000 packages on CRAN and nearly 1,300 more on Bioconductor, with more being added daily. In the past, building a package had the potential to be mystifying and complicated but that is no longer the case, especially when using Hadley Wickham's **devtools** package.

All packages submitted to CRAN (or Bioconductor) must follow specific guidelines, including the folder structure of the package, inclusion of `DESCRIPTION` and `NAMESPACE` files and proper help files.

30.1 Folder Structure

An R package is essentially a folder of folders, each containing specific files. At the very minimum there must be two folders, one called `R` where the included functions go, and the other called `man` where the documentation files are placed. It used to be that the documentation had to be be written manually, but thanks to **roxygen2** that is no longer necessary, as is seen in Section 30.3. Starting with R 3.0.0, CRAN is very strict in requiring that all files must end with a blank line and that code examples must be shorter than 105 characters.

In addition to the `R` and `man` folders, other common folders are `src` for compiled code such as C++ and FORTRAN, `data` for data that is included in the package and `inst` for files that should be available to the end user. No files from the other folders are available in a human-readable form (except the `INDEX`, `LICENSE` and `NEWS` files in the root folder) when a package is installed. Table 30.1 lists the most common folders used in an R package.

30.2 Package Files

The root folder of the package must contain at least a `DESCRIPTION` file and a `NAMESPACE` file, which are described in Sections 30.2.1 and 30.2.2. Other files like `NEWS`, `LICENSE` and `README` are recommended but not necessary. Table 30.2 lists commonly used files.

Table 30.1 **Folders used in R packages. While there are other possible folders, these are the most common.**

Folder	Explanation
R	Files containing R code. Filenames must end in .R, .S, .q, .r, or .s as an extension, with .r being the most common.
man	Documentation files ending in .Rd, one for each function in the R folder. These can be generated automatically using roxygen2.
src	Compiled code such as C/C++/FORTRAN.
data	Data included in the package.
inst	Files to be included in the installed package for the end user.
tests	Code that tests the functions in the R folder.

Table 30.2 **Files used in R packages. While there are other possible files these are the most common.**

File	Explanation
DESCRIPTION	Package information including dependencies.
NAMESPACE	List of functions exposed to end user and functions imported from other packages.
NEWS	What has been updated in each version.
LICENSE	Copyright information.
README	Basic description of package.

30.2.1 DESCRIPTION File

The DESCRIPTION file contains information about the package such as its name, version, author and other packages it depends on. The information is entered, each on one line, as Item1: Value1. Table 30.3 lists a number of fields that are used in DESCRIPTION files.

The Package field specifies the name of the package. This is the name that appears on CRAN and how users access the package.

Type is a bit archaic; it can be either Package or one other type, Frontend, which is used for building a graphical front end to R and will not be helpful for building an R package of functions.

Title is a short description of the package. It should actually be relatively brief and cannot end in a period. Description is a complete description of the package, which can be several sentences long but no longer than a paragraph.

Version is the package version and usually consists of three period-separated integers; for example, 1.15.2. Date is the release date of the current version.

The Author and Maintainer fields are similar but both are necessary. Author can be multiple people, separated by commas, and Maintainer is the person in charge, or

Table 30.3 **Fields in the DESCRIPTION file**

Field	Required	Explanation
Package	Yes	Name of package
Type	No	Just use Package
Title	Yes	Short description of package
Version	Yes	Current version: v.s.ss
Date	No	Latest build date
Author	Yes	Name of author
Maintainer	Yes	Author name and email address
Description	Yes	Complete description of package
License	Yes	License type
Depends	No	Comma separated list of packages to be loaded
Imports	No	Comma separated list of packages to use but not load
Suggests	No	Comma separated list of packages that are nice to have
Collate	No	List (no commas) of R files in the R directory in processing order
ByteCompile	No	If the package should be byte-compiled on installation

rather the person who gets complained to, and should be a name followed by an email address inside angle brackets (<>). An example is Maintainer: Jared P. Lander <packages@jaredlander.com>. CRAN is actually very strict about the Maintainer field and can reject a package for not having the proper format.

License information goes in the appropriately named License field. It should be an abbreviation of one of the standard specifications such as GPL-2 or BSD and the string 'file LICENSE' referring to the LICENSE file in the package's root folder.

Things get tricky with the Depends, Imports and Suggests fields. Often a package requires functions from other packages. In that case the other package, for example, **ggplot2**, should be listed in either the Depends or Imports field as a comma-separated list. If **ggplot2** is listed in Depends, then when the package is loaded so will **ggplot2**, and its functions will be available to functions in the package and to the end user. If **ggplot2** is listed in Imports, then when the package is loaded **ggplot2** will not be loaded, and its functions will be available to functions in the package but not the end user. Packages should be listed in one or the other, not both. Packages listed in either of these fields will be automatically installed from CRAN when the package is installed. If the package depends on a specific version of another package, then that package name should be followed by the version number in parentheses; for example, Depends: ggplot2 (>= 0.9.1). Packages that are needed for the examples in the documentation, vignettes or testing but are not necessary for the package's functionality should be listed in Suggests.

The Collate field specifies the R code files contained in the R folder.

A relatively new feature is byte-compilation, which can significantly speed up R code. Setting `ByteCompile` to `TRUE` will ensure the package is byte-compiled when installed by the end user.

The `DESCRIPTION` file from **coefplot** is shown next.

```
Package: coefplot
Type: Package
Title: Plots Coefficients from Fitted Models
Version: 1.2.4
Date: 2016-01-09
Author: Jared P. Lander
Maintainer: Jared P. Lander <packages@jaredlander.com>
Description: Plots the coefficients from model objects.  This very
        quickly shows the user the point estimates and
        confidence intervals for fitted models.
License: BSD_3_clause + file LICENSE
LazyLoad: yes
Depends:
    ggplot2 (>= 2.0.0)
Imports:
    plyr,
    reshape2,
    useful,
    stats,
    dplyr
Enhances:
    glmnet,
    maxLik,
    mfx
ByteCompile: TRUE
Packaged: 2016-01-09 05:16:05 UTC; Jared
Suggests:
    testthat,
    sandwich, lattice, nnet
RoxygenNote: 5.0.1
ByteCompile: TRUE
```

30.2.2 NAMESPACE File

The `NAMESPACE` file specifies which functions are exposed to the end user (not all functions in a package should be) and which other packages are imported into the `NAMESPACE`. Functions that are exported are listed as `export(multiplot)` and imported packages are listed as `import(plyr)`. Building this file by hand can be quite tedious, so fortunately **roxygen2** and **devtools** can, and should, build this file automatically.

R has three object-oriented systems: `S3`, `S4` and `Reference Classes`. `S3` is the oldest and simplest of the systems and is what we will focus on in this book. It consists of a number of generic functions such as **print**, **summary**, **coef** and **coefplot**. The generic functions exist only to dispatch object-specific functions. Typing **print** into the console shows this.

```
> print

standardGeneric for "print" defined from package "base"

function (x, ...)
standardGeneric("print")
<environment: 0x00000000095e1cb8>
Methods may be defined for arguments: x
Use  showMethods("print")  for currently available ones.
```

It is a single-line function containing the command `UseMethod("print")` that tells R to call another function depending on the class of the object passed. These can be seen with `methods(print)`. To save space we show only 20 of the results. Functions not exposed to the end user are marked with an asterisk (*). All of the names are `print` and the object class separated by a period.

```
> head(methods(print), n=20)

 [1] "print,ANY-method"            "print,bayesglm-method"
 [3] "print,bayespolr-method"      "print,diagonalMatrix-method"
 [5] "print,modelMatrix-method"    "print,sparseMatrix-method"
 [7] "print.aareg"                 "print.abbrev"
 [9] "print.acf"                   "print.AES"
[11] "print.agnes"                 "print.anova"
[13] "print.Anova"                 "print.anova.gam"
[15] "print.anova.lme"             "print.anova.loglm"
[17] "print.Anova.mlm"             "print.anova.rq"
[19] "print.aov"                   "print.aovlist"
```

When **print** is called on an object, it then calls one of these functions depending on the type of object. For instance, a `data.frame` is sent to **print.data.frame** and an `lm` object is sent to **print.lm**.

These different object-specific functions that get called by generic `S3` functions must be declared in the `NAMESPACE` in addition to the functions that are exported. This is indicated as `S3Method(coefplot, lm)` to say that **coefplot.lm** is registered with the **coefplot** generic function.

The `NAMESPACE` file from **coefplot** is shown next.

```
# Generated by roxygen2: do not edit by hand

S3method(buildModelCI,default)
S3method(coefplot,data.frame)
S3method(coefplot,default)
S3method(coefplot,glm)
S3method(coefplot,lm)
S3method(coefplot,logitmfx)
S3method(coefplot,rxGlm)
S3method(coefplot,rxLinMod)
S3method(coefplot,rxLogit)
S3method(extract.coef,maxLik)
export(buildModelCI)
export(buildModelCI.default)
export(coefplot)
export(coefplot.data.frame)
export(coefplot.default)
export(coefplot.glm)
export(coefplot.lm)
export(coefplot.logitmfx)
export(coefplot.rxGlm)
export(coefplot.rxLinMod)
export(coefplot.rxLogit)
export(extract.coef)
export(extract.coef.maxLik)
export(invlogit)
export(multiplot)
export(plotcoef)
export(position_dodgev)
import(ggplot2)
import(plyr)
import(reshape2)
import(useful)
```

Even with a small package like **coefplot**, building the `NAMESPACE` file by hand can be tedious and error prone, so it is best to let **devtools** and **roxygen2** build it.

30.2.3 Other Package Files

The `NEWS` file, which can be either plain text or Markdown, is for detailing what is new or changed in each version. The four most recent entries in the **coefplot** `NEWS` file are shown next. Notice how it is good practice to thank people who helped with or inspired the update. This file will be available to the end user's installation.

```
# Version 1.2.4
Patched to accommodate changes to ggplot2.

# Version 1.2.3
Can run coefplot on a data.frame that is properly setup like on
resulting from coefplot(..., plot=FALSE).

# Version 1.2.2
Support for glmnet models. Added tests.

# Version 1.2.1
In mulitplot there is now an option to reverse the order of the
legend so it matches the ordering in the plot.
```

The LICENSE file is for specifying more detailed information about the package's license and will be available to the end user's installation. CRAN has a strict policy about what can be written inside the license file. It must be excatly three lines with the first indicating the years of the copyright, the second specifies the copyright holder and the third lists the organization. The LICENSE file from **coefplot** is shown here.

```
YEAR: 2011-2017
COPYRIGHT HOLDER: Jared Lander
ORGANIZATION: Lander Analytics
```

The README file is purely informational and is not included in the end user's installation. Its biggest benefit may be for packages hosted on GitHub, where the README will be the information displayed on the project's home page. It is possible—even advisable—to write the README in RMarkdown and render it to Markdown before pushing to GitHub. This way code examples, and results, can be included in the online README.

It is best to create the RMarkdown README using the **use_readme_rmd** function from **devtools**, which will not only create the README.Rmd file but also create a Git hook requiring that if the README.Rmd has been modified, then the README.md Markdown file must be built.

The README file for **coefplot** is shown in the following listing. While it does not have code examples, it shows the general structure of the file and the necessary yaml header. It also includes badges to indicate its build, CRAN and test statuses.

```
---
output:
  md_document:
```

(Continues)

(Continued)

```
  variant: markdown_github
---

[![Travis-CI Build Status](https://travis-ci.org/jaredlander/
        coefplot.svg?branch=master)](https://travis-ci.org/
        jaredlander/coefplot)
[![CRAN_Status_Badge](http://www.r-pkg.org/badges/version/coefplot)](
        http://cran.r-project.org/package=coefplot)
[![Downloads from the RStudio CRAN mirror](http://cranlogs.r-pkg.org/
        badges/coefplot)](http://cran.rstudio.com/package=coefplot)

<!-- README.md is generated from README.Rmd. Please edit that file -->

```{r, echo = FALSE}
knitr::opts_chunk$set(
  collapse = TRUE,
  comment = "#>",
  fig.path = "README-"
)
```

Coefplot is a package for plotting the coefficients and standard
        errors from a variety of models. Currently lm, glm, glmnet,
        maxLik, rxLinMod, rxGLM and rxLogit are supported.

The package is designed for S3 dispatch from the functions coefplot
        and getModelInfo to make for easy additions of new models.

If interested in helping please contact the package author.
```

30.3 Package Documentation

A very strict requirement for R packages to be accepted by CRAN is proper documentation. Each exported function in a package needs its own .Rd file that is written in a LaTeX-like syntax. This can be difficult to write for even simple functions like the following one.

```
> simpleEx <- function(x, y)
+ {
+     return(x * y)
+ }
```

Even though it has only two arguments and simply returns the product of the two, it has a lot of necessary documentation, shown here.

```
\name{simpleEx}
\alias{simpleEx}
\title{within.distance}
\usage{simpleEx(x, y)}
\arguments{
    \item{x}{A numeric}
    \item{y}{A second numeric}
}
\value{x times y}
\description{Compute distance threshold}
\details{This is a simple example of a function}
\author{Jared P. Lander}
\examples{
    simpleEx(3, 5)
}
```

Rather than taking this two-step approach, it is better to write function documentation along with the function. That is, the documentation is written in a specially commented out block right above the function, as shown here.

```
> #' @title simpleEx
> #' @description Simple Example
> #' @details This is a simple example of a function
> #' @aliases simpleEx
> #' @author Jared P. Lander
> #' @export simpleEx
> #' @param x A numeric
> #' @param y A second numeric
> #' @return x times y
> #' @examples
> #' simpleEx(5, 3)
> simpleEx <- function(x, y)
+ {
+     return(x * y)
+ }
```

Running **document** from **devtools** will automatically generate the appropriate `.Rd` file based on the block of code above the function. The code is indicated by `#'` at the beginning of the line. Table 30.4 lists a number of commonly used **roxygen2** tags.

Every argument must be documented with a `@param` tag, including the dots (...), which are written as `\dots`. There must be an exact correspondence between `@param` tags and arguments; one more or less will cause an error.

Table 30.4 **Tags used in `roxygen2` documentation of functions**

Tag	Explanation
`@param`	The name of an argument and a short description
`@inheritParams`	Copies the `@param` tags from another function so that they do not need be to rewritten
`@examples`	Examples of the function being used
`@return`	Description of the object that is returned by the function
`@author`	Name of the author of the function
`@aliases`	Names by which a user can search for the function
`@export`	Lists the function as an export in the `NAMESPACE` file
`@import`	Lists a package as an import in the `NAMESPACE` file
`@seealso`	A list of other functions to see
`@title`	Title of help file
`@description`	Short description of the function
`@details`	Detailed information about the function
`@useDynLib`	Indicates that compiled source code will be used for the package
`@S3Method`	Declares functions that go with `S3` generic functions

It is considered good form to show examples of a function's usage. This is done on the lines following the `@examples` tag. In order to be accepted by CRAN all of the examples must work without error. In order to show, but not run, the examples wrap them in `\dontrun{...}`.

Knowing the type of object is important when using a function, so `@return` should be used to describe the returned object. If the object is a list, the `@return` tag should be an itemized list of the form

```
\item{name a}{description a}\item{name b}{description b}}
```

Help pages are typically arrived at by typing `?FunctionName` into the console. The `@aliases` tag uses a space-separated list to specify the names that will lead to a particular help file. For instance, using `@aliases coefplot plotcoef` will result in both `?coefplot` and `?plotcoef`, leading to the same help file.

In order for a function to be exposed to the end user, it must be listed as an export in the `NAMESPACE` file. Using `@export FunctionName` automatically adds `export(FunctionName)` to the `NAMESPACE` file. Similarly, to use a function from another package, that package must be imported and `@import PackageName` adds `import(PackageName)` to the `NAMESPACE` file.

When building functions that get called by generic functions, such as **coefplot.lm** or **print.anova**, the `@S3method` tag should be used. `@S3method GenericFunction Class` adds `S3method(GenericFunction,class)` to the `NAMESPACE` file. When using @S3method it is a good idea to also use @method with the same arguments. This is shown in the following function.

```
> #' @title print.myClass
> #' @aliases print.myClass
> #' @method print myClass
> #' @S3method print myClass
> #' @export print.myClass
> #' @param x Simple object
> #' @param \dots Further arguments to be passed on
> #' @return The top 5 rows of x
> print.myClass <- function(x, ...)
+ {
+     class(x) <- "list"
+     x <- as.data.frame(x)
+     print.data.frame(head(x, 5))
+ }
```

30.4 Tests

Testing code is a very important part of the pacage building process. It not only confirms that the code is behaving as designed but also warns when changes to code break functionality. Knowing when code changes break existing code can save hours of frustration and prevent production outages.

There are two main packages for writing tests: **RUnit** and **testthat**. They both have their benefits, though **testthat** has become more popular due to its integration with **devtools**.

When using **testthat** the `tests` folder contains a file called "testthat.R" and a folder called `testthat`, which holds the tests. Running **use_testthat** from **devtools** will set up that structure automatically and add **testthat** to the `Suggests` field of the `DESCRIPTION` file.

The "testthat.R" file is very simple. It loads the **testthat** package and the package we are testing. It also calls the **test_check** function to run the tests. The following is a sample file:

```
library(testthat)
library(ExamplePackage)

test_check("ExamplePackage")
```

The code that runs the actual tests go inside the `testthat` folder. A good rule of thumb is to test one function per file. Each file must be of the form `test-<file-name>.R`. These files can be automatically generated with **use_test** and specifying the name of the function to be tested.

```
> use_test('simpleEx')
```

This generates the file `test-simeEx.R` with the following contents.

```
context("simpleEx")

## TODO: Rename context
## TODO: Add more tests

test_that("multiplication works", {
  expect_equal(2 * 2, 4)
})
```

The first line is there to give us information about what we are testing. This will be useful when looking through the results of numerous tests so it should be long enough to make sense but not so long as to be a burden. The following lines are simply instructions telling us to flesh out the testing. Then there is an example test.

For any given function there should be multiple tests, each one testing a certain aspect of the function. Within a single test there should be multiple expectations about that aspect. If any expectation fails the entire test fails. For our example function some good tests are to check that it returns the proper data types, that the result is the correct length and that it should cause an error when necessary.

The first argument to **test_that** is a brief description of what is being tested. The second argument is the collection of expectations. Since we have multiple expectations, each of which is an expression, we encapsulate them in curly braces ({ and }). When executed interactively, the expectations and tests should return no result if there is not an error. However, if an expectation fails then an error is returned.

```
> library(testthat)
>
> test_that('Correct Answer', {
+     expect_equal(simpleEx(2, 3), 6)
+     expect_equal(simpleEx(5, 4), 20)
+     expect_equal(simpleEx(c(1, 2, 3), 3), c(3, 6, 9))
+     expect_equal(simpleEx(c(1, 2, 3), c(2, 4, 6)), c(2, 8, 18))
+ })
>
> test_that('Correct Type', {
+     expect_is(simpleEx(2, 3), 'numeric')
+     expect_is(simpleEx(2L, 3L), 'integer')
+     expect_is(simpleEx(c(1, 2, 3), c(2, 4, 6)), 'numeric')
+ })
>
> test_that('Correct length', {
+     expect_length(simpleEx(2, 3), 1)
+     expect_length(simpleEx(c(1, 2, 3), 3), 3)
+     expect_length(simpleEx(c(1, 2, 3), c(2, 4, 6)), 3)
+ })
```

Sometimes an error, or warning, should result from running a function. In this case we use **expect_error** or **expect_warning**, both of which will return no result if they capture an error or warning.

```
> test_that('Appropriate error or warning', {
+     expect_error(simpleEx(3, 'A'))
+     expect_equal(simpleEx(1:3, 1:2), c(1, 4, 3))
+     expect_warning(simpleEx(1:3, 1:2))
+ })
```

There are many expectations such as **expect_gte** for testing if a result is greater than or equal to a value, **expect_false** for expecting a `FALSE` result and **expect_named** to see if the result has names. The full set of expectations can be printed using **apropos**.

```
> apropos('expect_')

 [1] "expect_cpp_tests_pass"      "expect_equal"
 [3] "expect_equal_to_reference" "expect_equivalent"
 [5] "expect_error"               "expect_failure"
 [7] "expect_false"               "expect_gt"
 [9] "expect_gte"                 "expect_identical"
[11] "expect_is"                  "expect_length"
[13] "expect_less_than"           "expect_lt"
[15] "expect_lte"                 "expect_match"
[17] "expect_message"             "expect_more_than"
[19] "expect_named"               "expect_null"
[21] "expect_output"              "expect_output_file"
[23] "expect_s3_class"            "expect_s4_class"
[25] "expect_silent"              "expect_success"
[27] "expect_that"                "expect_true"
[29] "expect_type"                "expect_warning"
```

Rather than check one test at a time, we are more likely to check all the tests for a package at once, perhaps after making a change to ensure that the new code did not break the old code. This can be done with the **test** function in **devtools**.

```
> devtools::test()
```

Tests are also checked when checking the package with **check** or at the command line using `R CMD check`. The tests are checked again by CRAN before the package is accepted.

30.5 Checking, Building and Installing

Building a package used to require going to the command prompt and using commands like `R CMD check`, `R CMD build` and `R CMD INSTALL` (in Windows it is `Rcmd` instead of `R CMD`), which required being in the proper directory, knowing the correct options and other bothersome time wasters. Thanks to Hadley Wickham, this has all been made much easier and can be done from within the R console.

The first step is to make sure a package is properly documented by calling **document**. The first argument is the path to the root folder of the package as a string. (If the current working directory is the same as the root folder, then no arguments are even needed. This is true of all the **devtools** functions.) This builds all the necessary `.Rd` files, the `NAMESPACE` file and the `Collate` field of the `DESCRIPTION` file.

```
> devtools::document()
```

After the package is properly documented (and with tests written), it is time to check it. This is done using **check** with the path to the package as the first argument. This will make note of any errors or warnings that would prevent CRAN from accepting the package. If tests exist those will be checked as well. CRAN can be very strict, so it is essential to address all the issues. If submitting to CRAN, then the package should be against the current version of R on multiple operating systems and on R devel.

```
> devtools::check()
```

Building the package is equally simple using the **build** function, which also takes the path to the package as the first argument. By default it builds a `.tar.gz`, which is a collection of all the files in the package, that still needs to be built into a binary that can be installed in R. It is portable in that it can be built on any operating system. The `binary` argument, if set to `TRUE`, will build a binary that is operating system specific. This can be problematic if compiled source code is involved.

```
> devtools::build()
> devtools::build(binary=TRUE)
```

Other functions to help with the development process are **install**, which rebuilds and loads the package, and **load_all**, which simulates the loading of the package and `NAMESPACE`.

Another great function, not necessarily for the development process so much as for getting other people's latest work, is **install_github**, which can install an R package directly from a GitHub repository. There are analogous functions for installing from BitBucket (**install_bitbucket**) and Git (**install_git**) in general.

For instance, to get the latest version of **coefplot** the following code should be run. By the time of publication this may no longer be the the latest version.

```
> devtools::install_github(repo="jaredlander/coefplot",
+                          ref="survival")
```

Sometimes an older version of a package on CRAN is needed, which under normal circumstances that is hard to do without downloading source packages manually and building them. However, **install_version** was recently added to **devtools**, allowing a specific version of a package to be downloaded from CRAN, built and installed.

30.6 Submitting to CRAN

The best way to get a package out to the R masses is to have it on CRAN. Assuming the package passed the checks and tests using **check** and **test** from **devtools**, it is ready to be uploaded to CRAN using the new Web uploader (as opposed to using FTP) at `http://xmpalantir.wu.ac.at/cransubmit/`. The `.tar.gz` file is the one to upload. After submission, CRAN will send an email requiring confirmation that the package was indeed uploaded by the maintainer.

Alternatively, the **release** function from **devtools** will once again document, test, check and build the package and then upload it to CRAN. This process helps prevent little mistakes from slipping through. Any notes intended to be seen by the CRAN maintainers should be stored in `cran-comments.md` file in the root of the package. These notes should include information about where the package was tested and `R CMD check`. It would also be nice to include the words "thank you" somewhere, because the CRAN team puts in an incredible amount of effort despite not getting paid.

30.7 C++ Code

Sometimes R code is just not fast enough (even when byte-compiled) for a given problem and a compiled language must be used. R's foundation in C and links to FORTRAN libraries (digging deep enough into certain functions, such as **lm**, reveals that the underpinnings are written in FORTRAN) makes incorporating those languages fairly natural. **.Fortran** is used for calling a function written in FORTRAN and **.Call** is used for calling C and C++ functions.[1] Even with those convenient functions, knowledge of either FORTRAN or C/C++ is still necessary, as is knowledge of how R objects are represented in the underlying language.

Thanks to Dirk Eddelbuettel and Romain François, integrating C++ code has become much easier using the **Rcpp** package. It handles a lot of the scaffolding necessary to make C++ functions callable from R. Not only did they make developing R packages with C++ easier, but they also made running ad hoc C++ possible.

A number of tools are necessary for working with C++ code. First, a proper C++ compiler must be available. To maintain compatibility it is best to use `gcc`.

Linux users should already have `gcc` installed and should not have a problem, but they might need to install g++.

Mac users need to install Xcode and might have to manually select g++. The compiler offered on Mac generally lags behind the most recent version available, which has been known to cause some issues. Installing `gcc` via Homebrew might offer a more recent version.

Windows users should actually have an easy time getting started, thanks to RTools developed by Brian Ripley and Duncan Murdoch. It provides all necessary development tools, including `gcc` and `make`. The proper version, depending on the installed version of R, can be downloaded from `http://cran.r-project.org/bin/windows/Rtools/` and installed like any other program. It installs `gcc` and makes the Windows command prompt act more like a Bash terminal. If building packages from within R using

1. There is also a **.C** function, although despite much debate it is generally frowned upon.

devtools and RStudio (which is the best way now), then the location of `gcc` will be determined from the operating system's registry. If building packages from the command prompt, then the location of `gcc` must be put at the very beginning of the system PATH like `c:\Rtools\bin;c:\Rtools\gcc-4.6.3\bin;C:\Users\Jared\Documents\R\R-3.4.0\bin\x64`.

A LaTeX distribution is needed for building package help documents and vignettes. Table 27.1 lists the primary distributions for the different operating systems.

30.7.1 sourceCpp

To start, we build a simple C++ function for adding two `vectors`. Doing so does not make sense from a practical point of view because R already does this natively and quickly, but it will be good for illustrative purposes. The function will have arguments for two `vectors` and return the element-wise sum. The `// [[Rcpp::export]]` tag tells **Rcpp** that the function should be exported for use in R.

```
#include <Rcpp.h>
using namespace Rcpp;

// [[Rcpp::export]]
NumericVector vector_add(NumericVector x, NumericVector y)
{
    // declare the result vector
    NumericVector result(x.size());

    // loop through the vectors and add them element by element
    for(int i=0; i<x.size(); ++i)
    {
        result[i] = x[i] + y[i];
    }

    return result;
}
```

This function should be saved in a `.cpp` file (for example, `vector_add.cpp`) or as a `character` variable so it can be sourced using **sourceCpp**, which will automatically compile the code and create a new R function with the same name that when called, executes the C++ function.

```
> library(Rcpp)
> sourceCpp("vector_add.cpp")
```

Printing the function shows that it points to a temporary location where the compiled function is currently stored.

```
> vector_add

function (x, y)
.Primitive(".Call")(<pointer: 0x0000000032681a40>, x, y)
```

The function can now be called just like any other R function.

```
> vector_add(x=1:10, y=21:30)

 [1] 22 24 26 28 30 32 34 36 38 40

> vector_add(1, 2)

[1] 3

> vector_add(c(1, 5, 3, 1), 2:5)

[1] 3 8 7 6
```

JJ Allaire (the founder of RStudio) is responsible for **sourceCpp**, the `// [[Rcpp::export]]` shortcut and a lot of the magic that simplifies using C++ with R in general. **Rcpp** maintainer Dirk Eddelbuettel cannot stress enough how helpful Allaire's contributions have been.

Another nice feature of **Rcpp** is the syntactical sugar that allows C++ code to be written like R. Using sugar we can rewrite `vector_add` with just one line of code.

```
#include <Rcpp.h>
using namespace Rcpp;

// [[Rcpp::export]]
NumericVector vector_add(NumericVector x, NumericVector y)
{
    return x + y;
}
```

The syntactic sugar allowed two `vectors` to be added just as if they were being added in R.

Because C++ is a strongly typed language, it is important that function arguments and return types be explicitly declared using the correct type. Typical types are `NumericVector`, `IntegerVector`, `LogicalVector`, `CharacterVector`, `DataFrame` and `List`.

30.7.2 Compiling Packages

While **sourceCpp** makes ad hoc C++ compilation easy, a different tactic is needed for building R packages using C++ code. The C++ code is put in a `.cpp` file inside the `src` folder. Any functions preceded by `// [[Rcpp::export]]` will be converted into end user facing R functions when the package is built using **build** from **devtools**. Any **roxygen2** documentation written above an exported C++ function will be used to document the resulting R function.

The `vector_add` function should be rewritten using **roxygen2** and saved in the appropriate file.

```
# include <Rcpp.h>
using namespace Rcpp;

//' @title vector_add
//' @description Add two vectors
//' @details Adding two vectors with a for loop
//' @author Jared P. Lander
//' @export vector_add
//' @aliases vector_add
//' @param x Numeric Vector
//' @param y Numeric Vector
//' @return a numeric vector resulting from adding x and y
//' @useDynLib ThisPackage
// [[Rcpp::export]]
NumericVector vector_add(NumericVector x, NumericVector y)
{
    NumericVector result(x.size());

    for(int i=0; i<x.size(); ++i)
    {
        result[i] = x[i] + y[i];
    }

    return result;
}
```

The magic is that **Rcpp** compiles the code, and then creates a new `.R` file in the `R` folder with the corresponding R code. In this case it builds the following.

```
> # This file was generated by Rcpp::compileAttributes
> # Generator token: 10BE3573-1514-4C36-9D1C-5A225CD40393
>
> #' @title vector_add
> #' @description Add two vectors
> #' @details Adding two vectors with a for loop
> #' @author Jared P. Lander
> #' @export vector_add
> #' @aliases vector_add
> #' @param x Numeric Vector
> #' @param y Numeric Vector
```

```
> #' @useDynLib RcppTest
> #' @return a numeric vector resulting from adding x and y
> vector_add <- function(x, y) {
+     .Call('RcppTest_vector_add', PACKAGE = 'RcppTest', x, y)
+ }
```

It is simply a wrapper function that uses **.Call** to call the compiled C++ function.

Any functions that are not preceded by `// [[Rcpp::export]]` are available to be called from within other C++ functions, but not from R, using **.Call**. Specifying a name attribute in the export statement, like `// [[Rcpp::export(name="NewName"]]`, causes the resulting R function to be called that name. Functions that do not need an R wrapper function automatically built, but need to be callable using **.Call**, should be placed in a separate `.cpp` file where `// [[Rcpp::interfaces(cpp)]]` is declared and each function that is to be user accessible is preceded by `// [[Rcpp::export]]`.

In order to expose its C++ functions, a package's `NAMESPACE` must contain `useDynLib(PackageName)`. This can be accomplished by putting the `@useDynLib PackageName` tag in any of the **roxygen2** blocks. Further, if a package uses **Rcpp** the `DESCRIPTION` file must list **Rcpp** in both the `LinkingTo` and `Depends` fields. The `LinkingTo` field also allows easy linking to other C++ libraries such as **RcppArmadillo**, **bigmemory** and **BH** (Boost).

The `src` folder of the package must also contain `Makevars` and `Makevars.win` files to help with compilation. The following examples were automatically generated using **Rcpp.package.skeleton** and should be sufficient for most packages.

First the `Makevars` file:

```
## Use the R_HOME indirection to support installations of multiple R version
PKG_LIBS = `$(R_HOME)/bin/Rscript -e "Rcpp:::LdFlags()"`

## As an alternative, one can also add this code in a file 'configure'
##
##    PKG_LIBS=`${R_HOME}/bin/Rscript -e "Rcpp:::LdFlags()"`
##
##    sed -e "s|@PKG_LIBS@|${PKG_LIBS}|" \
##        src/Makevars.in > src/Makevars
##
## which together with the following file 'src/Makevars.in'
##
##    PKG_LIBS = @PKG_LIBS@
##
## can be used to create src/Makevars dynamically. This scheme is more
## powerful and can be expanded to also check for and link with other
## libraries.  It should be complemented by a file 'cleanup'
##
##    rm src/Makevars
##
```

(Continues)

(Continued)

```
## which removes the autogenerated file src/Makevars.
##
## Of course, autoconf can also be used to write configure files. This is
## done by a number of packages, but recommended only for more advanced
## users comfortable with autoconf and its related tools.
```

Now the `Makevars.win` file:

```
## Use the R_HOME indirection to support installations of multiple R version
PKG_LIBS = $(shell "${R_HOME}/bin${R_ARCH_BIN}/Rscript.exe" -e "Rcpp:::LdFlags()")
```

This just barely scratches the surface of **Rcpp**, but should be enough to start a basic package that relies on C++ code. Packages containing C++ code are built the same as any other package, preferably using **build** in **devtools**.

30.8 Conclusion

Package building is a great way to make code portable between projects and to share it with other people. A package purely built with R code only requires working functions that can pass the CRAN check using **check** and proper help files that can be easily built by including **roxygen2** documentation above functions and calling **document**. Building the package is as simple as using **build**. Packages with C++ should utilize **Rcpp**.

A
Real-Life Resources

One of the greatest aspects of R is the surrounding community, both online and in person. This includes Web resources like Twitter and Stack Overflow, meetups and textbooks.

A.1 Meetups

Meetup.com is a fantastic resource for finding like-minded people and learning experiences for just about anything including programming, statistics, video games, cupcakes and beer. They are so pervasive that as of early 2017, there were over 260,000 meetup groups in 184 countries. Data meetups draw particularly large crowds and usually take the format of socializing, a talk for 45 to 90 minutes, and then more socializing. Meetups are not only great for learning, but also for hiring or getting hired.

R meetups are very common, although some are starting to rebrand from R meetups to statistical programming meetups. Some popular meetups take place in New York, Chicago, Boston, Amsterdam, Washington, DC, San Francisco, Tel Aviv, London, Cleveland, Singapore and Melbourne. The talks generally show cool features in R, new packages or software or just an interesting analysis performed in R. The focus is usually on programming more than statistics. The New York Open Statistical Programming Meetup is the world's largest R and open stats meetup with over 8,000 members as of mid 2017. Table A.1 lists a number of popular meetups but it is an incredibly short list compared to how many meetups exist for R.

Machine Learning meetups are also good for finding presentations on R, although they will not necessarily be as focused on R. They are located in many of the same cities as R meetups and draw similar speakers and audiences. These meetups tend more toward the academic than focusing on programming.

The third core meetup type is Predictive Analytics. While they may seem similar to Machine Learning meetups, they cover different material. The focus is somewhere in between that of R and Machine Learning meetups. And yes, there is significant overlap in the audiences for these meetups.

Other meetup groups that might be of interest are data science, big data and data visualization.

Table A.1 R and related meetups

City	Group Name	URL
New York	New York Open Statistical Programming Meetup	`http://www.meetup.com/nyhackr/`
Washington, DC	Statistical Programming DC	`http://www.meetup.com/stats-prog-dc/`
Amsterdam	amst-R-dam	`http://www.meetup.com/amst-R-dam/`
Boston	Greater Boston useR Group (R Programming Language)	`http://www.meetup.com/Boston-useR/`
San Francisco	Bay Area useR Group (R Programming Language)	`http://www.meetup.com/R-Users/`
Chicago	Chicago R User Group (Chicago RUG) Data and Statistics	`http://www.meetup.com/ChicagoRUG/`
London	LondonR	`http://www.meetup.com/LondonR/`
Singapore	R User Group - Singapore (RUGS)	`http://www.meetup.com/R-User-Group-SG/`
Cleveland	Greater Cleveland R Group	`http://www.meetup.com/Cleveland-useR-Group/`
Melbourne	Melbourne Users of R Network (MelbURN)	`http://www.meetup.com/MelbURN-Melbourne-Users-of-R-Network/`
Connecticut	Connecticut R Users Group	`http://www.meetup.com/Conneticut-R-Users-Group/`[a]
New York	NYC Machine Learning Meetup	`http://www.meetup.com/NYC-Machine-Learning/`
Tel Aviv	Big Data & Data Science - Israel	`http://www.meetup.com/Big-Data-Israel/`

a. Though this is an incorrect spelling, it is the correct URL.

A.2 Stack Overflow

Sometimes when confronted with a burning question that cannot be solved alone, a good place to turn for help is Stack Overflow (`http://stackoverflow.com/`). Previously the R mailing list was the best, or only, online resource for help, but that has since been superseded by Stack Overflow.

The site is a forum for asking programming questions where both questions and answers are voted on by users and people can build reputations as experts. This is a very quick way to get answers for even difficult questions.

Common search tags related to R are `r`, `statistics`, `rcpp`, `ggplot2`, `shiny` and other statistics-related terms.

Many R packages these days are hosted on GitHub, so if a bug is found and confirmed, the best way to address it is not on Stack Overflow but on the GitHub issues list for the package.

A.3 Twitter

Sometimes just a quick answer is needed that would fit in 140 characters. In this case, Twitter is a terrific resource for R questions ranging from simple package recommendations to code snippets.

To reach the widest audience, it is important to use hash tags such `#rstats`, `#ggplot2`, `#knitr`, `#rcpp`, `#nycdatamafia` and `#statistics`.

Great people to follow are `@drewconway`, `@mikedewar`, `@harlanharris`, `@xieyihui`, `@hadleywickham`, `@jeffreyhorner`, `@revodavid`, `@eddelbuettel`, `@johnmyleswhite`, `@Rbloggers`, `@statalgo`, `@drob`, `@hspter`, `@JennyBryan`, `@ramnath_vaidya`, `@timelyportfolio`, `@ProbablePattern`, `@CJBayesian`, `@RLangTip`, `@cmastication`, `@pauldix`, `@nyhackr`, `@rstatsnyc` and `@jaredlander`.

A.4 Conferences

There are a number of conferences where R is either the focus or receives a lot of attention. There are usually presentations about or involving R, and sometimes classes that teach something specific about R.

The main one is the appropriately named useR! conference, which is a yearly event at rotating locations around the world. It alternates each year between Europe and the United States. It is organized by a local committee, so each year it has a different focus and theme. The Web site is at `http://www.r-project.org/conferences.html`. It is supported by the R Project.

The New York R Conference is a yearly conference organized by Lander Analytics and Work-Bench in New York City. It is two days of twenty-minute talks with no questions. The environment is meant to be open and convivial and emulate the New York Open Statistical Programming Meetup. The Web site is `http://www.rstats.nyc`.

R in Finance is a yearly conference that takes place in Chicago and is coorganized by Dirk Eddelbuettel. It is very quantitatively focused and heavy in advanced math. The Web site is at `http://www.rinfinance.com/`.

Other statistics conferences that are worth attending are the Joint Statistical Meetings organized by the American Statistical Association (`http://www.amstat.org/meetings/jsm.cfm`) and Strata New York (`http://strataconf.com/strata2013/public/content/home`).

Data Gotham is a new data science conference organized by some of the leaders of the data science community like Drew Conway and Mike Dewar. It went dormant for a few years but is being revitalized by Lander Analytics. The Web site is at `http://www.datagotham.com/`.

A.5 Web Sites

Being that R is an open source project with a strong community, it is only appropriate that there is a large ecosystem of Web sites devoted to it. Most of them are maintained by people who love R and want to share their knowledge. Some are exclusively focused on R and some only partially.

Besides `http://www.jaredlander.com/`, some of our favorites are R-Bloggers (`http://www.r-bloggers.com/`), htmlwidgets (`http://www.htmlwidgets.org/`), Rcpp Gallery (`http://gallery.rcpp.org/`), Revolution Analytics (`http://blog.revolutionanalytics.com/`), Andrew Gelman's site (`http://andrewgelman.com/`), John Myles White's site (`http://www.johnmyleswhite.com/`) and RStudio (`https://blog.rstudio.org/`).

A.6 Documents

Over the years, a number of very good documents have been written about R and made freely available.

An Introduction to R, by William N. Venables, David M. Smith and The R Development Core Team, was has been around since S, the precursor of R, and can be found at `http://cran.r-project.org/doc/manuals/R-intro.pdf`.

The R Inferno is a legendary document by Patrick Burns that delves into the nuances and idiosyncrasies of the language. It is available as both a printed book and a free PDF. Its Web site is `http://www.burns-stat.com/documents/books/the-r-inferno/`.

Writing R Extensions is a comprehensive treatise on building R packages that expands greatly on Chapter 30. It is available at `http://cran.r-project.org/doc/manuals/R-exts.html`.

A.7 Books

For a serious dose of statistics knowledge, textbooks offer a huge amount of material. Some are old fashioned and obtuse, while others are modern and packed with great techniques and tricks.

Our favorite statistics book—which happens to include a good dose of R code—is *Data Analysis Using Regression and Multilevel/Hierarchical Models* by Andrew Gelman and Jennifer Hill. The first half of the book is a good general text on statistics with R used for examples. The second half of the book focuses on Bayesian models using BUGS; the next edition is rumored to use STAN.

For advanced machine learning techniques, but not R code, Hastie, Tibshirani and Friedman's landmark *The Elements of Statistical Learning: Data Mining, Inference, and Prediction* details a number of modern algorithms and models. It delves deep into the underlying math and explains how the algorithms, including the Elastic Net, work.

Other books, not necessarily textbooks, have recently came out that are focused primarily on R. *Machine Learning for Hackers* by Drew Conway and John Myles White uses R as a tool in learning some basic machine learning algorithms. *Dynamic Documents with R and knitr* by Yihui Xie is an in-depth look at **knitr** and expands greatly on Chapter 27. Integrating C++ into R, discussed in Section 30.7 receives full treatment in *Seamless R and*

C++ Integration with Rcpp by Dirk Eddelbuettel. David Robinson and Julia Silge wrote *Text Mining with R* which covers modern techniques for analyzing text data in R.

A.8 Conclusion

Making use of R's fantastic community is an integral part of learning R. Person-to-person opportunities exist in the form of meetups and conferences. The best online resources are Stack Overflow and Twitter. And naturally there are a number of books and documents available both online and in bookstores.

B
Glossary

ACF	See autocovariance function
AIC	See Akaike Information Criterion
AICC	See Akaike Information Criterion Corrected
Akaike Information Criterion	Measure of model fit quality that penalizes model complexity
Akaike Information Criterion Corrected	Version of AIC with greater penalty for model complexity
Analysis of Variance	See ANOVA
Andersen-Gill	Survival analysis for modelling time to multiple events
ANOVA	Test for comparing the means of multiple groups. The test can only detect if there is a difference between any two groups; it cannot tell which ones are different from the others
Ansari-Bradley test	Nonparametric test for the equality of variances between two groups
AR	See autoregressive
ARIMA	Like an ARMA model but it includes a parameter for the number of differences of the time series data
ARMA	See Autoregressive Moving Average
`array`	Object that holds data in multiple dimensions
autocorrelation	When observations in a single variable are correlated with previous observations
autocovariance function	The correlation of a time series with lags of itself
Autoregressive	Time series model that is a linear regression of the current value of a time series against previous values

Autoregressive Moving Average	Combination of AR and MA models
average	While generally held to be the arithmetic mean, average is actually a generic term that can mean any number of measures of centrality, such as the mean, median or mode
Bartlett test	Parametric test for the equality of variances between two groups
BASH	A command line processor in the same vein as DOS; mainly used on Linux and MAC OS X, though there is an emulator for Windows
basis functions	Functions whose linear combination make up other functions
basis splines	Basis functions used to compose splines
Bayesian	Type of statistics where prior information is used to inform the model
Bayesian Information Criterion	Similar to AIC but with an even greater penalty for model complexity
Beamer	LaTeX document class for producing slide shows
Bernoulli distribution	Probability distribution for modelling the success or failure of an event
Beta distribution	Probability distribution for modelling a set of possible values on a finite interval
BIC	See Bayesian Information Criterion
Binomial distribution	Probability distribution for modelling the number of successful independent trials with identical probabilities of success
Bioconductor	Repository of R packages for the analysis of genomic data
BitBucket	Online Git repository
Boost	Fast C++ library
boosted tree	An extension of decision trees which fits successive trees on data that are iteratively reweighted to make the model stronger
Bootstrap	A process in which data are resampled repeatedly and a statistic is calculated for each resampling to form an empirical distribution for that statistic

Boxplot	A graphical display of one variable where the middle 50 percent of the data are in a box, there are lines reaching out to 1.5 times the interquartile range and dots representing outliers
BUGS	Probabilistic programming language specializing in Bayesian computations
byte-compilation	The process of turning human readable code into machine code that runs faster
C	A fast, low-level programming language; R is written primarily in C
C++	A fast, low-level programming language that is similar to C
Cauchy distribution	Probability distribution for the ratio of two Normal random variables
censored data	Data with unknown information such as the occurrence of an event after a cutoff time
`character`	Data type for storing text
chi-squared distribution	The sum of k squared standard normal distributions
chunk	Piece of R code inside a LaTeX or Markdown document
class	Type of an R object
classification	Determining the class membership of data
clustering	Partitioning data into groups
coefficient	A multiplier associated with a variable in an equation; in statistics this is typically what is being estimated by a regression
coefficient plot	A visual display of the coefficients and standard errors from a regression
Comprehensive R Archive Network	See CRAN
confidence interval	A range within which an estimate should fall a certain percentage of time
correlation	The strength of the association between two variables
covariance	A measure of the association between two variables; the strength of the relationship is not necessarily indicated

Cox proportional hazards	Model for survival analysis where predictors have a multiplicative effect on the survival rate
CRAN	The central repository for all things R
cross-validation	A modern form of model assessment where the data are split into *k* discrete folds and a model is repeatedly fitted on all but one and used to make predictions on the holdout fold
Data Gotham	Data science conference in New York
data munging	The process of cleaning, correcting, aggregating, joining and manipulating data to prepare it for analysis
data science	The confluence of statistics, machine learning, computer engineering, visualization and social skills
`data.frame`	The main data type in R, similar to a spreadsheet with tabular rows and columns
`data.table`	A high speed extension of `data.frames`
database	Store of data, usually in relational tables
Date	Data type for storing dates
DB2	Enterprise level database from IBM
Debian	Linux distribution
decision tree	Modern technique for performing nonlinear regression or classification by iteratively splitting predictors
degrees of freedom	For some statistic or distribution, this is the number of observations minus the number of parameters being estimated
density plot	Display showing the probability of observations falling within a sliding window along a variable of interest
deviance	A measure of error for generalized linear models
drop-in deviance	The amount by which deviance drops when adding a variable to a model; a general rule of thumb is that deviance should drop by two for each term added
DSN	Data source connection used to describe communication to a data source—often a database
dzslides	HTML5 slide show format
EDA	See exploratory data analysis

Elastic Net	New algorithm that is a dynamic blending of lasso and ridge regressions and is great for predictions and dealing with high dimensional datasets
Emacs	Text editor popular among programmers
ensemble	Method of combining multiple models to get an average prediction
Excel	The most commonly used data analysis tool in the world
expected value	Weighted mean
exploratory data analysis	Visually and numerically exploring data to get a sense for it before performing rigorous analysis
exponential distribution	Probability distribution often used to model the amount of time until an event occurs
F-test	Statistical test often used for comparing models, as with the ANOVA
F distribution	The ratio of two chi-squared distributions, often used as the null distribution in analysis of variance
`factor`	Special data type for handling character data as an integer value with character labels; important for including categorical data in models
fitted values	Values predicted by a model, mostly used to denote predictions made on the same data used to fit the model
`formula`	Novel interface in R that allows the specification of a model using convenient mathematical notation
FORTRAN	High-speed, low-level language; much of R is written in FORTRAN
FRED	Federal Reserve Economic Data
ftp	File transfer protocol
g++	Open source compiler for C++
GAM	See Generalized Additive Models
gamma distribution	Probability distribution for the time one has to wait for n events to occur
gamma regression	GLM for response data that are continuous, positive and skewed, such as auto insurance claims

gap statistic	Measure of clustering quality that compares the within-cluster dissimilarity for a clustering of the data with that of a bootstrapped sample of data
GARCH	See generalized autoregressive conditional heteroskedasticity
Gaussian distribution	See normal distribution
gcc	Family of open source compilers
Generalized Additive Models	Models that are formed by adding a series of smoother functions fitted on individual variables
Generalized Autoregressive Conditional Heteroskedasticity	Time series method that is more robust for extreme values of data
Generalized Linear Models	Family of regression models for non-normal response data such as binary and count data
geometric distribution	Probability distribution for the number of Bernoulli trials required before the first success occurs
Git	Popular version control standard
GitHub	Online Git repository
GLM	See Generalized Linear Models
Hadoop	Framework for distributing data and computations across a grid of computers
Hartigan's rule	Measure of clustering quality that compares the within-cluster sum of squares for a clustering of k clusters and one with $k + 1$ clusters
heatmap	Visual display where the relationship between two variables is visualized as a mix of colors
hierarchical clustering	Form of clustering where each observation belongs to a cluster, which in turn belongs to a larger cluster and so on until the whole dataset is represented
histogram	Display of the counts of observations falling in discrete buckets of a variable of interest
HTML	Hypertext Markup Language; used for creating Web pages
htmlwidgets	A collection of R packages that generates HTML and JavaScript code for interactive data display
hypergeometric distribution	Probability distribution for drawing k successes out of a possible N items, of which K are considered successes

hypothesis test	Test for the significance of a statistic that is being estimated
IDE	See Integrated Development Environment
indicator variables	Binary variables representing one level of a categorical variable; also called dummy variables
inference	Drawing conclusions on how predictors affect a response
integer	Data type that is only whole numbers, be they positive, negative or zero
Integrated Development Environment	Software with features to make programming easier
Intel Matrix Kernel Library	Optimized matrix algebra library
interaction	The combined effect of two or more variables in a regression
intercept	Constant term in a regression; literally the point where the best fit line passes through the y-axis; it is generalized for higher dimensions
interquartile range	The third quartile minus the first quartile
inverse link function	Function that transforms linear predictors to the original scale of the response data
inverse logit	Transformation needed to interpret logistic regression on the 0/1 scale; scales any number to be between 0 and 1
IQR	See interquartile range
Java	Low-level programming language
JavaScript	A Web-based scripting language that much of the modern Web is built upon
Joint Statistical Meetings	Conference for statisticians
JSM	See Joint Statistical Meetings
K-means	Clustering that divides the data into k discrete groups as defined by some distance measurement
K-medoids	Similar to K-means except it handles categorical data and is more robust to outliers
knitr	Modern package for interweaving R code with LaTeX or Markdown
lasso regression	Modern regression using an L1 penalty to perform variable selection and dimension reduction

LaTeX	High-quality typesetting program especially well suited for mathematical and scientific documents and books
`level`	A unique value in a `factor` variable
linear model	Model that is linear in terms of its coefficients
link function	Function that transforms response data so it can be modeled with a GLM
Linux	Open source operating system
`list`	Robust data type that can hold any arbitrary data types
log	The inverse of an exponent; typically the natural log in statistics
log-normal distribution	Probability distribution whose log is normally distributed
`logical`	Data type that takes on the values `TRUE` or `FALSE`
logistic distribution	Probability distribution used primarily for logistic regression
logistic regression	Regression for modelling a binary response
logit	The opposite of the inverse logit; transforms number between 0 and 1 to the real numbers
loop	Code that iterates through some index
MA	See moving average
Mac OS X	Apple's proprietary operating system
machine learning	Modern, computationally heavy statistics
Map Reduce	Paradigm where data is split into discrete sets, computed on, and then recombined in some fashion
Markdown	Simplified formatting syntax used to produce an elegant HTML document in a simple fashion
Matlab	Expensive commercial software for mathematical programming
`matrix`	Two-dimensional data type
matrix algebra	Algebra performed on matrices that greatly simplifies the math
maximum	Largest value in a set of data
mean	Mathematical average; typically either arithmetic (traditional average) or weighted

Term	Definition
mean squared error	Quality measure for an estimator; the average of the squares of the differences between an estimator and the true value
median	Middle number of an ordered set of numbers; when there are an even amount of numbers, the median is the mean of the middle two numbers
Meetup	A Web site that facilitates real life social interaction for any number of interests; particularly popular in the data field
memory	Also referred to as RAM, this is where the data that R analyzes is stored while being processed; this is typically the limiting factor on the size of data that R can handle
Microsoft Access	Lightweight database from Microsoft
Microsoft R	Commercial distribution of R developed by Microsoft designed to be faster, be more stable and scale better
Microsoft SQL Server	Enterprise level database from Microsoft
minimum	Smallest value in a set of data
Minitab	GUI based statistical package
missing data	A big problem in statistics, this is data that are not available to compute for any one of a number of reasons
MKL	See Intel Matrix Kernel Library
model complexity	Primarily how many variables are included in the model; overly complex models can be problematic
model selection	Process of fitting the optimal model
Moving Average	Time series model that is a linear regression of the current value of a time series against current and previous residuals
multicolinearity	When one column in a matrix is a linear combination of any other columns
multidimensional scaling	Projecting multiple dimensions into a smaller dimensionality
multinomial distribution	Probability distribution for discrete data that can take on any of k classes
multinomial regression	Regression for discrete response that can take on any of k classes

multiple comparisons	Doing repeated tests on multiple groups
multiple imputation	Advanced process to fill in missing data using repeated regressions
multiple regression	Regression with more than one predictor
MySQL	Open source database
`NA`	Value that indicates missing data
namespace	Convention where functions belong to specific packages; helps solve conflicts when multiple functions have the same name
natural cubic spline	Smoothing function with smooth transitions at interior breakpoints and linear behavior beyond the endpoints of the input data
negative binomial distribution	Probability distribution for the number of trials required to obtain *r* successes; this is often used as the approximate distribution for pseudo-Poisson regression
nonlinear least squares	Least squares regression (squared error loss) with nonlinear parameters
nonlinear model	Model where the variables do not necessarily have a linear relationship, such as decision trees and GAMs
nonparametric model	Model where the response does not necessarily follow the regular GLM distributions, such as normal, logistic or Poisson
normal distribution	The most common probability distribution that is used for a wide array of phenomena; the familiar bell curve
`NULL`	A data concept that represents nothingness
null hypothesis	The assumed true value in hypothesis tests
`numeric`	Data type for storing numeric values
NYC Data Mafia	Informal term for the growing prevalence of data scientists in New York City
NYC Open Data	Initiative to make New York City government data transparent and available
Octave	Open-source version of Matlab
ODBC	See Open Database Connectivity
Open Database Connectivity	Industry standard for communicating data to and from a database

`ordered factor`	Character data where one level can be said to be greater or less than another level
overdispersion	When data show more variability than indicated by the theoretical probability distribution
p-value	The probability, if the null hypothesis was correct, of getting an as extreme, or more extreme, result
PACF	See partial autocovariance function
paired t-test	Two sample t-test where every member of one sample is paired with a member of a second sample
PAM	See partitioning around medoids
pandoc	Software for easy conversion of documents between various formats such as Markdown, HTML, LaTeX and Microsoft Word
parallel	In the computational context, the running of multiple instructions simultaneously to speed computation
parallelization	The process of writing code to run in parallel
partial autocovariance function	The amount of correlation between a time series and lags of itself that is not explained by previous lags
partitioning around medoids	Most common algorithm for K-medoids clustering
PDF	Common document format, most often opened with Adobe Acrobat Reader
penalized regression	Form of regression where a penalty term prevents the coefficients from growing too large
Perl	Scripting language commonly used for text parsing
Poisson distribution	Probability distribution for count data
Poisson regression	GLM for response data that are counts, such as the number of accidents, number of touchdowns or number of ratings for a pizzeria
`POSIXct`	Date-time data type
prediction	Finding the expected value of response data for given values of predictors
predictor	Data that are used as an input into a model and explain and/or predict the response
prior	Bayesian statistics use prior information, in the form of distributions for the coefficients of predictors, to improve the model fit
Python	Scripted language that is popular for data munging

Q-Q plot	Visuals means of comparing two distributions by seeing if the quantiles of the two fall on a diagonal line
quantile	Numbers in a set where a certain percentage of the numbers are smaller than that quantile
quartile	The 25th quantile
Quasi-Poisson distribution	Distribution (actually the negative binomial) used for estimating count data that are overdispersed
R-Bloggers	Popular site from Tal Galili that aggregates blogs about R
R Console	Where R commands are entered and results are shown
R Core Team	Group of 20 prime contributors to R who are responsible for its maintenance and direction
R Enthusiasts	Popular R blog by Romain François
R in Finance	Conference in Chicago about using R for finance
RAM	See memory
Random Forest	Ensemble method that builds multiple decision trees, each with a random subset of predictors, and combines the results to make predictions
Rcmdr	GUI interface to R
Rcpp Gallery	Online collection of Rcpp examples
Rdata	File format for storing R objects on disk
regression	Method that analyzes the relationship between predictors and a response; the bedrock of statistics
regression tree	See decision tree
regular expressions	String pattern matching paradigm
regularization	Method to prevent overfitting of a model, usually by introducing a penalty term
residual sum of squares	Summation of the squared residuals
residuals	Difference between fitted values from a model and the actual response values
response	Data that are the outcome of a model and are predicted and/or explained by the predictors
ridge regression	Modern regression using an L2 penalty to shrink coefficients for more stable predictions
RSS	See residual sum of squares

RStudio	Powerful and popular open-source IDE for R
RTools	Set of tools needed in Windows for integrating C++, and other compiled code, into R
S	Statistical language developed at Bell Labs, which was the precursor to R
S3	Basic object type in R
S4	Advanced object type in R
s5	HTML5 slide show format
SAS	Expensive commercial scripting software for statistical analysis
scatterplot	Two-dimensional display of data where each point represents a unique combination of two variables
shapefile	Common file format for map data
Shiny	A framework that allows Web development and backend computations
shrinkage	Reducing the size of coefficients to prevent overfitting
simple regression	Regression with one predictor, not including the intercept
slideous	HTML5 slide show format
slidy	HTML5 slide show format
slope	Ratio of a line's rise and run; in regression this is represented by the coefficients
smoothing spline	Spline used for fitting a smooth trend to data
spline	Function f that is a linear combination of N functions (one for each unique data point) that are transformations of the variable x
SPSS	Expensive point-and-click commercial software for statistical analysis
SQL	Database language for accessing or inserting data
Stack Overflow	Online resource for programming questions
STAN	Next generation probabilistic programming language specializing in Bayesian computations
standard deviation	How far, on average, each point is from the mean
standard error	Measure of the uncertainty for a parameter estimate
Stata	Commercial scripting language for statistical analysis

stationarity	When the mean and variance of a time series are constant for the whole series
stepwise selection	Process of choosing model variables by systematically fitting different models and adding or eliminating variables at each step
Strata	Large data conference
survival analysis	Analysis of time to event, such as death or failure
SUSE	Linux distribution
SVN	Older version control standard
Sweave	Framework for interweaving R code with LaTeX; has been superseded by knitr
Systat	Commercial statistical package
t-statistic	Ratio where the numerator is the difference between the estimated mean and the hypothesized mean and the denominator is the standard error of the estimated mean
t-test	Test for the value of the mean of a group or the difference between the means of two groups
t distribution	Probability distribution used for testing a mean with a student t-test
tensor product	A way of representing transformation functions of predictors, possibly measured on different units
text editor	Program for editing code that preserves the structure of the text
TextPad	Popular text editor
time series	Data where the order and time of the data are important to their analysis
`ts`	Data type for storing time series data
two sample t-test	Test for the difference of means between two samples
Ubuntu	Linux distribution
UltraEdit	Popular text editor
uniform distribution	Probability distribution where every value is equally likely to be drawn
USAID Open Government	Initiative to make US Aid data transparent and available
useR!	Conference for R users

VAR	See Vector Autoregressive Model
variable	R object; can be data, functions, any object
variance	Measure of the variability, or spread, of the data
`vector`	A collection of data elements, all of the same type
Vector Autoregressive Model	Multivariate times series model
version control	Means of saving snapshots of code at different time periods for easy maintenance and collaboration
vim	Text editor popular among programmers
violin plot	Similar to a boxplot except that the box is curved, giving a sense of the density of the data
Visual Basic	Programming language for building macros, mostly associated with Excel
Visual Studio	IDE produced by Microsoft
Wald test	Test for comparing models
Weibull distribution	Probability distribution for the lifetime of an object
weighted mean	Mean where each value carries a weight allowing the numbers to have different effects on the mean
weights	Importance given to observations in data so that one observation can be valued more or less than another
Welch t-test	Test for the difference in means between two samples, where the variances of each sample can be different
white noise	Essentially random data
Windows	Microsoft's operating system
Windows Live Writer	Desktop blog publishing application from Microsoft
Xcode	Apple's IDE
xkcd	Web comic by Randall Munroe, beloved by statisticians, physicists and mathematicians
XML	Extensible Markup Language; often used to descriptively store and transport data
`xts`	Advanced data type for storing time series data

List of Figures

List of Tables

General Index

Symbols

A

B

C

D

X

Y

Z

Index of Functions

Symbols

A

B

C

D

E

F

G

H

I

J

K

L

M

N

O

T

V

W

Index of Packages

S

T

U

V

W

X

Index of People

Data Index

A

B

C

D

E

F

G

I

N

R

T

U

V

W

Register Your Product at informit.com/register

Access additional benefits and **save 35%** on your next purchase

- Automatically receive a coupon for 35% off your next purchase, valid for 30 days. Look for your code in your InformIT cart or the Manage Codes section of your account page.
- Download available product updates.
- Access bonus material if available.
- Check the box to hear from us and receive exclusive offers on new editions and related products.

InformIT.com—The Trusted Technology Learning Source

InformIT is the online home of information technology brands at Pearson, the world's foremost education company. At InformIT.com, you can:

- Shop our books, eBooks, software, and video training
- Take advantage of our special offers and promotions (informit.com/promotions)
- Sign up for special offers and content newsletter (informit.com/newsletters)
- Access thousands of free chapters and video lessons

Connect with InformIT—Visit informit.com/community

Addison-Wesley · Adobe Press · Cisco Press · Microsoft Press · Pearson IT Certification · Prentice Hall · Que · Sams · Peachpit Press